Lecture Notes in Artificial Intelligence 7884

Subseries of Lecture Notes in Computer Science

Osmar R. Zaïane Sandra Zilles (Eds.)

Advances in Artificial Intelligence

26th Canadian Conference
on Artificial Intelligence, Canadian AI 2013
Regina, SK, Canada, May 28-31, 2013
Proceedings

 Springer

Series Editors

Randy Goebel, University of Alberta, Edmonton, Canada
Jörg Siekmann, University of Saarland, Saarbrücken, Germany
Wolfgang Wahlster, DFKI and University of Saarland, Saarbrücken, Germany

Volume Editors

Osmar R. Zaïane
University of Alberta
Department of Computing Science
Edmonton, AB, Canada
E-mail: zaiane@cs.ualberta.ca

Sandra Zilles
University of Regina
Department of Computer Science
Regina, SK, Canada
E-mail: zilles@cs.uregina.ca

ISSN 0302-9743   e-ISSN 1611-3349
ISBN 978-3-642-38456-1   e-ISBN 978-3-642-38457-8
DOI 10.1007/978-3-642-38457-8
Springer Heidelberg Dordrecht London New York

Library of Congress Control Number: 2013937990

CR Subject Classification (1998): I.2.7, I.2, H.3, H.4, I.3-5, F.1

LNCS Sublibrary: SL 7 – Artificial Intelligence

Typesetting: Camera-ready by author, data conversion by Scientific Publishing Services, Chennai, India

Printed on acid-free paper

Springer is part of Springer Science+Business Media (www.springer.com)

Preface

We are delighted to present the proceedings of the 26th Canadian Artificial Intelligence Conference held for the first time in Regina, Saskatchewan, Canada, in co-location with the Canadian Graphics Interface Conference and the Canadian Conference on Computer and Robot Vision during May 28-31, 2013. This volume, published in the *Lecture Notes in Artificial Intelligence* series by Springer, contains the research papers presented at the conference. They were 32 research papers covering a variety of subfields of AI. In addition to these research papers thoroughly selected by the Program Committee, the technical program of the conference also encompassed two invited keynote speeches by eminent researchers, an Industry Track and a Graduate Student Symposium. The contributions from the Graduate Student Symposium are also included in these proceedings.

This year's conference continued the tradition of bringing together researchers from Canada and beyond to discuss and disseminate innovative ideas, methods, algorithms, principles, and solutions to challenging problems involving AI. We were thrilled to have prestigious invited speakers: Sheila McIlraith from the University of Toronto and Eric Xing from Carnegie Mellon University.

The papers presented at AI 2013 covered a variety of topics within AI. The topics included: information extraction, knowledge representation, search, text mining, social networks, temporal associations, etc. This wide range of topics bears witness to the vibrant research activities and interest in our community and the dynamic response to the new challenges posed by innovative types of AI applications.

We received 73 papers submitted from 17 countries including Australia, Belgium, Bosnia, Brazil, Canada, China, Denmark, Egypt, France, Germany, India, Iran, Nigeria, Saudi Arabia, Spain, Tunisia, and the USA. Each submission was rigorously reviewed by three to four reviewers. The Program Committee finally selected 17 regular papers and 15 short papers yielding an acceptance rate of 23% for regular papers and 43% overall. The eight contributions from the Graduate Student Symposium were selected from 14 submissions through a thorough reviewing process with a separate Program Committee.

We would like to express our most sincere gratitude to all authors of submitted papers for their contributions and to the members of the Program Committee and the external reviewers, who made a huge effort to review the papers in a timely and thorough manner. We gratefully acknowledge the valuable support of the executive committee of the Canadian Artificial Intelligence Association with whom we met regularly in order to put together a memorable program for this conference. We would also like to express our gratitude to Cory Butz and Atefeh Farzindar, the General Co-chairs of the AI/GI/CRV Conferences 2013, to Narjes Boufaden who organized the Industrial Track, and to Svetlana

Kiritchenko and Howard Hamilton who organized the Graduate Student Symposium. Thanks are due to Leila Kosseim and Diana Inkpen for their advice concerning the workflow of creating a conference program. We thank the members of the Program Committee of the Graduate Student Symposium: Ebrahim Bagheri, Julien Bourdaillet, Chris Drummond, Alistair Kennedy, Vlado Keselj, Adam Krzyzak, Guy Lapalme, Bradley Malin, Stan Matwin, Gordon McCalla, Martin Müller, David Poole, Fred Popowich, and Doina Precup. Acknowledgements are further due to the providers of the EasyChair conference management system; the use of EasyChair for managing the reviewing process and for creating these proceedings eased our work tremendously. Finally, we would like to thank our sponsors: The University of Regina, GRAND, the Alberta Innovates Centre for Machine Learning, iQmetrix, GB Internet Solutions, keatext, the University of Regina Alumni Association, SpringBoard West Innovations, Houston Pizza, and Palomino System Innovations.

March 2013

Osmar Zaïane
Sandra Zilles

Organization

Program Committee

Reda Alhajj	University of Calgary, Canada
Aijun An	York University, Canada
Xiangdong An	York University, Canada
John Anderson	University of Manitoba, Canada
Wolfgang Banzhaf	Memorial University of Newfoundland, Canada
Denilson Barbosa	University of Alberta, Canada
Andre Barreto	McGill University, Canada
Sabine Bergler	Concordia University, Germany
Karsten Berns	University of Kaiserslautern, Germany
Giuseppe Carenini	University of British Columbia, Germany
Yllias Chali	University of Lethbridge, Canada
Colin Cherry	National Research Council Canada
Cristina Conati	University of British Columbia, Canada
Joerg Denzinger	University of Calgary, Canada
Ralph Deters	University of Saskatchewan, Canada
Michael Fleming	University of New Brunswick, Canada
Gosta Grahne	Concordia University, Canada
Kevin Grant	University of Lethbridge, Canada
Marek Grzes	University of Waterloo, Canada
Howard Hamilton	University of Regina, Canada
Robert Hilderman	University of Regina, Canada
Robert C. Holte	University of Alberta, Canada
Michael C. Horsch	University of Saskatchewan, Canada
Frank Hutter	University of British Columbia, Canada
Vlado Keselj	Dalhousie University, Canada
Ziad Kobti	University of Windsor, Canada
Grzegorz Kondrak	University of Alberta, Canada
Leila Kosseim	Concordia University, Canada
Anthony Kusalik	University of Saskatchewan, Canada
Laks V.S. Lakshmanan	University of British Columbia, Canada
Marc Lanctot	Maastricht University, The Netherlands
Guy Lapalme	Université de Montréal, Canada
Kate Larson	University of Waterloo, Canada
Levi Lelis	University of Alberta, Canada
Carson K. Leung	University of Manitoba, Canada
Daniel Lizotte	University of Waterloo, Canada

Cristina Manfredotti   Pierre and Marie Curie University, France
Stan Matwin   Dalhousie University, Canada
Sheila McIlraith   University of Toronto, Canada
Martin Memmel   German Research Center for Artificial
                              Intelligence, Germany
Robert Mercer   University of Western Ontario, Canada
Evangelos Milios   Dalhousie University, Canada
Shamima Mithun   Concordia University, Canada
Gabriel Murray   University of the Fraser Valley, Canada
Jeff Orchard   University of Waterloo, Canada
Gerald Penn   University of Toronto, Canada
Lourdes Peña-Castillo   Memorial University of Newfoundland,
                              Canada
David Poole   University of British Columbia, Canada
Fred Popowich   Simon Fraser University, Canada
Doina Precup   McGill University, Germany
Michael Richter   University of Kaiserslautern, Germany
Rafael Schirru   German Research Center for
                              Artificial Intelligence, Germany
Armin Stahl   German Research Center for
                              Artificial Intelligence, Germany
Ben Steichen   University of British Columbia, Canada
Csaba Szepesvári   University of Alberta, Canada
Peter van Beek   University of Waterloo, Canada
Paolo Viappiani   Centre National de la Recherche
                              Scientifique, France
Asmir Vodenčarević   University of Paderborn, Germany
Osmar Zaïane   University of Alberta, Canada
Harry Zhang   University of New Brunswick, Canada
Sandra Zilles   University of Regina, Canada

Additional Reviewers

Abnar, Afra   Hudson, Jonathan
Agrawal, Ameeta   Jiang, Fan
Baumann, Stephan   Kardan, Samad
Cao, Peng   Makonin, Stephen
Delpisheh, Elnaz   Moghaddam, Samaneh
Dimkovski, Martin   Mousavi, Mohammad
Dosselmann, Richard   Nicolai, Garrett
Esmin, Ahmed   Obradovic, Darko
Gurinovich, Anastasia   Onet, Adrain
Havens, Timothy   Patra, Pranjal
Hees, Jörn   Rabbany, Reihaneh

Roth-Berghofer, Thomas
Salameh, Mohammad
Samuel, Hamman
Sanden, Chris
Sturtevant, Nathan
Tanbeer, Syed

Thompson, Craig
Tofiloski, Milan
Trabelsi, Amine
van Seijen, Harm
Zier-Vogel, Ryan

Table of Contents

Long Papers

Short Papers

Contributions From the Graduate Student Symposium

Logo Recognition Based on the Dempster-Shafer Fusion of Multiple Classifiers

Mohammad Ali Bagheri[1], Qigang Gao[1], and Sergio Escalera[2]

[1] Faculty of Computer Science, Dalhousie University, Halifax, Canada
[2] Computer Vision Center, Campus UAB, Edifici O, 08193, Bellaterra, Spain
Dept. Matemática Aplicada i Análisi, Universitat de Barcelona, Gran Via de les
Corts Catalanes 585, 08007, Barcelona, Spain

Abstract. The performance of different feature extraction and shape description methods in trademark image recognition systems have been studied by several researchers. However, the potential improvement in classification through feature fusion by ensemble-based methods has remained unattended. In this work, we evaluate the performance of an ensemble of three classifiers, each trained on different feature sets. Three promising shape description techniques, including Zernike moments, generic Fourier descriptors, and shape signature are used to extract informative features from logo images, and each set of features is fed into an individual classifier. In order to reduce recognition error, a powerful combination strategy based on the Dempster-Shafer theory is utilized to fuse the three classifiers trained on different sources of information. This combination strategy can effectively make use of diversity of base learners generated with different set of features. The recognition results of the individual classifiers are compared with those obtained from fusing the classifiers' output, showing significant performance improvements of the proposed methodology.

Keywords: Logo recognition, ensemble classification, Dempster-Shafer fusion, Zernike moments, generic Fourier descriptor, shape signature.

1 Introduction

The research of document image processing has received great attention in recent years because of its diverse applications, such as digital libraries, online shopping, and office automation systems. An important problem in the field of document image processing is the recognition of graphical items, such as trademarks and company logos. Logos are mainly used by companies and organizations to identify themselves on documents. Given an image segment from a document image and a logo database, the task of logo recognition is to find whether the image segment corresponds to a logo in the database. The successful recognition of logos facilitates automatic classification of source documents, which is considered a key strategy for document image analysis and retrieval.

Logo analysis in document images involves two main steps: (1) detecting the probable logo from a document image; (2) classifying the detected logo candidate

O. Zaïane and S. Zilles (Eds.): Canadian AI 2013, LNAI 7884, pp. 1–12, 2013.

segment into one of the learned logos in the database [25]. The first step is referred to as logo detection, while the second is usually called logo recognition. In this work, we focus on the logo recognition phase.

From the machine learning point of view, logo recognition is considered a multi-class classification task since each logo category is considered a separate target class. In this view, the classification system involves two main stages: the selection and/or extraction of informative features and the construction of a classification algorithm. In such a system, a desirable feature set can greatly simplify the construction of a classification algorithm, and a powerful classification algorithm can work well even with a low discriminative feature set.

In the last decade, active research has been conducted on logo recognition. Most of the research work has focused on providing a framework for logo recognition by the extraction of informative features [8] or the analysis of image structures [1]. The classification algorithm is usually used as a black box tool.

In this work, we aim to enhance the recognition efficiency of logo images by augmenting the classification stage. Here, we evaluate the performance of an ensemble of three classifiers, each trained on different feature sets extracted from three shape description techniques. Three promising shape desciptors, including Zernike moments, generic Fourier descriptors (GFD), and shape signature based on centroid distance are used to extract an informative set of features from logo images. Then, each set of features is fed into a base classifier and fused by the Demspter-Shafer based combination method. The classification results of the individual classifiers are compared with those obtained from fusing the classifiers' output. The experimental results show that this strategic combination of shape description techniques can significantly improve the recognition accuracy.

The contribution of this work is two-fold: (1) the application of the ensemble approach to address a challenging image vision classification problem; (2) improving the recognition performance by utilizing a combination strategy that is appropriate for fusing different sources of information. This strategy can effectively make use of diversity of base classifiers trained on different set of features.

The rest of this work is organized as follows: Section 2 first provides a brief review of the ensemble classification approaches and then explains the Dempster-Shafer fusion of ensemble classifiers. In Section 3, the proposed logo classification framework is explained in detail. The experimental results on a well-know logo dataset are reported in Section 4. Finally, Section 5 states the conclusions of the paper.

2   Multiple Classifier Systems

Combining multiple classifiers to achieve higher accuracy is one of the most active research areas in the machine learning community [7]. It is known under various names, such as multiple classifier systems, classifier ensemble, committee of classifiers, mixture of experts, and classifier fusion. Multiple classifier systems can generate more accurate classification results than each of the individual classifiers [22]. In such systems, the classification task can be solved by integrating different classifiers, leading to better performance. However, the ensemble approach depends on the assumption that single classifiers' errors are uncorrelated, which is known as classifier diversity. The intuition is that if each classifier makes different errors, then the total errors can be reduced by an appropriate combination of these classifiers.

The design process of a multiple classifier system generally involves two steps [22]: the collection of an ensemble of classifiers and the design of the combination rule. These steps are explained in detail in the next subsections.

2.1 Creating an Ensemble of Classifiers

There are three general approaches to creating an ensemble of classifiers in state-of-the-art research, which can be considered as different ways to achieve diversity. The most straightforward approach is using different learning algorithms for the base classifiers or variations of the parameters of the base classifiers e.g. different initial weights or different topologies of a series of neural network classifiers. Another approach, which has been getting more attention in the related literature, is to use different training sets to train base classifiers. Such sets are often obtained from the original training set by resampling techniques, such as the procedures presented in Bagging and AdaBoost [10].

The third approach, which is employed in this work for classification of logo images, is to train the individual classifiers with datasets that consist of different feature subsets, or so-called ensemble feature selection [21]. While traditional feature selection algorithms seek to find an optimal subset of features, the goal of ensemble feature selection is to find different feature subsets to generate accurate and diverse classifiers. The Random subspace method (RMS) proposed by Hu in [12] is one early algorithm that builds an ensemble by randomly choosing the feature subsets. More recently, different techniques based on this approach have been proposed.

2.2 Design of a Combination Rule

Once a set of classifiers are generated, the next step is to construct a combination function to merge their outputs, which is also called decision optimization. The most straightforward strategy is the simple majority voting, in which each classifier votes on the class it predicts, and the class receiving the largest number of votes is the ensemble decision. Other strategies for combination function include weighted majority voting, sum, product, maximum and minimum, fuzzy integral, decision templates, and the Dempster-Shafer (DS) based combiner [16],[17].

Inspired by the Dempster-Shafer (DS) theory of evidence [6], a combination method is proposed in [24], which is commonly known as the Dempster-Shafer fusion method. By interpreting the output of a classifier as a measure of evidence provided by the source that generated the training data, the DS method fuses an ensemble of classifiers. Here, we skip the details of how this originated from DS theory and will explain the DS fusion algorithm in the following subsection.

Dempster-Shafer Fusion Method. Let $x \in R^n$ be a feature vector and $\Omega = \{\omega_1, \omega_2, \ldots, \omega_c\}$ be the set of class labels. Each classifier h_i in the ensemble $H = \{h_1, h_2, \ldots, h_L\}$ outputs c degrees of support. Without loss of generality, we can assume that all c degrees are in the interval $[0, 1]$. The support that classifier h_i, gives to the hypothesis that $\mathbf{x}$ comes from class ω_j is denoted by $d_{i,j}(x)$. Clearly, the larger the support, the more likely the class label ω_j. The L classifier outputs for a particular instance $\mathbf{x}$ can be organized in a decision profile, $DP(x)$, as the following matrix [17]:

$$DP(x) = \begin{pmatrix} d_{1,1}(x) & \cdots & d_{1,j}(x) & \cdots & d_{1,c}(x) \\ \vdots & & \vdots & & \vdots \\ d_{i,1}(x) & \cdots & d_{i,j}(x) & \cdots & d_{i,c}(x) \\ \vdots & & \vdots & & \vdots \\ d_{L,1}(x) & \cdots & d_{L,j}(x) & \cdots & d_{L,c}(x) \end{pmatrix}$$

The Dempster-Shafer fusion method uses decision profile to find the overall support for each class and subsequently labels the instance $\mathbf{x}$ in the class with the largest support. In order to obtain the ensemble decision based on DS fusion method, first, the c decision templates, $DT_1, \ldots, DT_c$, are built from the training data. Roughly speaking, decision templates are the most typical decision profile for each class ω_j. For each test sample, $\mathbf{x}$, the DS method compare the decision profile, $DP(x)$, with decision templates. The closest match will label $\mathbf{x}$. In order to predict the target class of each test sample, the following steps are performed [17][24]:

1. Build Decision Templates: For $j = 1, \ldots, c$, calculate the means of the decision profiles for all training samples belonging to ω_j. Call the mean a decision template of class ω_j, DT_j.

$$DT_j = \frac{1}{N_j} \sum_{z_k \in \omega_j} DP(z_k) \tag{1}$$

where N_j in the number of training samples belong to ω_j.

2. Calculate the Proximity: Let DT_j^i denote the ith row of the decision template DT_j, and D_i the output of the ith classifier, that is, the ith row of the decision profile $DP(x)$. Instead of similarity, we now calculate proximity Φ, between DT_j^i and the output of classifier D_i for the test sample x:

$$\Phi_{j,i}(x) = \frac{(1 + \|DT_j^i - D_i(x)\|)^{-1}}{\sum_{k=1}^{c} (1 + \|DT_j^i - D_i(x)\|)^{-1}} \tag{2}$$

where $\|.\|$ is a matrix norm.

3. Compute Belief Degrees: Using Eq. (2), calculate for each class $j = 1, \ldots, c$ and for each classifier $i = 1, \ldots, L$, the following belief degrees, or evidence, that the ith classifier is correctly identifying sample $\mathbf{x}$ into class ω_j:

$$b_j(D_i(x)) \frac{\Phi_{j,i}(x) \prod_{k \neq j}(1 - \Phi_{k,i}(x))}{1 - \Phi_{j,i}(x)[1 - \prod_{k \neq j}(1 - \Phi_{k,i}(x))]} \tag{3}$$

4. Final Decision Based on Class Support: Once the belief degrees are achieved for each source (classifier), they can be combined by Dempster's rule of combination, which simply states that the evidences (belief degree) from each source should be multiplied to obtain the final support for each class:

$$\mu_j(x) = K \prod_{i=1} b_j(D_i(x)), \; j = 1, \ldots, c \tag{4}$$

where K is a normalizing constant ensuring that the total support for ω_j from all classifiers is 1. The DS combiner gives a preference to class with largest $\mu_j(x)$.

3   Framework of the Proposed Logo Recognition System

As mentioned earlier, this work focuses on the second step of logo analysis: logo recognition. The problems of image segmentation and logo detection are beyond the scope of this work. Figure 1 shows the framework of our logo recognition system. In the followings, we describe the main phases of the framework.

The logo image database we used is the MPEG7 dataset[1]. This dataset consists of $C = 70$ classes with 20 instances per class, which represents a total of 1400 object images. Figure 2 shows a few of samples for some categories of this dataset. This dataset has been widely used as the benchmark dataset for logo classification and retrieval [9], [19], [23].

3.1   Preprocessing

An effective classification system should be invariant to the translation, rotation, and scaling (TRS) of logo images. Generally, there are two approaches to achieve the invariance property. The first one is to use shape descriptors that are naturally invariant to TRS. The second approach is to employ some preprocessing steps before using shape description techniques in order to provide TRS invariance.

Here, we used three shape description techniques: shape signature based on centroid distance, Zernike moments, and generic Fourier descriptor (GFD). The shape signature descriptor has the desirable properties of translation, rotation, and scaling invariance. However, the Zernike moments are invariant only to the rotation, and are not invariant to scaling and translation. Similarly, generic

---

[1] MPEG7 Repository dataset: `http://www.cis.temple.edu/~latecki/`

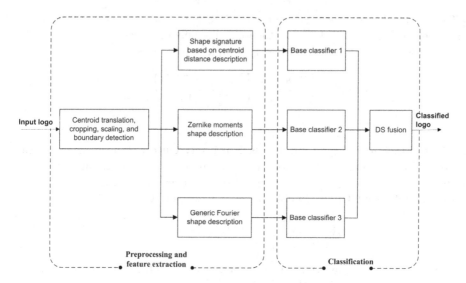

Fig. 1. The framework of the proposed logo classification system based on the Dempster-Shafer fusion of multiple classifiers

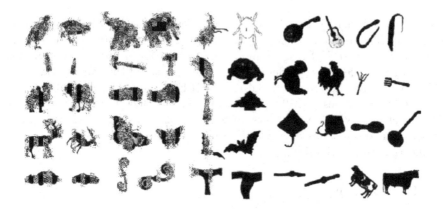

Fig. 2. Some examples of labeled images in the MPEG7 dataset

Fourier descriptor is not natural translation invariant. Therefore, for effective usage of Zernike moments and generic Fourier descriptor in the logo classification framework, the input images need to be normalized for scale and translation.

In the preprocessing phase, translation invariance is achieved by finding the geometrical centroid, (x_0, y_0), of the image and shifting the origin to the centroid of every image. For scale invariance, we create a circular image by superposing a circle centred at the geometrical centroid, with a radius equal to the distance between the centroid and the outermost pixel of the logo image. Finally, we scale the circular image to a square of size 256×256 pixels.

3.2 Feature Extraction of Logo Images

Some researchers have studied the problem of logo recognition by applying different feature extraction methods such as algebraic and differential invariants [8],[14], edge direction histogram [4],[14], Zernike and pseudo-Zernike moments [15],[18], string-matching techniques [5], template matching [14], and wavelet features [20].

In this work, we employed three different image description techniques:

- **Shape Signatures Based on Centroid Distance:** A shape signature, $z(u)$, is a 1-D function representing 2-D areas or boundaries, which can be a unique descriptor of a shape. Shape signatures are mostly used as an input vector to the Fourier Descriptor (FD). Zhang and Lu studied different FD methods for image retrieval and showed that FDs derived from centroid distance perform better than FDs derived from other shape signatures in terms of overall performance [27]. Thus, we used centroid distance-based shape signatures in this work.
- **Zernike Moments (ZM):** ZMs are observed to outperform many moment-based shape descriptors, such as geometric moments, Legendre moments, and pseudo-ZMs [28]. The superiority of ZMs is mainly due to the fact that their basis functions are orthogonal. Therefore, Zernike moments can describe an image with no redundancy or overlap of information between the moments [13]. Here, logo images are mapped onto a set of complex Zernike polynomials and the first 4-order Zernike moments are computed. The reader is referred to [13] for a more detailed description of the ZM computation.
- **Generic Fourier Descriptors (GFD):** The GFD is extracted from spectral domain by applying the 2D Fourier transform on polar-raster sampled shape images [26]. The process of employing GFD is similar to the conventional FD:

$$GFD(\rho, \phi) = \sum_r \sum_i f(r, \theta_i) exp[-j2\pi(\frac{r}{R}\rho + \frac{2\pi i}{T}\phi)]$$

where $0 \leq r \leq R$ and $\theta_i = i(2\pi/T)(0 \leq i \leq T)$; $0 \leq \rho \leq R$, $0 \leq \phi \leq T$. R and T are the radial and angular resolutions, respectively and $f(x, y)$ is the binary image function [26].

4 Classification Results

In this stage, we aim to classify different logo images based on DS fusion of individual classifiers. The classification performance is obtained by means of stratified 10-fold cross-validation over 10 runs. In order to improve the reliability of the results, the experiments are conducted using different numbers of classes, i.e. different numbers of logo categories and using two classification algorithms, including 1) Support Vector Machine with the Gaussian Kernel and 2) Multilayer Perceptron (MLP) with 10 nodes in the hidden layer. For SVM implementation, we use the LIBSVM package (version 3.1) developed by Chang and Lin [3], tuning Kernel parameters via cross-validation.

The summary of the results are reported in Table 1 and Table 2 for SVM and MLP as the base learners. These tables show the classification accuracy of individual classifiers and the one achieved by the Dempster-Shafer fusion of them.

Table 1. Classification accuracy of single and fused classifiers using SVM as the base learner

| # classes | Single classifier trained only on | | | DS fusion |
	GFD	Zernike moments	Shape signature	
10	98.30	98.50	97.20	99.20
20	95.65	96.10	95.20	98.15
30	94.33	93.23	92.30	96.47
40	93.93	93.73	91.65	96.78
50	92.66	91.94	90.36	95.72
60	92.43	92.50	90.30	96.07
70	91.79	91.57	89.36	95.37

Table 2. Classification accuracy of single and fused classifiers using MLP as the base learner

| # classes | Single classifier trained only on | | | DS fusion |
	GFD	Zernike moments	Shape signature	
10	89.50	94.00	90.90	98.70
20	81.65	84.95	79.30	96.60
30	66.80	67.13	63.43	93.57
40	54.03	55.58	53.85	90.55
50	47.98	46.86	45.16	88.44
60	40.97	40.52	39.72	86.30
70	34.59	34.84	35.64	84.91

It is important to note the outperformance of the fused results in comparison with the individual classifier. This improvement is clearer when the number of classes of the datasets is increased. In that case, the inter-class variability is reduced, and thus, it is easier to confuse patterns from different classes.

As an additional analysis, we compare classification results of merging classifiers by different combination methods. Figure 3 and Figure 4 show the classification accuracy of individual classifiers and ensemble systems by different fusion methods. Considered combination methods include fusion by majority voting, maximum, sum, minimum, average, naive-Bayes, and the Dempster-Shafer fusion method.

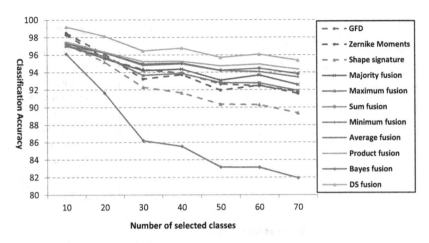

Fig. 3. Classification accuracy of single and fused classifiers by different combination methods using SVM as the base learner

As these figure show, the best classification accuracy is achieved by means of the Dempster-Shafer fusion method. The general results for different numbers of classes are summarized below:

- In these experiments, the three descriptors used show similar performance in terms of classification accuracy on the MPEG-7 dataset.
- The classification results reveal the critical role of the combination method. As Figure 3 and 4 show, only using diverse classifiers is not enough to improve the classification performance of the ensemble system. If the combination method does not properly make use of the ensemble diversity, then no benefit arises from fusing multiple classifiers [2]. For example, the commonly used majority voting combination method does not make significant use of the diversity among ensemble classifiers in these experiments. Therefore, the classification accuracy obtained by fusing the classifiers' outputs can be even worse than the one achieved by single classifiers trained only with one set of shape descriptors. On the other hand, the Dempster-Shafer fusion method has significantly improved the classification performance.
- The ensemble system, even by using a poor fusion method, generally per- forms better than the base classifiers. This finding confirms the philosophy of the ensemble systems: combining the outputs of several learners can reduce the risk of an unfortunate selection of a poorly performing learner.
- The MLP classification accuracy of individual classifiers dramatically de- creases as the number of classes increase. This finding is mainly due to the fact that MLP classifiers solve the whole multi-class classification problem concurrently. Therefore, it is more difficult to separate a large number of classes. However, the case for SVM is different. The SVM algorithm solves the multi-class problem by decomposing it into several smaller binary prob- lems using the one-versus-one scheme. It has been shown that this approach,

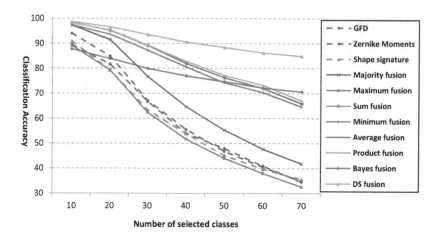

Fig. 4. Classification accuracy of single and fused classifiers by different combination methods using MLP as the base learner

known as class binarization, achieves better classification performance compared to the approach that aims to solve the whole multiclass problem at once [11].

5   Conclusions

In this work, we evaluated the performance of an ensemble of three classifiers, each trained on different feature sets. Three efficient shape description methods, including shape signature, Zernike moments, and generic Fourier descriptors, were used to extract informative features from logo images and each set of features was fed into an individual classifier. In order to reduce recognition error, the Dempster-Shafer combination theory was employed to fuse the three classifiers trained on different sources of information. The classification results of the individual classifiers were compared with those obtained from fusing the classifiers by the Dempster-Shafer combination method.

Generally speaking, using ensemble methods for the classification of logo images is effective, though different combination methods would show different performances, and even some combination of base classifiers and ensemble methods would deteriorate the performance of the best single classifier. However, as demonstrated by our experiments, by using the DS fusion method, the classification performance was significantly increased compared with single classifiers trained by a specific set of features.

Acknowledgment. The authors would like to thank Ms. Fatemeh Yazdanpanah for her kind help in conducting experiments.

References

1. Alajlan, N., Kamel, M., Freeman, G.: Multi-object image retrieval based on shape and topology. Signal Processing: Image Communication 21(10), 904–918 (2006)
2. Brodley, A., Lane, T.: Creating and exploiting coverage and diversity. In: Proceedings of AAAI 1996 Workshop on Integrating Multiple Learned Models, Portland, OR, pp. 8–14 (1996)
3. Chang, C.C., Lin, C.J.: Libsvm: A library for support vector machines. ACM Transactions on Intelligent Systems and Technology 2(3), 1–27 (2011)
4. Ciocca, G., Schettini, R.: Content-based similarity retrieval of trademarks using relevance feedback. Pattern Recognition 34(8), 1639–1655 (2001)
5. Cortelazzo, G., Mian, G., Vezzi, G., Zamperoni, P.: Trademark shapes description by string-matching techniques. Pattern Recognition 27(8), 1005–1018 (1994)
6. Dempster, A.: Upper and lower probabilities induced by multivalued mappings. Annals of Mathematical Statistics 38(2), 325–339 (1967)
7. Dietterich, T.G.: Machine learning research: Four current directions. Artificial Intell. Mag. 18(4), 97–136 (1997)
8. Doermann, D., Rivlin, E., Weiss, I.: Applying algebraic and differential invariants for logo recognition. Machine Vision and Applications 9(2), 73–86 (1996)
9. Escalera, S., Fornés, A., Pujol, O., Lladós, J., Radeva, P.: Circular blurred shape model for multiclass symbol recognition. IEEE Transactions on Systems, Man, and Cybernetics, Part B: Cybernetics 41(2), 497–506 (2011)
10. Freund, Y., Schapire, R.: A decision-theoretic generalization of on-line learning and an application to boosting. Journal of Computer and System Sciences 55(1), 119–139 (1997)
11. García-Pedrajas, N., Ortiz-Boyer, D.: An empirical study of binary classifier fusion methods for multiclass classification. Information Fusion 12(2), 111–130 (2011)
12. Ho, T.K.: The random subspace method for constructing decision forests. IEEE Transactions on Pattern Analysis and Machine Intelligence 20, 832–844 (1998)
13. Hwang, S., Kim, W.: A novel approach to the fast computation of zernike moments. Pattern Recognition 39(11), 2065–2076 (2006)
14. Jain, A., Vailaya, A.: Shape-based retrieval: A case study with trademark image databases. Pattern Recognition 31(9), 1369–1390 (1998)
15. Kim, Y., Kim, W.: Content-based trademark retrieval system using a visually salient feature. Image and Vision Computing 16(12-13), 931–939 (1998)
16. Kittler, J., Hatef, M., Duin, R., Matas, J.: On combining classifiers. IEEE Transactions on Pattern Analysis and Machine Inteligence 20(3), 226–239 (1998)
17. Kuncheva, L.I.: Combining Pattern Classifiers: Methods and Algorithms. Wiley, New York (2004)
18. Li, S., Lee, M., Pun, C.: Complex zernike moments features for shape-based image retrieval. IEEE Transactions on Systems, Man and Cybernetics, Part A: Systems and Humans 39(1), 227–237 (2009)
19. Mohd Anuar, F., Setchi, R., Lai, Y.K.: Trademark image retrieval using an integrated shape descriptor. Expert Systems with Applications (2012), http://dx.doi.org/10.1016/j.eswa.2012.07.031
20. Neumann, J., Samet, H., Soffer, A.: Integration of local and global shape analysis for logo classification. Pattern Recognition Letters 23, 1449–1457 (2002)
21. Opitz, D.W.: Feature selection for ensembles (1999)
22. Polikar, R.: Ensemble based systems in decision making. IEEE Circuits and Systems Magazine 6(3), 21–45 (2006)

23. Qi, H., Li, K., Shen, Y., Qu, W.: An effective solution for trademark image retrieval by combining shape description and feature matching. Pattern Recognition 43(6), 2017–2027 (2010)
24. Rogova, G.: Combining the results of several neural network classifiers. Neural Networks 7, 777–781 (1994)
25. Seiden, S., Dillencourt, M., Irani, S., Borrey, R., Murphy, T.: Logo detection in document images. In: International Conference on Imaging Science, Systems, and Technology, Las Vegas, Nevada, pp. 446–449 (1997)
26. Zhang, D., Lu, G.: Shape-based image retrieval using generic fourier descriptor. Signal Processing: Image Communication 17(10), 825–848 (2002)
27. Zhang, D., Lu, G.: Study and evaluation of different fourier methods for image retrieval. Image and Vision Computing 23(1), 33–49 (2005)
28. Zhang, D., Lu, G.: Review of shape representation and description techniques. Pattern Recognition 37(1), 1–19 (2004)

d-Separation: Strong Completeness of Semantics in Bayesian Network Inference

Cory J. Butz[1], Wen Yan[1], and Anders L. Madsen[2,3]

[1] Department of Computer Science, University of Regina, Canada
{butz,yanwe111}@cs.uregina.ca
[2] HUGIN EXPERT A/S, Aalborg, Denmark
anders@hugin.com
[3] Aalborg University, Department of Computer Science, Denmark

Abstract. It is known that d-separation can determine the minimum amount of information needed to process a query during exact inference in discrete Bayesian networks. Unfortunately, no practical method is known for determining the semantics of the intermediate factors constructed during inference. Instead, all inference algorithms are relegated to denoting the inference process in terms of potentials. In this theoretical paper, we give an algorithm, called *Semantics in Inference* (SI), that uses d-separation to denote the semantics of every potential constructed during inference. We show that SI possesses four salient features: polynomial time complexity, soundness, completeness, and strong completeness. SI provides a better understanding of the theoretical foundation of Bayesian networks and can be used for improved clarity, as shown via an examination of Bayesian network literature.

1 Introduction

In [12], Pearl advocated the restoration of probabilistic methods in artificial intelligence systems and explored the possibility of representing and manipulating probabilistic knowledge in graphical forms, latter called Bayesian networks. When recounting the development of Bayesian networks, Pearl [14] states that perhaps [12] made its greatest immediate impact through the notion of *d-separation*. As a method for deciding which conditional independence relations are implied by the directed acyclic graph of a Bayesian network, d-separation provides the semantics needed for defining and characterizing Bayesian networks. Observe that Pearl emphasizes the importance of d-separation with respect to Bayesian network modeling. With respect to inference, Pearl only states that d-separation can determine the minimum information needed for answering a query posed to a Bayesian network. No claim has ever been made that d-separation can also provide semantics during Bayesian network inference.

Koller and Friedman [8] state that it is interesting to consider the semantics of the potentials constructed during inference. They mention that sometimes the probabilities are defined with respect to the joint distribution, but not at other times. As no practical algorithm exists for deciding the semantics of inference, all inference algorithms denote the intermediate factors constructed during inference

O. Zaïane and S. Zilles (Eds.): Canadian AI 2013, LNAI 7884, pp. 13–24, 2013.

as potentials. Potentials have no constraints [8] meaning they do not have clear physical interpretation [4].

In this theoretical paper, we present *Semantics in Inference* (SI), an algorithm for denoting semantics during exact inference in discrete Bayesian networks. SI works by introducing the notion of evidence normal form to organize how each potential was constructed. SI then decides semantics of the potential by performing one d-separation test. Formal properties of the SI algorithm are obtained, namely, polynomial time complexity, soundness, completeness, and strong completeness. SI can be utilized for clarity of exposition in Bayesian network literature, since the semantics of potentials can now be articulated.

2   Inference

Here we consider only discrete Bayesian networks. $U = \{v_1, v_2, \ldots, v_n\}$ is a finite set of random variables and each $v_i \in U$ can take a value from a finite domain, $dom(v_i)$. Given $X \subseteq U$, $dom(X)$ is the Cartesian product of $dom(v_i)$, $v_i \in X$. A *potential* on $dom(X)$ is a function ψ on $dom(X)$ such that $\psi(x) \geq 0$ for each $x \in dom(X)$, and at least one $\psi(x)$ is positive. For brevity, we refer to ψ as a mapping on X rather than $dom(X)$. A potential p on U that sums to 1 is called a *joint probability distribution* on U, denoted $p(U)$. A *conditional probability table* (CPT) for X given disjoint Y, denoted $\psi(X|Y)$, is a potential on XY that sums to 1, for each configuration $y \in dom(Y)$. The *unity-potential* $1(v_i)$ for v_i is a function 1 mapping every element of $dom(v_i)$ to one. The unity-potential for a non-empty set $X = \{v_1, v_2, \ldots, v_k\}$ of variables, denoted $1(X)$, is defined as $1(X) = 1(v_1) \cdot 1(v_2) \cdots 1(v_k)$. For simplified notation, we may write $\{v_1, v_2, \ldots, v_k\}$ as $v_1, v_2, \ldots, v_k$.

A *Bayesian network* [13] is a pair (B, C). B denotes a directed acyclic graph with vertex set U and C is a set of *conditional probability tables* (CPTs) $\{p(v_i | P(v_i)) \mid i = 1, 2, \ldots, n\}$, where $P(v_i)$ denotes the parents (immediate predecessors) of $v_i \in B$. The product of CPTs in C is a joint probability distribution $p(U)$. For example, the directed acyclic graph in Figure 1 is called the *extended student Bayesian network* (ESBN) [8]. We give CPTs in Table 1, where only binary variables are used in examples, and probabilities not shown can be obtained by definition. By the above,

$$p(U) \; = \; p(c) \cdot p(d|c) \cdot p(i) \cdot p(g|d, i) \cdots p(h|g, j). \tag{1}$$

We say X and Z are *conditionally independent* [16] given Y in $p(U)$, denoted $I_p(X, Y, Z)$, if given any $x \in dom(X)$, $y \in dom(Y)$, for all $z \in dom(Z)$: $p(x|y, z) = p(x|y)$, whenever $p(y, z) > 0$, where $X, Y, Z \subseteq U$.

Pearl [12] gave a method, called d-separation, for determining those independencies encoded in a directed acyclic graph. The following is the definition of d-separation based on [8]. In a Bayesian network B, a trail (an undirected path) $v_1, v_2, \ldots, v_n$ is *active* given Y, if: (i) whenever we have a v-structure $v_{i-1} \rightarrow v_i \leftarrow v_{i+1}$, then v_i or one of its descendants are in Y; (ii) no other node along the trail is in Y. Note that if v_1 or v_n are in Y the trail is not active.

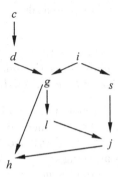

Fig. 1. The directed acyclic graph of ESBN

Table 1. CPTs for the ESBN in Figure 1

c	$p(c)$		c	d	$p(d\|c)$		d	i	g	$p(g\|d,i)$
0	0.20		0	0	0.40		0	0	0	0.90
			1	0	0.70		0	1	0	0.20
i	$p(i)$						1	0	0	0.50
0	0.75		g	l	$p(l\|g)$		1	1	0	0.40
			0	0	0.30					

g	j	h	$p(h\|g,j)$		i	s	$p(s\|i)$		s	l	j	$p(j\|s,l)$
0	0	0	0.25		0	0	0.40		0	0	0	0.10
0	1	0	0.65		1	0	0.80		0	1	0	0.60
1	0	0	0.50						1	0	0	0.45
1	1	0	0.85						1	1	0	0.50

We say that X and Z are *d-separated* given Y in B, denoted $I_B(X,Y,Z)$, if there is no active trail between any variable $v \in X$ and $v' \in Z$ given Y.

In inference, $p(X|E = e)$ is the most common query type, which are useful for many reasoning patterns, including explanation, prediction, intercausal reasoning, and many more [8]. Here, X and E are disjoint subsets of U, and E is observed taking value e. We describe a basic algorithm for computing $p(X|E = e)$, called *variable elimination* (VE), first put forth in [17]. We do not consider alternative approaches to inference such as conditioning [6] and join tree propagation [1,2,10]. Inference involves the elimination of variables. Algorithm 1, called *sum-out* (SO), eliminates a single variable v from a set Φ of *potentials* [8], and returns the resulting set of potentials. The algorithm *collect-relevant* simply returns those potentials in Φ involving variable v.

Algorithm 1. SO(v,Φ)
Ψ = collect-relevant(v,Φ)
ψ = the product of all potentials in Ψ
$\tau = \sum_v \psi$
return $(\Phi - \Psi) \cup \{\tau\}$

SO uses Lemma 1, which means that potentials not involving the variable being eliminated can be ignored.

Lemma 1. [15] *If ψ_1 is a potential on W and ψ_2 is a potential on Z, then the marginalization of $\psi_1 \cdot \psi_2$ onto W is the same as ψ_1 multiplied with the marginalization of ψ_2 onto $W \cap Z$, where $W, Z \subseteq U$.*

The *evidence potential* for $E = e$, denoted $1(E = e)$, assigns probability 1 to the single value e of E and probability 0 to all other values of E. Hence, for a variable v observed taking value λ and $v \in \{v_i\} \cup P(v_i)$, the product $p(v_i|P(v_i)) \cdot 1(v = \lambda)$ keeps only those configurations agreeing with $v = \lambda$.

Algorithm 2, taken from [8], computes $p(X|E = e)$ from a discrete Bayesian network B. VE calls SO to eliminate variables one by one. More specifically, in Algorithm 2, Φ is the set C of CPTs for B, X is a list of query variables, E is a list of observed variables, e is the corresponding list of observed values, and σ is an elimination ordering for variables $U - XE$, where XE denotes $X \cup E$.

Algorithm 2. VE(Φ, X, E, e, σ)
Multiply evidence potentials with appropriate CPTs
While σ is not empty
    Remove the first variable v from σ
    Φ = sum-out(v, Φ)
$p(X, E = e)$ = the product of all potentials $\psi \in \Phi$
return $p(X, E = e)/\sum_X p(X, E = e)$

As in [8], suppose the observed evidence for the ESBN is $i = 1$ and $h = 0$ and the query is $p(j|h = 0, i = 1)$. The weighted-min-fill algorithm [8] can yield $\sigma = (c, d, l, s, g)$. VE first incorporates the evidence:

$$\psi(i = 1) = p(i) \cdot 1(i = 1),$$
$$\psi(d, g, i = 1) = p(g|d, i) \cdot 1(i = 1),$$
$$\psi(i = 1, s) = p(s|i) \cdot 1(i = 1),$$
$$\psi(g, h = 0, j) = p(h|g, j) \cdot 1(h = 0).$$

To eliminate c, the SO algorithm computes

$$\psi(d) = \sum_c p(c) \cdot p(d|c).$$

SO computes the following to eliminate d

$$\psi(g, i = 1) = \sum_d \psi(d) \cdot \psi(d, g, i = 1).$$

To eliminate l,

$$\psi(g, j, s) = \sum_l p(l|g) \cdot p(j|l, s).$$

SO computes the following when eliminating s,

$$\psi(g, i = 1, j) = \sum_s \psi(i = 1, s) \cdot \psi(g, j, s). \tag{2}$$

For g, SO can compute:

$$\sum_g \psi(g, i = 1, j) \cdot \psi(g, i = 1) \cdot \psi(g, h = 0, j)$$

$$= \sum_g \psi(g, i = 1, j) \cdot \psi(g, h = 0, i = 1, j) \tag{3}$$

$$= \psi(h = 0, i = 1, j).$$

Next, VE multiplies all remaining potentials as

$$p(h = 0, i = 1, j) = \psi(i = 1) \cdot \psi(h = 0, i = 1, j).$$

Finally, VE answers the query by

$$p(j|h = 0, i = 1) = \frac{p(h = 0, i = 1, j)}{\sum_j p(h = 0, i = 1, j)}.$$

3 Understanding Semantics

We review the current limited understanding of semantics in inference.

Kjaerulff and Madsen [7] suggest that in working with probabilistic networks it is convenient to denote distributions as potentials. In fact, the use of potentials is built into the standard inference algorithms (see the SO and VE algorithms, for instance). For example, suppose query $p(j)$ is posed to the ESBN [8]. Even without evidence being considered, the initial step of VE is to regard CPTs as potentials, i.e., $p(U)$ is factorized as

$$p(U) = \psi(c) \cdot \psi(c, d) \cdot \psi(i) \cdots \psi(g, h, j). \tag{4}$$

By comparing (1) and (4), it is clear that semantics are destroyed even before the CPTs in computer memory are modified. The notation used for potentials does not convey the semantic meaning of the probabilities comprising the potential.

Darwiche [6] ascribes meaning during inference by representing each potential by what we will call evidence expanded form, except that products involving evidence potentials are taken. Let ψ be any potential constructed by VE. The *evidence expanded form* of ψ, denoted $F(\psi)$, is the unique expression defining how ψ was built using the multiplication and marginalization operators on the Bayesian network CPTs together with any appropriate evidence potentials.

For example, consider potential $\psi(g, i = 1, j)$ in (2). $F(\psi(g, i = 1, j))$, the evidence expanded form, can be easily obtained in a recursive manner as follows:

$$\sum_s \psi(i = 1, s) \cdot \psi(g, j, s)$$

$$= \sum_s \psi(i = 1, s) \cdot \left(\sum_l (p(l|g) \cdot p(j|l, s))\right)$$

$$= \sum_s ((p(s|i) \cdot 1(i = 1)) \cdot \left(\sum_l (p(l|g) \cdot p(j|l, s))\right)). \tag{5}$$

Henceforth, parentheses are understood and may not be shown. Unfortunately, the expanded form by itself does not directly articulate semantics.

By semantics, we mean that a CPT $\psi(X|Y)$ constructed by VE's manipulation of Bayesian network CPTs is not necessarily equal to the CPT $p(X|Y)$ obtained from the defined joint probability distribution $p(U)$. For instance, it can be verified that in the ESBN,

$$p(h|g, j) \cdot \sum_d p(g|d, i) \cdot \sum_c p(c) \cdot p(d|c) \tag{6}$$

produces the CPT $\psi(g, h|i, j)$ in Table 2 (left). In contrast, the CPT $p(g, h|i, j)$ built from the joint distribution $p(U)$ in (1) is shown in Table 2 (right).

Table 2. (left) CPT $\psi(g, h|i, j)$ built by (6). (right) CPT $p(g, h|i, j)$ built from $p(U)$ in (1).

i	j	g	h	$\psi(g, h|i, j)$	i	j	g	h	$p(g, h|i, j)$
0	0	0	0	0.1890	0	0	0	0	0.1960
0	0	0	1	0.5670	0	0	0	1	0.5880
0	0	1	0	0.1220	0	0	1	0	0.1080
0	1	0	0	0.4914	0	1	0	0	0.4762
0	1	0	1	0.2646	0	1	0	1	0.2564
0	1	1	0	0.2074	0	1	1	0	0.2272
1	0	0	0	0.0680	1	0	0	0	0.0846
1	0	0	1	0.2040	1	0	0	1	0.2537
1	0	1	0	0.3640	1	0	1	0	0.3309
1	1	0	0	0.1768	1	1	0	0	0.1518
1	1	0	1	0.0952	1	1	0	1	0.0817
1	1	1	0	0.6188	1	1	1	0	0.6515

Semantics in inference are not well understood. In their comprehensive and highly recommended text, Koller and Friedman [8] consider the semantics of potential $\psi(b, c, d)$ built by eliminating variable a from the Bayesian network B in Figure 2 (left):

$$\psi(b, c, d) = \sum_a p(a) \cdot p(b|a) \cdot p(d|a, c). \tag{7}$$

Koller and Friedman [8] incorrectly state

$$p(b, d|c) \neq \psi(b, c, d). \tag{8}$$

While this claim is almost always true, there are a few exceptions to refute it. For one counter-example, eliminating variable a using the CPTs in Table 3 yields:

$$p(b, d|c) = \psi(b, c, d). \tag{9}$$

Koller and Friedman [8] also state it must necessarily be the case that

$$p'(b, d|c) = \psi(b, c, d), \tag{10}$$

where $p'(U)$ is defined by a *different* Bayesian network B' - the one given in Figure 2 (right). Our objective is to stipulate semantics in the *current* Bayesian network B - the one on which inference is being conducted.

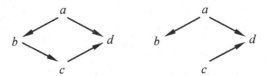

Fig. 2. Bayesian networks B (left) and B' (right)

Table 3. Exceptional CPTs for B in Figure 2 (left)

a	$p(a)$	a	b	$p(b\|a)$	b	c	$p(c\|b)$	a	c	d	$p(d\|a,c)$
0	0.2	0	0	0.4	0	0	0.5	0	0	0	0.5
		1	0	0.9	1	0	0.5	0	1	0	0.5
								1	0	0	0.5
								1	1	0	0.5

4   CPT Structure

It is instructive to review that, when evidence is not considered, each potential built by VE is a CPT.

A *topological ordering* [8] is an ordering $\prec$ of the variables in a Bayesian network B so that for every arc (v_i, v_j) in B, v_i precedes v_j in $\prec$. For example, $c \prec d \prec i \prec g \prec s \prec l \prec j \prec h$ is a topological ordering of the directed acyclic graph in Figure 1, but $d \prec c \prec i \prec g \prec h \prec l \prec j \prec s$ is not.

Recall this feature of Bayesian networks,

$$p(U) = \prod_{v_i \in U} p(v_i | P(v_i)).$$

This can be established by showing

$$1 = \sum_{U} \prod_{v_i \in U} p(v_i | P(v_i)).$$

More generally, we have the following two lemmas.

Lemma 2. [3] *Consider a Bayesian network (B, C) on U. Given any non-empty subset X of U, $\prod_{v_i \in X} p(v_i | P(v_i))$ is a CPT $\psi(X | P(X))$, where $P(X) = (\cup_{v_i \in X} P(v_i)) - X$.*

Lemma 3. [3] *When evidence is not considered, each potential constructed by VE is a CPT.*

Lemma 3 can be seen as first applying Lemma 1 on the evidence expanded form of a potential built by VE, keeping in mind $E = \emptyset$, and then applying Lemma 2.

For example, consider the potential ψ built by (6), which is already in evidence expanded form. By applying Lemma 1,

$$\sum_d \sum_c p(h | g, j) \cdot p(g | d, i) \cdot p(c) \cdot p(d | c). \tag{11}$$

By Lemma 2,

$$\psi(g, h | i, j) \;=\; \sum_d \sum_c \psi(c, d, g, h | i, j),$$

Thus, the potential ψ built by (6) is, in fact, a CPT $\psi(g, h | i, j)$, in Table 2 (left).

5   Denoting Semantics

The evidence expanded form $F(\psi)$ of any potential ψ constructed by VE is in evidence normal form, if $F(\psi)$ is written as

$$\gamma \cdot N,$$

where γ is the product of 1 and all evidence potentials in $F(\psi)$, and N is the same factorization as $F(\psi)$ except without products involving evidence potentials.

Recall $\psi(g, h = 0, i = 1, j)$ in (3). The evidence expanded form $F(\psi)$ is

$$p(h | g, j) \cdot 1(h = 0) \cdot \sum_d p(g | d, i) \cdot 1(i = 1) \cdot \sum_c p(c) \cdot p(d | c), \tag{12}$$

and the evidence normal form $\gamma \cdot N$ is

$$1(h = 0, i = 1) \;\cdot\; p(h | g, j) \cdot \sum_d p(g | d, i) \cdot \sum_c p(c) \cdot p(d | c), \tag{13}$$

namely, $\gamma = 1(h = 0, i = 1)$ and N is (6).

Lemma 4. *The evidence expanded form $F(\psi)$ of any potential ψ constructed by VE always can be equivalently written in normal form, i.e., $F(\psi) = \gamma \cdot N$.*

Proof. Since evidence variables are never marginalized in VE, the claim follows from Lemma 1.

Observe that, by Lemma 3, N in evidence normal form is a CPT. We may denote evidence normal form $\gamma \cdot N$ simply as N with evidence γ understood, since γ only serves to select configurations of N agreeing with the evidence. We now turn to denoting semantics.

To understand when $N = p(X|Y)$ in evidence normal form, some terminology is needed. A *path* from v_1 to v_n is a sequence $v_1, v_2, \ldots, v_n$ with arcs (v_i, v_{i+1}) in B, $i = 1, \ldots, n-1$. With respect to a variable v_i, we define three sets: (i) the ancestors of v_i, denoted $A(v_i)$, are those variables having a path to v_i; (ii) the descendants of v_i, denoted $D(v_i)$, are those variables to which v_i has a path; and, (iii) the children of v_i are those variables v_j such that arc (v_i, v_j) is in B. The ancestors of a set $X \subseteq U$ are defined as $A(X) = (\cup_{v_i \in X} A(v_i)) - X$. The descendants $D(X)$ are defined similarly. $I_B(X, Y, Z)$ means an independence statement $I(X, Y, Z)$ [13] holds in B by d-separation, where $X, Y, Z \subseteq U$.

We now give the *Semantics in Inference* (SI) algorithm, which uses d-separation to denote the semantics of any potential ψ built by VE on B. Each potential ψ constructed by VE is represented in evidence normal form $\psi(X|Y)$. If the semantics of B ensure the $\psi(X|Y) = p(X|Y)$, then ψ is denoted as $p_B(X|Y)$; otherwise, it is denoted as $\phi_B(X|Y)$. S is the set of variables marginalized in $F(\psi)$. $A(XS)$ and $D(XS)$ are computed from the *transitive closure*, denoted T, of B [5].

Algorithm 3. $\text{SI}(\psi)$
Compute the evidence expanded form $F(\psi)$ of ψ
Compute the evidence normal form $\gamma \cdot N$ of $F(\psi)$
Compute the CPT structure $\psi(X|Y)$ of N
Compute $Z = A(XS) \cap D(XS)$
Compute $X_1 = X \cap P(Z)$
if $I_B(X_1, \emptyset, Y)$ holds in B by d-separation
    return $p_B(X|Y)$
else
    return $\phi_B(X|Y)$

Recall $\psi(g, i = 1, j)$ in (2). The evidence expanded form is (5). Its evidence normal form $\gamma \cdot N$ is $\gamma = 1(i = 1)$ and $N = \psi(j|g, i)$. Now $X = \{j\}$, $Y = \{g, i\}$ and $S = \{l, s\}$. By the transitive closure T of the ESBN, $A(XS) = \{c, d, i, g\}$ and $D(XS) = \{h\}$. Hence, $Z = \emptyset$, $P(Z) = \emptyset$, and $X_1 = \emptyset$. Trivially, $I_B(X_1, \emptyset, Y)$ holds. Thus, SI denotes $\psi(g, i = 1, j)$ in (2) as $p_B(j|g, i = 1)$.

Now consider $\psi(g, h = 0, i = 1, j)$ in (3). The evidence expanded form is (12). The evidence normal form $\gamma \cdot N$ is (13). Here $N = \psi(g, h|i, j)$, as seen in (6). With $X = \{g, h\}$, $Y = \{i, j\}$ and $S = \{c, d\}$, from T on the ESBN we have $A(\{c, d, g, h\}) = \{i, j, l, s\}$ and $D(\{c, d, g, h\}) = \{j, l\}$. Thus, $Z = \{j, l\}$, giving $P(Z) = \{g, s\}$ and $X_1 = \{g\}$. Now, $I_B(X_1, \emptyset, Y)$ does not hold. Thereby, SI denotes $\psi(g, h = 0, i = 1, j)$ in (3) as $\phi_B(g, h = 0|i = 1, j)$.

In this example, there is a path from $XS = \{c, d, g, h\}$ to XS through $Z = \{j, l\}$, starting at $X_1 = \{g\}$. With $X_1 = \{g\}$ and $Y = \{i, j\}$, we focus on $I_B(g, \emptyset, ij)$. Note that when deciding semantics of $\psi(X|Y)$, the independence to be tested is $I_B(X_1, \emptyset, Y)$ and not $I_B(XS, Y, A(XSY))$. In Figure 2 (left), $I_B(abd, c, \emptyset)$ holds, but $p(b, d|c) \neq \psi(b, d|c)$ in (8) is possible.

6   Theoretical Foundation

We present four salient features of SI. Only the proofs of time complexity and strong completeness are shown due to space considerations.

Theorem 1. *Let ψ be any potential built by VE during exact inference in a discrete Bayesian network with n variables. Then the time complexity of the SI algorithm to determine the semantics of ψ is $O(n^3)$.*

Proof. As ψ may require $n-1$ multiplications and n marginalizations, computing $F(\psi)$ takes $2n$ steps. The normal form $\gamma \cdot N$ can be decided in linear time, as can the CPT structure $\psi(X|Y)$ of N. The transitive closure T of the directed acyclic graph can be computed in $O(n^3)$ [5]. Let XS be a set of k variables, $1 \leq k \leq n$. Then $A(XS)$ and $D(XS)$ each can be computed in $O(k \cdot n)$. Now Z and X_1 each can be computed in $O(n^2)$. Testing $I_B(X_1, \emptyset, Y)$ is linear in the size of B [6]. Thus, the semantics of ψ can be determined by SI in $O(n^3)$.

Theorem 2. *In a Bayesian network (B, C) defining a joint distribution $p(U)$, suppose VE computes a potential ψ whose evidence normal form is $\gamma \cdot N$. If SI denotes the semantics of N as $p_B(X|Y)$, then $N = p(X|Y)$.*

Theorem 2 guarantees that if SI denotes the semantics of a VE potential ψ as $\gamma \cdot p_B(X|Y)$, then

$$\psi \; = \; \gamma \cdot p(X|Y).$$

Recall potential $\psi(g, i = 1, j)$ in (2). As illustrated in Table 4, Theorem 2 ensures that $\psi(g, i = 1, j)$ is equal to $p(j|g, i = 1)$, since SI denotes it as $p_B(j|g, i = 1)$.

Table 4. Potential $\psi(g, i = 1, j)$ in (2) is $p(j|g, i = 1)$

			i	g	j	$p_B(j|g, i=1)$
			1	0	0	0.457
$\psi(g, i=1, j)$	$=$	$p(j|g, i=1) =$	1	0	1	0.543
			1	1	0	0.334
			1	1	1	0.666

With respect to inference, the question of completeness is this. Can SI determine the semantics of every VE potential defined with respect to the joint distribution? The answer is no.

Theorem 3. *In a Bayesian network B on U, suppose VE computes a potential ψ whose evidence normal form is $\gamma \cdot N$. If SI denotes the semantics of N as $\phi_B(X|Y)$, there exists a set C of CPTs for B defining a joint distribution $p(U)$ such that $N \neq p(X|Y)$.*

Theorem 3 states that whenever SI indicates that a potential is not defined with respect to the joint distribution, then this is true for at least one set of CPTs for the given Bayesian network. Recall once again $\psi(g, h = 0, i = 1, j)$ in (3), which SI denotes as $\phi_B(g, h = 0, l|i = 1, j)$. With respect to $p(U)$ defined by the CPTs in Table 1, we have

$$\psi(g, h = 0, i = 1, j) \neq p(g, h = 0|i = 1, j).$$

However, Theorem 3 can be made significantly stronger.

Lemma 5. [11] *Except for a measure zero set in the space of all joint distributions $p(U)$ defined by all discrete Bayesian networks (B, C), the independencies satisfied by $p(U)$ are precisely those satisfied by d-separation in B.*[1]

Lemma 5 says that for nearly all choices C of CPTs for a Bayesian network B defining $p(U)$, d-separation perfectly characterizes the independencies in $p(U)$, i.e., for $X, Y, Z \subseteq U$,

$$I_p(X, Y, Z) \iff I_B(X, Y, Z).$$

Theorem 4. *Except for a measure zero set in the space of all joint distributions $p(U)$ defined by all discrete Bayesian networks (B, C), for any potential ψ built by VE,*

$$\psi = \gamma \cdot p(X|Y) \iff \text{SI denotes } \psi \text{ as } p_B(X|Y),$$

where $\gamma \cdot N$ is the evidence normal form of ψ.

Proof. ($\Rightarrow$) Suppose VE constructs a ψ whose evidence normal form is $\gamma \cdot N$ and whose semantics are defined with respect to $p(U)$. By contraposition, suppose SI denotes N as $\phi_B(X|Y)$. By SI, $I_B(X_1, \emptyset, Y)$ does not hold. Then, by Lemma 5, $I_p(X_1, \emptyset, Y)$ does not hold in essentially all possible $p(U)$ defined over B. It follows that for each such $p(U)$,

$$\gamma \cdot p(X|Y) \neq \gamma \cdot N.$$

A contradiction to our initial assumption. Therefore, SI correctly denotes the potential ψ as $p_B(X|Y)$.

($\Leftarrow$) Follows directly from Theorem 2.

Let B be any Bayesian network. Theorem 4 states that for nearly all choices C of CPTs for B, the SI algorithm correctly denotes the semantics of potentials constructed by VE during exact inference on B.

---

[1] A set has measure zero if it is infinitesimally small relative to the overall space [8].

7   Conclusion

We extend d-separation's role from determining the minimum amount of information needed to answer a query $p(X|E = e)$ [12] to also giving the semantics of the potentials constructed when answering $p(X|E = e)$. Our results contribute to a deeper understanding of Bayesian networks, since semantics of VE's intermediate factors are now articulated with respect to the joint distribution. The main result (Theorem 4) showed that our SI algorithm correctly denotes the semantics of inference in nearly all Bayesian networks. Future work will include applying the results here to differential semantics in Bayesian networks [6,9].

References

1. Butz, C.J., Hua, S., Konkel, K., Yao, H.: Join Tree Propagation with Prioritized Messages. Networks 55(4), 350–359 (2010)
2. Butz, C.J., Konkel, K., Lingras, P.: Join Tree Propagation Utilizing both Arc Reversal and Variable Elimination. Int. J. Approx. Reasoning 52(7), 948–959 (2011)
3. Butz, C.J., Yan, W., Lingras, P., Yao, Y.Y.: The CPT Structure of Variable Elimination in Discrete Bayesian Networks. In: Ras, Z.W., Tsay, L.S. (eds.) Advances in Intelligent Information Systems. SCI, vol. 265, pp. 245–257. Springer, Heidelberg (2010)
4. Castillo, E., Gutiérrez, J., Hadi, A.: Expert Systems and Probabilistic Network Models. Springer, New York (1997)
5. Cormen, T.H., Leiserson, C.E., Rivest, R.L., Stein, C.: Introduction to Algorithms. MIT Press, Cambridge (2009)
6. Darwiche, A.: Modeling and Reasoning with Bayesian Networks. Cambridge University Press, New York (2009)
7. Kjaerulff, U.B., Madsen, A.L.: Bayesian Networks and Influence Diagrams. Springer, New York (2008)
8. Koller, D., Friedman, N.: Probabilistic Graphical Models: Principles and Techniques. MIT Press, Cambridge (2009)
9. Madsen, A.L.: A Differential Semantics of Lazy AR Propagation. In: 21st Conference on Uncertainty in Artificial Intelligence, pp. 364–371. Morgan Kaufmann, San Mateo (2005)
10. Madsen, A.L.: Improvements to Message Computation in Lazy Propagation. Int. J. Approximate Reasoning 51(5), 499–514 (2010)
11. Meek, C.: Strong Completeness and Faithfulness in Bayesian Networks. In: 11th Conference on Uncertainty in Artificial Intelligence, pp. 411–418. Morgan Kaufmann, San Mateo (1995)
12. Pearl, J.: Fusion, Propagation and Structuring in Belief Networks. Artif. Intell. 29, 241–288 (1986)
13. Pearl, J.: Probabilistic Reasoning in Intelligent Systems: Networks of Plausible Inference. Morgan Kaufmann, San Francisco (1988)
14. Pearl, J.: Belief Networks Revisited. Artif. Intell. 59, 49–56 (1993)
15. Shafer, G.: Probabilistic Expert Systems. SIAM, Philadelphia (1996)
16. Wong, S.K.M., Butz, C.J., Wu, D.: On the Implication Problem for Probabilistic Conditional Independency. IEEE Trans. Syst. Man Cybern. A 30(6), 785–805 (2000)
17. Zhang, N.L., Poole, D.: A Simple Approach to Bayesian Network Computations. In: 7th Canadian Conference on Artificial Intelligence, pp. 171–178. Springer, New York (1994)

Detecting Health-Related Privacy Leaks in Social Networks Using Text Mining Tools

Kambiz Ghazinour[1,3], Marina Sokolova[1,2,3], and Stan Matwin[1,4]

[1] School of Electrical Engineering and Computer Science, University of Ottawa
{kghazino,sokolova}@uottawa.ca, stan@site.uottawa.ca
[2] Faculty of Medicine, University of Ottawa
[3] Electronic Health Information Lab, CHEO Research Institute
[4] Faculty of Computer Science, Dalhousie University

Abstract. In social media, especially in social networks, users routinely share personal information. In such sharing, they might inadvertently reveal some personal health information, an essential part of their private information. In this work, we present a tool for detection of personal health information (PHI) in a social network site, MySpace. We analyze the PHI with the use of two well-known medical resources MedDRA and SNOMED. We introduce a new measure – Risk Factor of Personal Information – that assesses a possibility of a term to disclose personal health information. We synthesize a profile of a potential PHI leak in a social network, and we demonstrate that this task benefits from the emphasis on the MedDRA and SNOMED terms.

Keywords: Medical electronic dictionaries, Personal health information, Social networks, Machine Learning.

1 Introduction

Four technologies: privacy preserving data mining, information leakage prevention, risk assessment and social network analysis are relevant to personal health information (PHI) posted on public communication hubs (e.g. blogs, forums, and online social networks).

PHI relates to the physical or mental health of the individual, including information that consists of the health history of the individual's family, and information about the health care provider. We differentiate terms revealing PHI from medical terms that convey health information which is not necessarily personal, and terms that despite their appearance have no medical meaning (e.g. "I have pain in my chest" vs. "I feel your pain").

In this work, we focus on analysis/development of methods protecting privacy of personal health information in online social networks. We believe the online social networks' growth and the general public involvement makes social networks an excellent candidate for health information privacy research. Other means of social media are left for future work.

Our ultimate goal is to develop tools that could detect and, if necessary, protect personal health information that might be unknowingly revealed by users of social

O. Zaïane and S. Zilles (Eds.): Canadian AI 2013, LNAI 7884, pp. 25–39, 2013.

networks. In this paper we find empirical support that shows personal health information is disclosed in social network, and furthermore we show how two existing electronic linguistic medical resources help detect personal health information in messages retrieved from a social network[1]. These resources are the Medical Dictionary for Regulatory Activities (MedDRA) [12] and the Systematized Nomenclature of Medicine (SNOMED) [18], two well-established medical dictionaries used in biomedical text mining. We use machine learning to validate the importance of the terms detected by these two medical dictionaries in revealing health information and analyse the results of MedDRA and SNOMED.

In Section 2 we present background and briefly discuss related work in the area of personal health information and social networks. Section 3 describes current computational linguistic resources used in medical research. Section 4 explains our empirical study and Section 5 discusses our findings and introduces the *Risk Factor of Personal Information* and contributions of this study. Section 6 discusses how we use machine learning to validate our hypotheses. Section 7 concludes the paper and gives future research directions.

2   Related Background

2.1   Personal Health Information in Social Networks

Emergence of social networks, weblogs and other online technologies, has given people more opportunities to share their personal information. Such sharing might include disclosing personal identifiable information (PII) (e.g., names, address, dates) and personal health information (PHI) (e.g., symptoms, treatments, medical care) among other factors of personal life. In fact, 19%-28% of all Internet users participate in medical online forums, health-focused groups and communities and visit health-dedicated web sites [1,14]. A recent study [9] had demonstrated a real-world example of cross-site information aggregation that resulted in disclosing PHI. A target patient has profiles on two online medical social networking sites. By comparing the attributes from both profiles, the adversary can link the two with high confidence. Furthermore, the attacker can use the attribute values to get more profiles of the target through searching the Web and other online public data sets. Medical information including lab test results was identified by aggregating and associating five profiles gathered by an attacker, including the patient's full name, date of birth, spouse's name, home address, home phone number, cell phone number, two email addresses, and occupation.

2.2   Protection of Personal Health Information

Uncontrolled access to health information could lead to privacy compromise, breaches of trust, and eventually harm. [23] proposed a role prediction model to protect the electronic medical records (EMR) and privacy of the patients. As another

---

[1] www.eecs.uottawa.ca/~stan/PHI2013data.txt

example, [13] study the privacy protection state laws and technology limitations with respect to the electronic medical records.

However, protection of personal health information in contents of social networks did not receive as much attention. In part, this is due to the lack of resources appropriate for detection and analysis of PHI in informally written messages posted by the users [19]. The currently available resources and tools were designed to analyse PHI in more structured and contrived text of electronic health records [21].

2.3   Previous Work

Some studies analyzed personal health information disseminated in blogs written by healthcare professionals/doctors [8]. However, these studies did not analyze large volumes of texts. Thus, the published results may not have sufficient generalization power, [7, 17]. In [10], the authors manually analyze 3500 messages posted on seven sub-boards of a UK peer moderated online infertility support group and the results of this study show that online support groups can provide a unique and valuable avenue through which healthcare professionals can learn more about the needs and experiences of patients.

In a recent study [3], Carroll et al. described experiments in the use of distributional similarity for acquiring lexical information from notes typed by primary care physicians who were general practitioners. They also present a novel approach to lexical acquisition from 'sensitive' text, which does not require the text to be manually anonymized. This enables the use of much larger datasets, compared to the situation where the sentences need to be manually anonymized and large datasets cannot be examined.

There is a considerable body of work that compares the practices of two popular social networking sites (Facebook and MySpace) related to trust and privacy concerns of their users, as well as self-disclosure of personal information and the development of new relationships [5]. In [15] the dissemination of health information through social networks is studied. The authors reviewed Twitter status updates mentioning antibiotic(s) to determine overarching categories and explore evidence of misunderstanding or misuse of antibiotics. Most of the work use only in-house lists of medical terms [19], each built for specific purposes, but do not use existing electronic resources of medical terms designed for analysis of text from biomedical domain. However, those resources are general and in need of evaluation with respect to their applicability to the PHI extraction from social networks. The presented work fills this gap.

In most of the above cases, the authors analyze text manually and do not use automated text analysis. In contrast, in this work, we want to develop an automated method for mining and analysis of personal health information.

3   Electronic Resources of Medical Terminology

Biomedical information extraction and text classification have a successful history of method and tool development, including deployed information retrieval systems, knowledge resources and ontologies [22]. However, these resources are designed to analyze knowledge-rich biomedical literature. For example, GENIA is built for the microbiology domain. Its categories include DNA-metabolism, protein metabolism, and cellular process. Another resource, Medical Subjects Heading (MeSH), is a controlled vocabulary thesaurus, which terms are informative to experts but might not be in use by the general public. The Medical Entities Dictionary (MED) is an ontology containing approximately 60,000 concepts, 208,000 synonyms, and 84,000 hierarchies. This powerful lexical and knowledge resource is designed with medical research vocabulary in mind. Unified Medical Language System (UMLS) has 135 semantic types and 54 relations that include organisms, anatomical structures, biological functions, chemicals, etc. Specialized ontology BioCaster was built for surveillance of traditional media. It helps to find disease outbreaks and predict possible epidemic threats. All these sources would require considerable modification before they could be used for analysis of messages posted on public Web forums.

Table 1. A sample of the MedDRA hierarchy and their labels

Category Label	Main category	First level sub-category	Second level sub-category
10	Biliary disorders		
10-1		Biliary neoplasms	
10-1-1			Biliary neoplasms benign

3.1 MedDRA and Its Use in Text Data Mining

The Medical Dictionary for Regulatory Activities (MedDRA) is an international medical classification for medical terms and drugs terminology used by medical professionals and industries. The standard set of MedDRA terms enables these users to exchange and analyze their medical data in a unified way. MedDRA has a hierarchical structure with 83 main categories in which some have up to five levels of sub-categories. MedDRA contains more than 11,400 nodes which are instances of medical terms, symptoms, etc. Table 1 shows a sample of the MedDRA hierarchy.

Since its appearance nearly a decade ago, MedDRA has been used by the research community to analyze the medical records provided or collected by health care professionals: e.g. [11] use MedDRA in their study to evaluate patient reporting of adverse drug reactions to the UK 'Yellow Card Scheme'. In another study [20] use MedDRA to group adverse reactions and drugs derived from reports were extracted from the World Health Organization (WHO) global ICSR database that originated from 97 countries from 1995 until February 2010.

The above examples show that the corpus on which MedDRA tested is generally derived from patients' medical history or other medical descriptions found in the structured medical documents that are collected or disclosed by healthcare professionals. Content and context of those documents considerably differ from those of messages written by the social network users. In our study, we aim to evaluate the usefulness of MedDRA in detection of PHI disclosed on social networks.

3.2   SNOMED and Its Use in Text Data Mining

Another internationally recognized classification scheme is the Systematized Nomenclature of Medicine Clinical Terms (SNOMED CT) maintained by the International Health Terminology Standards Development Organization. Although SNOMED is considered the most comprehensive clinical health care terminology classification system, it is primarily used in standardization of electronic medical records [2].

Medical terms in SNOMED are called *concepts*. A concept is indicating of a particular meaning. Each concept has a unique *id* that with which it is referred to. A concept has a *description* which is a string used to represent a concept. It is used to explain what the concept is about. *Relationship* is a tuple of *(object – attribute – value)* connecting two concepts through an attribute.

Same as MedDRA, SNOMED has also a hierarchical structure. The root node, *SNOMED Concept*, has 19 direct children which Figure 1 shows 10 of them from *Clinical finding* to *Record artifact*. As illustrated in Figure 1, one of the nodes, *procedure*, has 27 branches including but not limited to, *administrative procedures* (e.g. Medical records transfer), *education procedures* (e.g. Low salt diet education) and other procedures.

Among 353,154 instances of all 19 main branches we decided to only sub-select *procedures* and *clinical findings* (that encompasses *diseases* and *disorders*). These branches have more medical meanings than for instance the *Environment or geographical location* node which covers name of the cities, provinces/states, etc. The *clinical findings* node has 29,724 sub-nodes (19,349 *diseases and disorders*, 10,375 *findings*) and the node *procedure* has 15,078 sub-nodes. So in total we have selected 44,802 nodes out of 353,154.

Fig. 1. A sample of the SNOMED hierarchy

Clinical finding		Administrative procedure
Procedure	⇨	Blood bank procedure
Observable entity		Community health procedure
Body structure		Environmental care procedure
Organism		Educational procedure
Event		:
Environment or geographical location		:
:		
:		Procedure by intent
Physical object		Procedure by method
Qualifier value		Procedure by priority
Record artifact		Procedure by site

Table 2. MedDRA and SNOMED hierarchical structure

Dictionary	# total nodes	# unique sub-selected nodes	average depth level
MedDRA	11,400	8,561	3
SNOMED	353,154	44,802	6

Table 2 depicts a brief comparison between MedDRA and SNOMED hierarchical structure. It shows that SNOMED covers a larger set of terms and has deeper hierarchical levels compared to MedDRA.

3.3   Benefits of Using MedDRA and SNOMED

We believe referring to MedDRA and SNOMED as well-funded, well-studied and reliable sources has two benefits:

1. We introduce a field of new text applications (the posts, weblogs and other information sources directly written by individuals) which extend the use for MedDRA and SNOMED. These medical dictionaries were previously used only for the health information collected by healthcare professionals.
2. Since MedDRA and SNOMED have well-formed hierarchical structures, by examining them against the posts on the social network site, we should be able to identify which terms and branches in the MedDRA and SNOMED are used to identify PHI and which branches are not, hence can be pruned out. These operations should result in a more concise and practical dictionary that can be used on detecting PHI disclosed in diverse textual environments.

4   Empirical Study

In this research, we examined the amount of PHI disclosed by individuals on an online social network site, MySpace. Unlike previous research work, introduced in Section 2, the presence of PHI was detected through the use of the medical terminologies of MedDRA and SNOMED.

In our empirical studies we examined posts and comments publicly available on MySpace. We sorted and categorize the terms used in both MedDRA and SNOMED, and found in MySpace, based on the frequency of use and if they reveal PHI or not. We also studied the hierarchy branches that are used and the possibility of pruning the unused branches (if there exists any). Based on the hierarchical structures of MedDRA and SNOMED, the deeper we traversed down the branches, the more explicit the medical terms get and the harder the pruning phase is.

4.1   MySpace Data

MySpace is an online social networking site that people can share their thoughts, photos and other information on their profile or general bulletin, i.e., posts posted on to a "bulletin board" for everyone on a MySpace user's friends list to see. There have been several research publications on use of MySpace data in text data mining, but none of them analyzed disclosure of personal health information in posted messages [6, 16].

We obtained the MySpace data set from the repository of training and test data released by the workshop Content Analysis for the Web 2.0 (CAW 2009). The data creators stated that those datasets intended to comprehend a representative sample of what can be found in web 2.0. Our corpus was collected from more than 11,800 posts on MySpace. In the text pre-processing phase we eliminated numbers, prepositions and stop words. We also performed stemming which converted all the words to their stems (e.g. hospital, hospitals and hospitalized are treated the same).

4.2   Data Annotation

We manually reviewed 11,800 posts on MySpace to see to what extent those medical terms are actually revealing personal health information on MySpace. The terms were categorized into 3 groups:

- PHI: terms revealing personal health information.
- HI: medical terms that address health information (but not necessarily personal).
- NHI: terms with non-medical meaning.

To clarify this let's see the following examples:

The word *lung* which assumed to be a medical term appears in the following three sentences we got from our MySpace corpus: "...they are promoting cancer awareness particularly lung cancer..." which is a medical term but does not reveal any personal health information. "... I had a rare condition and half of my lung had to be removed..." this is clearly a privacy breach and "...I saw a guy chasing someone and screaming at the top of his lungs..." which carries no medical value. In this manner we have manually analyzed and performed manual labeling based on the annotator's judgment whether the post reveals information about the person who wrote it, or discloses information about other individuals that make them identifiable. We acknowledge that there might be cases where the person might be identified with a high probability in posts that mention "...my aunt..., my roommate". For simplicity in this research we categorize those posts as HI where the post has medical values but does not reveal a PHI. Table 3 shows some more examples of PHI, HI and NHI.

4.3   MedDRA Results

To assess MedDRA's usability for PHI detection, we performed two major steps:

1) We labeled the MedDRA hierarchy in a way that the label of each node reflects to what branch it belongs too. The result is corpus-independent.
2) We did a uni-gram and bi-gram (a contiguous sequence of one term from a given sequence of text or speech) comparison between the terms that appear in MySpace and the words detected by MedDRA. The result is corpus-dependent.

Table 3. Examples of terms found on MySpace which are PHI, HI and NHI

Term	PHI	HI	NHI
Fraction	...got a huge bump on my forehead, fractured my nose.	I wish the driver would've died as well instead of just suffering a fractured leg	The few people who did vote would be so fractured among the different parties.
Laser	For me the laser treatment had unpleasant side-effects.	I know someone who had laser surgery to remove the hair from his chest.	...with a laser writes something on a flower stem.
Allergic	I'm allergic to cigarette smoke	...the allergy is a valid reason.	I'm allergic to bullets!

After the execution of the step 1, MedDRA's main categories are labeled from 1 to 83 and for those with consequent sub-categories, the main category number is followed by a hyphen (-) and the sub-category's number (e.g. Biliary disorders (10) and its sub categories Biliary neoplasms (10-1) and Biliary neoplasms benign (10-1-1)).

After the execution of the step 2, there are 87 terms that appear both in MedDRA and in the MySpace corpus. A subset of them is illustrated in Table 4.

There are also identical terms that appear under different categories and increase the ambiguity of the term. For instance, *nausea* appears under categories *acute pancreatitis*, and *gastrointestinal nonspecific symptoms and therapeutic procedures*, so when *nausea* appears in a post, it is not initially clear which category of the MedDRA hierarchy has been used, and the text needs further semantic processing.

4.4   SNOMED Results

SNOMED leaves are very specific and have many more medical terms compared to MedDRA. Our manual analysis has shown that the general public uses less technical and therefore more general terms when they discuss personal health and medical conditions. Hence, we expect that for SNOMED's less granular terms appear more often in MySpace data than its more specific terms.

The structure of SNOMED is organized as follows: The root node has 19 sub-nodes. One of the sub-nodes is *procedure* that itself has 27 sub-nodes. Out of those sub-nodes is called *procedure by method* which has 134 sub-nodes. *Counseling* is one of those 134 sub-nodes and itself has 123 sub-nodes. Another node among the 134 sub-nodes is *cardiac pacing* that has 12 sub-nodes which are mostly leaves of the hierarchy.

In extreme cases there might be nodes that are located 11 levels deep down the hierarchy. For example the following shows the hierarchy associated with the node *hermaphroditism*. Each '>' symbol can be interpreted as 'is a...':

> *Hermaphroditism > Disorder of endocrine gonad > Disorder of reproductive system > Disorder of the genitourinary system > Disorder of pelvic region > Finding of pelvic structure > Finding of trunk structure > Finding of body region > Finding by site > Clinical finding > SNOMED Concept*

It is cumbersome to understand how many of the 44,802 nodes that we have sub-selected from SNOMED are leaves and how many are intermediate nodes; however, 66 nodes out of 44,802 appeared in MySpace, of which 9 were leaves (see Table 5).

Table 6 shows some terms. The number of times the term appears in SNOMED, and whether it is a leaf in the hierarchy. We can see that except *jet lag* and *dizzy*, the other terms do not reveal PHI. Even in the example *dizzy,* the appearance as PHI compared to the number of times they appear as HI and NHI is trivial.

Table 4. A subset of terms detected by MedDRA that appear in MySpace

Terms	PHI	HI	NHI	Terms	PHI	HI	NHI
depression	18	114	0	Dizzy	2	7	2
injury	9	12	0	Overdose	2	6	0
swell	4	2	1	Thyroid	2	0	0
concussions	3	0	0	Asthma	1	1	0

Table 5. A subset of terms detected by SNOMED that appear in MySpace

Terms	PHI	HI	NHI	Terms	PHI	HI	NHI
Sick	44	1	135	Fracture	3	3	1
Pain	17	3	141	Dizzy	2	7	2
infection	5	33	0	insomnia	2	6	0
Swell	4	2	1	thyroid	2	0	0

Table 6. Terms of MySpace detected by SNOMED leaf nodes

Term	Level of hierarchy	PHI	HI	NHI	# freq. in SNOMED
phlebotomy	3	0	3	0	2
histology	4	0	2	0	2
jet lag	4	1	0	0	1
domestic abuse	5	0	1	0	1
physic assault/abuse	5	0	5	0	1
black out	6	0	1	1	1
dizzy	6	2	7	2	2
hematological	6	0	1	0	1
Papillary conjunctivitis	6	0	1	0	2

This result indicates that although SNOMED has a deep hierarchical structure, one should not traverse all the nodes and branches to reach leaf nodes to be able to detect PHI terms. In contrast, we hypothesize that branches can be pruned to reduce the PHI detection time and still achieve an acceptable result. We leave this as potential future work.

5   Risk Factor of Personal Information

Due to the semantic ambiguity of the terms we had to manually examine the given context to see whether the terms were used for describing medical concepts or not. For instance, the term adult in the post "...today young people indifferent to the adult world..." has no medical meaning.

We aimed to find whether the terms were used for revealing PHI or HI. Although some terms like *surgery* and *asthma* have strictly (or with high certainty) medical meaning, some terms may convey different meanings depending on where or how they are used. For instance, the word heart has two different meanings in "...heart attack..." and "...follow your own heart...".

We also measured the ratio of the number of times that the term was used in MySpace and the number of times that revealed PHI. We called the ratio the Risk Factor of Personal Information (RFPI). In other words, for a term t, $RFPI_t$ is:

$$RFPI_t = number\ of\ times\ t\ reveals\ PHI\ /\ number\ of\ times\ t\ appears\ in\ a\ text$$

Table 7 illustrates the top RFPI terms from MedDRA and SNOMED that often reveal PHI. There is an overlap between the top most used terms of MedDRA and SNOMED with highest RFPI. These are terms that prone to the number of times they appear in data(*concussions, thyroid, hypothermia, swell, ulcer and fracture*).

Furthermore, according to our studies although the words *sick* and *pain* appear numerous times and reveal personal health information but their RFPI is relatively low and might not be as privacy revealing as words like *fracture* or *thyroid*.

For example *sick* in the sentence "...I am sick and tired of your attitude..." or "...the way people were treated made me sick..." clearly belong to the NHI group and does not carry any medical information. Or in the case of term *pain*, the sentences "...having a high-school next to your house is going to be a pain..." or "...I totally feel your pain!..." belong to NHI group as well.

In total, we found 127 terms that appear in MySpace and in both dictionaries. 87 terms in MySpace are captured by MedDRA and 66 terms are captured by SNOMED. (There are 26 common terms that appear in both dictionaries. Although SNOMED is a larger dataset compared to MedDRA, since its terms are more specific, fewer terms are appeared in SNOMED. Thus, we consider MedDRA to be more useful for PHI detection.

Table 7. Top terms detected by MedDRA and SNOMED that have highest RFPI

a)MedDRA

term	PHI	HI	NHI	RFPI
concussions	3	0	0	1.00
thyroid	2	0	0	1.00
disoriented	1	0	0	1.00
hyperthyroid	1	0	0	1.00
hypothermia	1	0	0	1.00
liposuction	1	0	0	1.00
swell	4	2	1	0.57
asthma	1	1	0	0.50
ulcer	1	1	0	0.50
injury	9	12	0	0.43
fracture	3	3	1	0.43

b)SNOMED

term	PHI	HI	NHI	RFPI
concussions	3	0	0	1.00
thyroid	2	0	0	1.00
bipolar disorder	1	0	0	1.00
hypothermia	1	0	0	1.00
jet lag	1	0	0	1.00
motion sickness	1	0	0	1.00
shoulder pain	1	0	0	1.00
stab wound	1	0	0	1.00
tetanus	1	0	0	1.00
thyrotoxicosis	1	0	0	1.00
swell	4	2	1	0.57

Table 8. Percentage of the sentences that are detected by these two dictionaries in each group

Dictionary	%PHI	%HI	%NHI
SNOMED	16.5	28.5	55
MEDDRA	11.5	60	28.5

Figure 2 illustrates the number of sentences (not terms) in MySpace for each category of PHI, HI and NHI that are detected by MedDRA, SNOMED and the union of them. Table 8 demonstrates that although SNOMED detects more PHI terms compared to MedDRA, since it also detects more NHI terms (false positive) as well, it is less useful compared to MedDRA. In addition, the summation of both PHI and HI

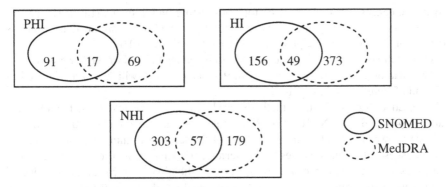

Fig. 2. HI, PHI and NHI sentences detected by each dictionary and their intersection

in MedDRA is greater than its equivalent in SNOMED which is another reason why MedDRA seems more useful than SNOMED.

In brief, since MedDRA covers a broader and more general area it detects more HI than SNOMED. In contrast, although SNOMED detects slightly more PHI and HI, it is less trustable than MedDRA since it also detects far more NHI. Hence, the precision of detection is much lower.

Although there are not many sentences (9 sentences out of almost 1000 sentences) that reveal PHI as a result of engaging in a conversation that initially contained HI, there is always the possibility that existence of HI sentences is more likely to result in PHI detection compared to the sentences that contain NHI.

As shown in Figure 2, the amount of terms that are detected by both MedDRA and SNOMED (their intersection in the Venn diagram) is not impressive and that is why these two dictionaries cannot be used interchangeably.

6   Learning the Profile of PHI Disclosure

We approach the task of detecting PHI leaks as acquiring a profile of what "language" is characteristic of this phenomenon happening in posts on health-related social networks. This can be achieved if a profile of the occurrence of this phenomenon is acquired, and Machine Learning, or more specifically text classification, is a natural technique to perform this acquisition. We studied the classification of the sentences under two categories of PHI-HI and NHI using Machine Learning methods.

Hypothesis. Focusing on terms from MedDRA and SNOMERD results in a better performing profiling than the straightforward method of bag of words.
Experiment. Our experiment consists of the following two parts:

6.1   Part I, Standard Bag of Words Model

We vectorized each of the 976 sentence detected by the two medical dictionaries. In these sentences, there are 1865 distinct terms. After removing the words with the same roots and deleting the symbols and numerical terms, 1669 unique words have been identified. Next, generating a standard Bag of Words document representation and the sentences are vectorized (0 for not existing and 1 for existing). Hence, we have vectors of 1669 attributes that are either 0 or 1 and one more attribute which is the label of the sentence, the privacy class (0 = NHI, 1 = PHI-HI).

After each vector is labeled accordingly to be either PHI-HI or NHI, we perform a bi-classification and train our model with 976 sentences of which 425 are labeled as NHI and 551 are labeled as PHI-HI.

We used two classification methods used most often in text classification, i.e., Naive Bayes (NB), KNN (IBK) in Weka based on the privacy class (0 = NHI, 1 = PHI-HI) shown in the left column of Table 9. Hence, our training data set would be the sentences with binary values of the terms appearing in them or not. Due to our small set data, we performed five by two cross-validation. In each fold, our collected data set were randomly partitioned into two equal-sized sets in which one was the training set which was tested on the other set. Then we calculated the average and variance of those five iterations for the privacy class. Table 9 shows the results of this 5X2-fold cross validation.

6.2   Part II, Special Treatment for Medical Terms

We took the vectors resulting from Part I and focused on the terms belong to the following three groups by weighting them stronger in the bag of words than the remaining words.

    a)   List of pronouns or possessive pronouns/family members/relatives (e.g. I, my, his, her, their, brother, sister, father, mother, spouse, wife, husband, ex-husband, partner, boyfriend, girlfriend, etc.).

    b)   Medical term detected by MedDRA and SNOMED.

    c)   Other medical terms that their existence in a sentence may result in a sentence to be a PHI or HI. Terms such as *hospital, clinic, insurance, surgery*, etc.

For group (a) and group (c) we associate weight 2 (one level more than the regular terms that are presented by 1 as an indication that the terms exist in the sentence). For group (b) which are the terms detected by the SNOMED and MEDRA and have higher value for us we associate weight with value of 3. So unlike the vectors in Part I which consist of 0s or 1s, in this part we have vectors of 0s, 1s, 2s and 3s. In fact, the values for weights are arbitrary and finding the right weights would be the task of optimization of the risk factor. We ran the experiment with values of 0s, 1s, 2s and 4s and we got the same results shown in Table 9.

Table 9. Two classification methods on the privacy class with and without medical terms

Classification		Privacy Class (Part I)	Privacy Class (Part II)
NB	Correctly classified%	75.51	85.75
	Mean absolute error	0.25	0.15
KNN (k=2)	Correctly classified%	74.48	86.88
	Mean absolute error	0.27	0.13

Next, we used the same two classification methods and performed five by two cross-validation. The results are shown in Table 9 in the right column.

Comparing the results from Part I and II show that there is almost 10% improvement in detecting sentences that reveal health information using the terms detected by MedDRA and SNOMED which confirms our hypothesis. The results are statistically significant [4], with the *p-value* of .95.

7   Conclusion and Future Work

In this research work, we developed tools that detect and, if necessary, protect personal health information that might be unknowingly revealed by users of social networks. In this paper we found empirical support for this hypothesis, and furthermore we showed how two existing electronic medical resources MedDRA and SNOMED help to detect personal health information in messages retrieved from a social network.

In our work we labeled the MedDRA and SNOMED hierarchy in a way that the label of each node reflects to what branch it belongs to. Next, we did a uni-gram and bi-gram comparisons between the terms appear on the MySpace corpus and the words appear on MedDRA and SNOMED. Comparing the number of terms captured by these two medical dictionaries, it suggests that MedDRA covers more general terms and seems more useful than SNOMED that has more detailed and descriptive nodes. Performing a bi-classification on the vectors resulted from the sentences labeled as PHI-HI and NHI support our hypotheses. We used two common classification methods to validate our hypothesis and analyse the results of MedDRA and SNOMED. Our experiments demonstrated that using the terms detected by MedDRA and SNOMED helps us to better identify sentences in which people reveal health information.

Future directions include analysis of Precision and Recall and analysis of words which tend to correlate but not perfectly match with the terms in medical dictionaries (e.g. in the sentence "I had my bell rung in the hockey game last night" words bell rung could indicate a concussion). Use methods such as Latent Dirichlet Allocation (LDA) might be a good approach. Furthermore, testing our model on different posts on other social networks such as Facebook or Twitter would be a good research experiment. We want to compare the terms that appear in MedDRA and SNOMED and evaluate their RFPI values.

An interesting project would be to develop a user interface or an application plug in to the current social networks such as MySpace, Facebook and Twitter that warns the user about revealing PHI when they use these potentially privacy violating words that we have introduced in this research.

Another potential future work could investigate use of more advanced NLP tools, beyond the lexical level and identifying some semantic structures in which those terms are involved and lead to health-related privacy violation.

Acknowledgement. The authors thank NSERC for the funding of the project and anonymous reviewers for many helpful comments.

References

[1] Balicco, L., Paganelli, C.: Access to health information: going from professional to public practices. In: 4th Int. Conference on Information Systems and Economic Intelligence, SIIE 2011 (2011)

[2] Campbell, J., Xu, J., Wah Fung, K.: Can SNOMED CT fulfill the vision of a compositional terminology? Analyzing the use case for Problem List. In: AMIA Annual Symposium Proc. 2011, pp. 181–188 (2011)

[3] Carroll, J., Koeling, R., Puri, S.: Lexical Acquisition for Clinical Text Mining Using Distributional Similarity. In: Gelbukh, A. (ed.) CICLing 2012, Part II. LNCS, vol. 7182, pp. 232–246. Springer, Heidelberg (2012)

[4] Dietterich, T.G.: Approximate Statistical Tests for Comparing Supervised Classification Learning Algorithms. Neural Computation 10(7), 1895–1924 (1998)

[5] Dwyer, C., Hiltz, S.R., Passerini, K.: Trust and privacy concern within social networking sites: A comparison of Facebook and MySpace. In: Proceedings of the Thirteenth Americas Conference on Information Systems, Keystone, Colorado, August 09-12 (2007)

[6] Grace, J., Gruhl, D., Haas, K., Nagarajan, M., Robson, C., Sahoo, N.: Artist ranking through analysis of on-line community comments (2007),
http://domino.research.ibm.com/library/cyberdig.nsf/
papers/E50790E50756F371154852573870068A371154852573870184/
$File/rj371154852573810421.pdf

[7] Kennedy, D.: Doctor blogs raise concerns about patient privacy,
http://www.npr.org/templates/story/story.php?storyId=88163567
(accessed June 13, 2012)

[8] Lagu, T., Kaufman, E., Asch, D., Armstrong, K.: Content of Weblogs Written by Health Professionals. Journal of General Internal Medicine 23(10), 1642–1646 (2008)

[9] Li, F., Zou, X., Liu, P., et al.: New threats to health data privacy. BMC Bioinformatics 12, S7 (2011)

[10] Malik, S., Coulson, N.: Coping with infertility online: an examination of self-help mechanisms in an online infertility support group. Patient Educ. Couns 81, 315–318 (2010)

[11] McLernon, D.J., Bond, C.M., Hannaford, P.C., Watson, M.C., Lee, A.J., Hazell, L., Avery, A.: Adverse drug reaction reporting in the UK: a retrospective observational comparison of Yellow Card reports submitted by patients and healthcare professionals. Drug Saf. 33(9), 775–788 (2010)

[12] MedDRA Maintenance and Support Services Organization,
http://www.meddramsso.com (accessed January 1, 2013)

[13] Miller, A.R., Tucker, C.: Privacy protection and technology adoption: The case of electronic medical records. Management Science 55(7), 1077–1093 (2009)

[14] Renahy, E.: Recherche bd'infomation en matiere de sante sur INternet: determinants, practiques et impact sur la sante et le recours aux soins, Paris 6 (2008)

[15] Scanfeld, D., Scanfeld, V., Larson, E.: Dissemination of health information through social networks: Twitter and antibiotics. American Journal of Infection Control 38(3), 182–188 (2010)

[16] Shani, G., Chickering, D.M., Meek, C.: Mining recommendations from the web. In: RecSys 2008: Proceedings of the 2008 ACM Conference on Recommender Systems, pp. 35–42 (2008)

[17] Silverman, E.: Doctor Blogs Reveal Patient Info & Endorse Products. Pharmalot (2008), http://www.pharmalot.com/2008/07/doctor-blogs-reveal-patient-info-endorse-products/ (December 15, 2009)

[18] Systematized Nomenclature of Medicine, http://www.ihtsdo.org/snomed-ct/ (accessed January 1, 2013)

[19] Sokolova, M., Schramm, D.: Building a patient-based ontology for mining user-written content. In: Recent Advances in Natural Language Processing, pp. 758–763 (2011)

[20] Star, K., Norén, G.N., Nordin, K., Edwards, I.R.: Suspected adverse drug reaction reported for children worldwide: an exploratory study using VigiBase. Drug Saf. 34, 415–428 (2011)

[21] Yeniterzi, R., Aberdeen, J., Bayer, S., Wellner, B., Clark, C., Hirschman, L., Malin, B.: Effects of personal identifier resynthesis on clinical text de-identification. J. Am. Med. Inform. Assoc. 17(2), 159–168 (2010)

[22] Yu, F.: High Speed Deep Packet Inspection with Hardware Support'- Technical Report No. UCB/EECS-2006-156 (2006), http://www.eecs.berkeley.edu/Pubs/TechRpts/2006/EECS-2006-156.html

[23] Zhang, W., Gunter, C.A., Liebovitz, D., Tian, J., Malin, B.: Role prediction using electronic medical record system audits. In: AMIA 2011 Annual Symposium, pp. 858–867. American Medical Informatics Association (2011)

Move Pruning and Duplicate Detection

Robert C. Holte

Computing Science Department
University of Alberta
Edmonton, AB Canada T6G 2E8
rholte@ualberta.ca

Abstract. This paper begins by showing that Burch and Holte's move pruning method is, in general, not safe to use in conjunction with the kind of duplicate detection done by standard heuristic search algorithms such as A*. It then investigates the interactions between move pruning and duplicate detection with the aim of elucidating conditions under which it is safe to use both techniques together. Conditions are derived under which simple interactions cannot possibly occur and it is shown that these conditions hold in many of the state spaces commonly used as research testbeds. Unfortunately, these conditions do not preclude more complex interactions from occurring. The paper then proves two conditions that must hold whenever move pruning is not safe to use with duplicate detection and discusses circumstances in which each of these conditions might not hold, i.e. circumstances in which it would be safe to use move pruning in conjunction with duplicate detection.

1 Introduction

Burch and Holte [1,2] introduced a generalization of the method for eliminating redundant operator sequences introduced by Taylor and Korf [3,4], proved its correctness, and showed that it could vastly reduce the size of a depth-first search tree in spaces containing short cycles or transpositions.[1] Both methods work by pruning moves, i.e. disallowing ("pruning") the use of an operator ("move") after a specific sequence of operators has been executed. Burch and Holte also showed that move pruning could not, in general, be safely used in conjunction with transposition tables [5]; i.e. there is a risk, if move pruning is used together with transposition tables, that all optimal paths from start to goal will be eliminated.

A*, breadth-first search, and many other search algorithms use a duplicate detection strategy that is simpler than the transposition tables considered by Burch and Holte. Such algorithms simply record each state that is generated and its distance from the start state. If the state is generated again by a path that is cheaper than the recorded distance, the distance is updated and the state is "re-opened" with a priority based on the new distance. Otherwise the

---

[1] A "transposition" occurs when there are two distinct paths leading from one state to another.

O. Zaïane and S. Zilles (Eds.): Canadian AI 2013, LNAI 7884, pp. 40–51, 2013.

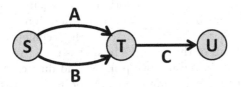

Fig. 1. Example in which duplicate detection and move pruning interact to produce erroneous behaviour

new path to the state is ignored. I shall refer to this as "duplicate detection" in the remainder of this paper.

Burch and Holte did not discuss whether it was safe to use their move pruning in conjunction with duplicate detection, but it was observed by Malte Helmert (personal communication) that it is not. Figure 1 shows the typical situation in which a problem arises. A and B are operators or operator sequences that are not redundant with each other in general, but happen to produce the same state, T, when applied to state S. AC and BC are the only two paths from S to U, and move pruning determines that AC is redundant with BC and decides to prohibit C from being applied after A. However, the search generates T via path A first, and records this fact using the usual backpointer method found in A* implementations. When the search later generates T via path B it notices that T has already been generated by a path of the same cost and therefore ignores B. Since the only recorded path from S to T is A, move pruning prevents C from being applied to T and state U is never reached.

To see that it is possible for such A, B, and C to exist, with AC and BC being redundant with each other but A and B not being redundant with one another, here is a very simple example (also due to Malte Helmert) presented in the notation of the PSVN language (see [2]). A state in this example is described by three state variables and is written as a vector of length three. The value of each variable is either 0, 1, or 2. The operators are written in the form $LHS \rightarrow RHS$ where LHS is a vector of length three defining the operator's preconditions and RHS is a vector of length three defining its effects. The LHS may contain variable symbols (X_i in this example); when the operator is applied to a state, the variable symbols are bound to the value of the state in the corresponding position. Preconditions test either that the state has a specific value for a state variable or that the value of two or more state variables are equal (this is done by having the same X_i occur in all the positions that are being tested for equality). For example, operator A below can only be applied to states whose first state variable has the value 0. The following operators behave like A, B, and C in Figure 1 when applied to state $S = \langle 0, 1, 1 \rangle$:

$A : \langle 0, X_1, X_2 \rangle \rightarrow \langle 1, 1, X_2 \rangle$
$B : \langle 0, X_3, X_4 \rangle \rightarrow \langle 1, X_3, 1 \rangle$
$C : \langle 1, X_5, X_6 \rangle \rightarrow \langle 2, 1, 1 \rangle$

A and B are not redundant with one another, in general, but both can be applied to state $S = \langle 0, 1, 1 \rangle$ and doing so produces the same state, $T = \langle 1, 1, 1 \rangle$.

This example motivates the study reported in this paper, whose aim is to understand the interactions between move pruning and duplicate detection and to identify conditions under which it is safe to use the techniques together. There are two main contributions of this paper. The first is to derive conditions under which simple interactions between move pruning and duplicate detection, such as the one depicted in Figure 1, cannot possibly occur. It turns out that these conditions hold in many of the state spaces commonly used as research testbeds (Rubik's Cube, TopSpin, etc.). Unfortunately, these conditions do not preclude more complex interactions from occurring. The second contribution of the paper is to derive conditions that must hold whenever there is a deleterious interaction between move pruning and duplicate detection.

1.1 Motivation

The motivation for adding move pruning to a system that does duplicate detection is computational—move pruning is faster than duplicate detection. This is because with duplicate detection a state must be generated and looked up in a data structure to determine if it is a duplicate. Move pruning saves the time needed for duplicate detection because it avoids generating states when it knows (by analysis in a preprocessing step) the resulting state will be a duplicate. For example, the depth-first search system used in Burch and Holte's experiments [1] did "parent pruning," an elementary form of duplicate detection, and they reported that using move pruning to achieve the same effect as parent pruning was more than twice as fast as doing parent pruning by explicit duplicate detection. In addition, if suboptimal paths to a state are generated before optimal ones, duplicate detection will involve updating the data structure that stores the distance-from-start information. This can be relatively expensive—updating the priority queue used by A*, for example. Move pruning will avoid some of these updates by not generating some of the suboptimal paths at all.

On the other hand, duplicate detection is useful to add to a system that does move pruning because move pruning, in general, is incomplete: it only detects short sequences that are redundant (in the current implementation move pruning considers all and only sequences of length L or less) and it only detects "universal" redundancy, as opposed to "serendipitous" redundancy, as illustrated in the example above, where sequences A and B are redundant when applied to certain states but are not redundant in general. Duplicate detection is complete, unless there is not enough memory to store all the generated states.

The final motivation for undertaking this study is that it applies much more broadly than just to systems that use Burch and Holte's method for automatic move pruning. When problem domains with many obvious redundancies, such as TopSpin and Rubik's Cube, are coded by hand, the researchers writing the code manually do a simple version of the move pruning that Burch and Holte have automated. For example, here is a detailed description of the standard move pruning done by hand for Rubik's Cube [6]:

> Since twisting the same face twice in a row is redundant, ruling out such moves reduces the branching factor to 15 after the first move. Furthermore,

twists of opposite faces of the cube are independent and commutative. For example, twisting the front face, then twisting the back face, leads to the same state as performing the same twists in the opposite order. Thus, for each pair of opposite faces we arbitrarily chose an order, and forbid moves that twist the two faces consecutively in the opposite order.

These are precisely the kinds of redundant operator sequences that Burch and Holte's method detects automatically. The correctness of the move pruning done manually has never been questioned, but the problem illustrated in Figure 1 applies regardless of whether the move pruning was inferred by an automatic method or by hand. Thus it brings into question the correctness of the standard encodings of testbeds such as Rubik's Cube and TopSpin if they are used in a system that does duplicate detection. In fact, I have verified that the manually encoded move pruning in the IDA* code written in my research group for TopSpin results in non-optimal solutions being produced if it is used in A*.

2 Essential Theory by Burch and Holte [1]

This section defines terminology and repeats the key theoretical ideas from [1].

The empty sequence is denoted ε. If A is a finite operator sequence then $|A|$ denotes the length of A (the number of operators in A, $|\varepsilon| = 0$), $cost(A)$ is the sum of the costs of the operators in A ($cost(\varepsilon) = 0$), $pre(A)$ is the set of states to which A can be applied, and $A(s)$ is the state resulting from applying A to state $s \in pre(A)$. I assume the cost of each operator is non-negative. A prefix of A is a nonempty initial segment of A ($A_1...A_k$ for $1 \leq k \leq |A|$) and a suffix is a nonempty final segment of A ($A_k...A_{|A|}$ for $1 \leq k \leq |A|$).

Operator sequence B is redundant with operator sequence A if (i) the cost of A is no greater than the cost of B, and, for any state s that satisfies the preconditions of B, both of the following hold: (ii) s satisfies the preconditions of A, and (iii) applying A and B to s leads to the same end state. Formally,

Definition 1. *Operator sequence B is "redundant" with operator sequence A iff the following conditions hold:*

(**R1**) $cost(B) \geq cost(A)$
(**R2**) $pre(B) \subseteq pre(A)$
(**R3**) $s \in pre(B) \Rightarrow B(s) = A(s)$

We write $B \geq A$ to denote that B is redundant with A.

Let $\mathcal{O}$ be a total ordering on operator sequences. $B >_{\mathcal{O}} A$ indicates that B is greater than A according to $\mathcal{O}$. $\mathcal{O}$ has no intrinsic connection to redundancy so it can easily happen that $B \geq A$ according to Definition 1 but $B <_{\mathcal{O}} A$.

Definition 2. *A total ordering on operator sequences $\mathcal{O}$ is "nested" if $\varepsilon <_{\mathcal{O}} A$ for all $A \neq \varepsilon$ and $B >_{\mathcal{O}} A$ implies $XBY >_{\mathcal{O}} XAY$ for all $A, B, X,$ and Y.*

Definition 3. *Given a nested ordering $\mathcal{O}$, for any pair of states s, t define $min(s, t)$ to be the least-cost path from s to t that is smallest according to $\mathcal{O}$ ($min(s, t)$ is undefined if there is no path from s to t).*

Theorem 1. *Let $\mathcal{O}$ be any nested ordering on operator sequences and B any operator sequence. If there exists an operator sequence A such that $B \geq A$ according to Definition 1 and $B >_{\mathcal{O}} A$, then B does not occur as a consecutive subsequence in $min(s, t)$ for any states s, t.*

As noted by Burch and Holte [1], from Theorem 1 it immediately follows that a move pruning system that restricts itself to pruning only operator sequences B that are redundant with some operator sequence A **and** greater than A according to a fixed nested ordering will be "safe", i.e. it will not eliminate all the least-cost paths between any pair of states. In Burch and Holte's implementation of move pruning, all operator sequences of length L or less are generated in an order defined by a fixed nested ordering, and each newly generated sequence is tested for redundancy against all the non-redundant sequences generated before it.

3   Conditions Precluding Simple Interactions

I will call the situation depicted in Figure 1 a "simple" interaction between duplicate detection and move pruning, by which I mean the interaction takes place between two optimal paths, AC and BC, that have a common suffix (C). In this section I derive commonly occurring conditions under which simple interactions cannot possibly happen. Throughout the rest of the paper I assume there is a fixed nested ordering on operator sequences, $\mathcal{O}$, used for move pruning.

Because AC and/or BC can be longer than the sequences that move pruning considers, define A' to be the suffix of A, B' to be the suffix of B, and C' to be the prefix of C such that move pruning determines that $A'C' \geq B'C'$ and $A'C' >_{\mathcal{O}} B'C'$. The latter implies $A' >_{\mathcal{O}} B'$. This, together with the fact that A' is not pruned by move pruning (A' is fully executed) implies that $A' \not\geq B'$.

Thus, a simple interaction requires an interesting situation: $A'C' \geq B'C'$ but $A' \not\geq B'$. There are natural conditions in which this combination is impossible because $(A'C' \geq B'C') \Rightarrow (A' \geq B')$ for all sequences $A', B',$ and C'. To derive such conditions, recall that the definition of $X \geq Y$ has three requirements:

    (R1) $cost(X) \geq cost(Y)$
    (R2) $pre(X) \subseteq pre(Y)$
    (R3) $s \in pre(X) \Rightarrow X(s) = Y(s)$

In order to derive conditions under which $(A'C' \geq B'C') \Rightarrow (A' \geq B')$ we need to consider each of these in turn.

(R1) We require conditions under which $(cost(A'C') \geq cost(B'C')) \Rightarrow (cost(A') \geq cost(B'))$. In fact, no special conditions are needed, this is always true because the cost of an operator sequence is the sum of the costs of the operators in the sequence and operator costs are non-negative.

(R2) We require conditions under which $(pre(A'C') \subseteq pre(B'C')) \Rightarrow (pre(A') \subseteq pre(B'))$. This is often not true, but it certainly holds if $pre(XY) = pre(X)$ for all operator sequences X and Y (with X non-empty). There are at least two commonly occurring conditions in which this is true:

- operators have no preconditions (every operator is applicable to every state) as in Rubik's Cube;
- the precondition of any sequence is the precondition of the first operator in the sequence (because the preconditions of the next operator in the sequence are guaranteed by the effects and unchanged preconditions of the operators preceding it), as in the sliding-tile puzzles with one blank.

(R3) We require conditions under which $(s \in pre(A'C') \Rightarrow A'C'(s) = B'C'(s)) \Rightarrow (t \in pre(A') \Rightarrow A'(t) = B'(t))$. This follows if both of the following hold:

- $pre(XY) = pre(X)$ for all operator sequences X and Y (with X non-empty), the same condition discussed in connection with (R2); and
- all operators are 1-to-1 $((op(x)=op(y)) \Rightarrow (x=y))$.

The two conditions listed under (R3) are thus sufficient to prevent simple interactions from occurring. These conditions hold in many commonly used state spaces: the sliding-tile puzzle with one blank, Rubik's Cube, Scanalyzer [7], Top-Spin, and the Pancake puzzle. In all such spaces, simple interactions between move pruning and duplicate detection cannot occur.

Unfortunately, simple interactions are not the only way that move pruning and duplicate detection can interact deleteriously, i.e., the situation in Figure 1 is not a necessary condition for move pruning to be unsafe in conjunction with duplicate detection. Figure 2 gives an example based on an actual run of A* on $(10, 4)$-TopSpin[2] when move pruning is applied to sequences of length 4 or less. The start state is at the top of the figure, the goal state is at the bottom. Move pruning eliminates all but two of the optimal paths from start to goal; those two paths are labelled J (the leftmost path) and M (the rightmost path) in the figure; the individual operators in a path are indicated by a subscript (e.g. J_2 is the second operator in path J).

Three additional paths $(K, L, \text{and } N)$ are shown because they play a role in preventing J and M from being fully executed even though they themselves cannot be fully executed because of move pruning. The move pruning that eliminates K, L, and N is shown in the figure by an X through operators K_6, L_5, and N_6. The reasons for these are as follows. Move pruning detects that $N_5N_6 \geq J_5J_6$ and therefore prevents N_6 from being executed after N_5. It also detects that $K_3...K_6 \geq L_3...L_6$ and therefore prevents K_6 from being executed after $K_3...K_5$. Similarly, it detects that $L_2...L_5 \geq M_2...M_5$ and therefore prevents L_5 from being executed after $L_2...L_4$. These can all be seen in the figure as paths of length 4 or less that branch apart at some particular state and later rejoin.

---

[2] In $(10, 4)$-TopSpin there are 10 tokens (numbers 0 to 9) in a circle and there are operators that reverse the order of any 4 adjacent tokens. Because only the cyclic order matters and not the absolute location within the circle, in the figure a state is written as a vector with token 9 always placed at the end.

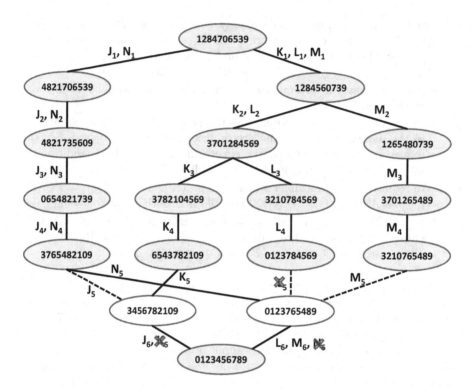

Fig. 2. Example from $(10, 4)$-TopSpin of move pruning and duplicate detection interacting to prevent the goal (bottom node) from being reached from the start (top node) by an optimal path

The effects of duplicate detection are shown by drawing the edges entering the two states just above the goal as either solid or broken. A solid edge indicates the path by which the state was first generated; a broken edge indicates an alternative path to the state that is generated later (or not at all in the case of L_5). For example, state 3456782109 is first generated by path K (operator K_5) and is later generated by path J (operator J_5). Since the path $J_1...J_5$ is not cheaper than the first path to generate the state ($K_1...K_5$), it is ignored. Similarly, $M_1...M_5$ is not cheaper than the first path to generate state 0123765489 ($N_1...N_5$), so it too is ignored.

What makes this fundamentally different than Figure 1 is that the path (K) that blocks J because of duplicate detection is not itself blocked by J because of move pruning, it is blocked by a different optimal path (L, which in turn is blocked by M because of move pruning). Likewise, the path (N) that blocks M because of duplicate detection is not itself blocked by M because of move pruning, it is blocked by a different optimal path (J). As I will show next, this represents the general situation in which move pruning and duplicate detection interact deleteriously.

4   Necessary Conditions for Move Pruning to Be Unsafe

In this section I state and prove conditions that must hold if move pruning is unsafe to use in conjunction with duplicate detection. The importance of identifying these "necessary" conditions is that one can then consider whether there are specific circumstances in which one or more of the necessary conditions are guaranteed not to hold. Move pruning is safe to use in such circumstances.

Let S be the start state, U any state that is reachable from S (U is the goal state), and BC any optimal path from S to U that contains no operator sequence considered redundant by move pruning (BC must exist because of Theorem 1).

Let Alg be a search algorithm that does neither duplicate detection nor move pruning and has the following properties.

- Alg enumerates the paths (operator sequences) emanating from S in a fixed sequence, thereby imposing a total order on the paths ($p_1 <_{Alg} p_2$ denotes that path p_1 is enumerated by Alg before path p_2).
- If operator sequence p_1 is a prefix of operator sequence p_2 then $p_1 <_{Alg} p_2$.
- Alg is able to prove the optimality of any optimal path it generates.[3]

When Alg is used in conjunction with move pruning, the resulting system is called Alg^{MP}. The effect of move pruning is to remove paths from Alg's enumeration sequence but not to change the order of those that remain. Path p_1 will be removed by move pruning if and only if it is determined that there exists another path p_2 such that $p_1 \geq p_2$ and $p_1 >_{\mathcal{O}} p_2$. Note that every such Alg^{MP} generates BC.

When Alg is used in conjunction with duplicate detection, the resulting system is called Alg_{DD}. The effect of duplicate detection is to remove paths from Alg's enumeration sequence but not to change the order of those that remain. For a given start state S, duplicate detection removes path p_1 if and only if there exists a prefix of p_1, p' (possible p_1 itself), and a path p_2 such that $p'(S) = p_2(S)$, $p' >_{Alg} p_2$, and p_2 is not itself eliminated by duplicate detection. A* and breadth-first search are examples of such Alg_{DD} search algorithms.

When Alg is used in conjunction with both move pruning and duplicate detection, the resulting system is called Alg_{DD}^{MP}.

I assume that the elimination of paths from Alg's enumeration sequence (whether by move pruning or duplicate detection or both) does not adversely affect its ability to prove a path it generates is optimal among the paths that remain in the enumeration sequence. This is true of A* and breadth-first search.

Definition 4. *We say move pruning is "safe" to use in conjunction with duplicate detection if, for any algorithm Alg with the properties stated above, Alg_{DD}^{MP} generates an optimal path from S to U for all states S and all states U that are reachable from S.*

---

[3] Algorithms such as A* and breadth-first search accomplish this by enumerating all paths that might be cheaper than the current cheapest path from S to U.

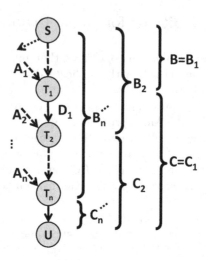

Fig. 3. Illustration of the Proof of REQ-2

In other words, move pruning is unsafe to use in conjunction with duplicate detection only if some Alg_{DD}^{MP} fails to generate any optimal path from S to U. In particular if move pruning is unsafe, Alg_{DD}^{MP} will fail to generate BC. From this fact, I will now derive necessary conditions for move pruning to be unsafe to use in conjunction with duplicate detection.

Theorem 2. *Let Alg be any search algorithm with the properties stated above. Then Alg_{DD}^{MP} can only fail to generate BC if both of the following conditions hold:*

REQ-1. *There exists a state $T_1 = B(S)$ on $BC(S)$ and an alternative path A_1 from S to T_1 such that $cost(A_1) = cost(B)$ and T_1 was generated by Alg_{DD}^{MP} via A_1 prior to being generated via B (i.e. $A_1 <_{Alg_{DD}^{MP}} B$).*

REQ-2. *There exists a state T_n on $BC(S)$, an alternative path A_n from S to T_n, and a suffix C_n of C such that $C_n(T_n) = U$, A_nC_n is an optimal path from S to U, and move pruning prohibits C_n from being applied after A_n.*

Proof of REQ-1. This is necessary because if no such T_1 and A_1 existed duplicate detection would not affect the generation of BC, which contradicts the premise that Alg_{DD}^{MP} fails to generate BC. A_1 cannot be cheaper than B because B is part of an optimal path to U and is therefore an optimal path to T_1.   □

Proof of REQ-2. Figure 3 depicts the key ideas needed to prove this. A_1 here is as in REQ-1 and sequences B and C from above are renamed here B_1 and C_1. As the proof proceeds, they are replaced by A_i, B_i, and C_i for larger values of i, with B_i increasing in length as i increases and C_i decreasing in length. In all cases $BC = B_iC_i$, $A_i(S) = B_i(S) = T_i$ and A_iC_i is an optimal path from S to U. D_i is the operator subsequence in BC that leads from T_i to T_{i+1}.

$A_1 C_1$ is an optimal path from S to U, why did Alg_{DD}^{MP} not generate it in full? Either because $A_1 C_1$ was eliminated by move pruning or because it was eliminated by duplicate detection. If it was eliminated by move pruning then we are done, with $n = 1$ ($T_n = T_1$, $A_n = A_1$, and $C_n = C_1 = C$). If it was eliminated by duplicate detection then there must be a state T_2 later in the $BC(S)$ sequence and alternative path A_2 from S to T_2 such that $cost(A_2) = cost(A_1 D_1)$ and T_2 was generated by Alg_{DD}^{MP} via A_2 prior to being generated via $A_1 D_1$. Let C_2 be the suffix of C such that $C_2(T_2) = U$ and B_2 be the prefix of BC such that $B_2(S) = T_2$. Now repeat this reasoning for the path $A_2 C_2$, which is an optimal path from S to U. If it was eliminated because of move pruning we are done with $n = 2$, and if it was eliminated because of duplicate detection, there must exist a T_3, A_3, B_3, and C_3 such that T_3 is later in the $BC(S)$ sequence than T_2, etc. This generates a sequence of states $T_1, T_2, ...$, each later in the $BC(S)$ sequence than the one before, and therefore there must be a final state in this sequence, T_n, with a corresponding A_n, B_n, and C_n, with $A_n C_n$ being an optimal path from S to U. This path was not executed and it cannot have been eliminated by duplicate detection (because if it were there would be a T_{n+1}), therefore it must have been eliminated because move pruning did not allow C_n to be executed after A_n.   □

Because both of these requirements are necessary for move pruning to be unsafe, if one of them does not hold, move pruning is safe to use in conjunction with duplicate detection. The remainder of this section considers each of them in turn.

4.1   Discussion of REQ-1

REQ-1 states that there must be an alternative optimal path, A_1, to $T_1 = B(S)$ that is generated before B. This could fail to hold in at least three different ways. First, it would fail to hold if there was only one path to each of the states on $BC(S)$ (namely, the path that is part of BC). This would happen, for example, if move pruning eliminated all alternative paths, as it does in the Arrow Puzzle [2]. In such cases, no duplicate is ever generated so duplicate detection is obviously safe to use with move pruning. Secondly, it would fail to hold if there were alternative paths to one or more states $T_1 = B(S)$ generated prior to B, but all of them were suboptimal. This is not impossible; for example, it would happen if there was a unique shortest path from S to each state in the state space.

The third way that REQ-1 could fail to hold, and perhaps the most interesting from a practical point of view, is that there are indeed alternative optimal paths to a state $T_1 = B(S)$ but none of them is generated before B. For example, consider the special case depicted in Figure 1, where $A_1 = A$ is generated before B (i.e. $A <_{Alg} B$) but $A >_\mathcal{O} B$. In other words there is a disagreement between how Alg orders the sequences and how they are ordered by $\mathcal{O}$. If the two orderings $>_\mathcal{O}$ and $>_{Alg}$ were chosen so that such a disagreement did not occur then the special case depicted in Figure 1 could not arise. Whether this can be done in practice, and whether it solves the general problem and not just the special case depicted in Figure 1 are open problems at present.

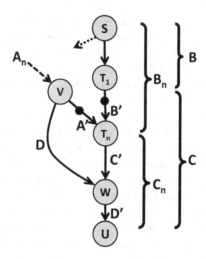

Fig. 4. General Case for Requirement 2

4.2   Discussion of REQ-2

REQ-2 says that there must exist optimal paths $A_n C_n$ and $B_n C_n$ such that move pruning prohibits C_n from being executed after A_n but allows it after B_n. This is very similar to the special case depicted in Figure 1, but with one important difference. In the special case, $C = C_n$ is prohibited after $A = A_n$ because of $B = B_n$, i.e., $AC \geq BC$. In the general case we are now considering we do not require $AC \geq BC$, we just require that AC is redundant with some path.

Let A' be the suffix of A_n and C' be the prefix of C_n such that $A'C'$ is the sequence within $A_n C_n$ that move pruning determines to be redundant with some other sequence D. There are two possibilities for D. The first possibility, which is what we saw in Figure 1, is that D is part of BC, i.e., there exists a suffix B' of B_n such that $A'C' \geq B'C'$ and $A'C' >_O B'C'$. Circumstances in which this cannot possibly happen have been discussed in Sections 3 and 4.1 above.

The other possibility for D is shown in Figure 4. Here D is a sequence entirely distinct from BC. In this case, we have another optimal path from S to U—one that follows A_n to state V, then executes D, which leads to state W on the BC path from which the goal is reached by sequence D'. This, in fact, is precisely what we saw in the TopSpin example in Figure 2. In that example optimal solution J was blocked by duplicate detection by another sequence, K, which in turn was blocked by move pruning by a sequence, L, that had nothing in common with J.

There is, however, one special circumstance in which REQ-2 cannot possibly occur and therefore move pruning is safe to use in conjunction with duplicate detection, and that is if move pruning is restricted to considering only sequences of length 1, i.e. redundancy among individual operators considered in a fixed order. If this restriction is imposed, move pruning cannot prohibit C_n from being executed after A_n but allow it after B_n since no "history" is taken into account.

5   Summary and Conclusions

This paper has investigated the interactions between move pruning and duplicate detection with the aim of elucidating conditions under which it is safe to use both techniques together. I have derived conditions under which simple interactions cannot possibly occur and shown that these conditions hold in many of the state spaces commonly used as research testbeds (Rubik's Cube, TopSpin, etc.). Unfortunately, these conditions do not preclude more complex interactions from occurring and an example was given where A* fails to find an optimal solution in TopSpin because of these more complex interactions. I then derived conditions that must hold whenever there is a deleterious interaction between move pruning and duplicate detection.

Acknowledgements. Thanks to Malte Helmert for his insights. I gratefully acknowledge the financial support of Canada's Natural Sciences and Engineering Research Council (NSERC).

References

1. Burch, N., Holte, R.C.: Automatic move pruning revisted. In: Proceedings of the 5th Symposium on Combinatorial Search, SoCS (2012)
2. Burch, N., Holte, R.C.: Automatic move pruning in general single-player games. In: Proceedings of the 4th Symposium on Combinatorial Search, SoCS (2011)
3. Taylor, L.A.: Pruning duplicate nodes in depth-first search. Technical Report CSD-920049, UCLA Computer Science Department (1992)
4. Taylor, L.A., Korf, R.E.: Pruning duplicate nodes in depth-first search. In: AAAI, pp. 756–761 (1993)
5. Reinefeld, A., Marsland, T.A.: Enhanced iterative-deepening search. IEEE Trans. Pattern Anal. Mach. Intell. 16(7), 701–710 (1994)
6. Korf, R.E.: Finding optimal solutions to Rubik's Cube using pattern databases. In: AAAI, pp. 700–705 (1997)
7. Helmert, M., Lasinger, H.: The scanalyzer domain: Greenhouse logistics as a planning problem. In: ICAPS, pp. 234–237 (2010)

Protocol Verification in a Theory of Action

Aaron Hunter[1], James P. Delgrande[2], and Ryan McBride[2]

[1] British Columbia Institute of Technology
[2] Simon Fraser University
Burnaby, BC

Abstract. Cryptographic protocols are usually specified in an informal language, with crucial elements of the protocol left implicit. We suggest that this is one reason that such protocols are difficult to analyse, and are subject to subtle and nonintuitive attacks. We present an approach for formalising and analysing cryptographic protocols in the situation calculus, in which all aspects of a protocol must be explicitly specified. We provide a declarative specification of underlying assumptions and capabilities, such that a protocol is translated into a sequence of actions to be executed by the principals, and a successful attack is an executable plan by an intruder that compromises the goal of the protocol. Our prototype verification software takes a protocol specification, translates it into a high-level situation calculus (Golog) program, and outputs any attacks that can be found. We describe the structure and operation of our prototype software, and discuss performance issues.

1 Introduction

A cryptographic protocol is a formalised sequence of messages exchanged between agents, where parts of the messages are protected using encryption. These protocols are used for many purposes, including authentication and secure information exchange. Protocols are typically specified in a quasi-formal language, as illustrated in the following example:

The Challenge-Response Protocol

1. $A \rightarrow B : \{N_A\}_{K_{AB}}$
2. $B \rightarrow A : N_A$

The goal is for agent A to determine whether B is alive on the network. The first step is for A to send B the message N_A encrypted with a shared key K_{AB}. The symbol N_A stands for a *nonce*, a random number assumed to be new to the network. The second step is for B to send A the message N_A unencrypted. The claim is that since only A and B have K_{AB}, and assuming that K_{AB} is indeed secure, then N_A could only have been decrypted by B, and so B must be alive. However, the protocol is flawed; here is an attack in which an intruder I masquerades as B and initiates a round of the protocol with A, thereby obtaining the decrypted nonce:

O. Zaïane and S. Zilles (Eds.): Canadian AI 2013, LNAI 7884, pp. 52–63, 2013.

An Attack on the Challenge-Response Protocol

1. $A \to I_B : \{N_A\}_{K_{AB}}$
1.1 $I_B \to A : \{N_A\}_{K_{AB}}$
1.2 $A \to I_B : N_A$
2. $I_B \to A : N_A$

This example is simplistic, but it illustrates the type of problems that arise in protocol verification. Even though protocols are generally short, they are notoriously difficult to prove correct. As a result, many different formal approaches have been developed for protocol verification. However, often these approaches are difficult to apply for anyone other than the original developers [3]. Part of the problem is that there is no clear agreement on what exactly constitutes an attack [1], which leaves one with considerable ambiguity about the status of a protocol when no attack is found. Moreover, as we later discuss, the language for specifying a protocol is highly ambiguous, and much information is left implicit.

Our thesis is that all aspects of a protocol need to be explicitly specified. The main contribution of this paper is the introduction of a declarative theory for protocol specification, including message passing between agents on a possibly compromised network, expressed in terms of a situation calculus (SitCalc) theory. A protocol is *compiled* into a sequence of actions for agents to execute. These actions may be interleaved with others, and indeed the framework allows simultaneous runnings of multiple protocols. The intention of an intruder is to construct a *plan* such that the goal of the protocol, in a precise sense, is thwarted. A protocol is secure when no such plan is possible. The approach is flexible, and significantly more general than previous approaches. We have used this approach to implement a verification tool that finds attacks by translating protocol specifications into Golog programs.

2   Motivation

2.1   Background

The standard intruder model used in cryptographic protocol verification is the so-called Dolev-Yao intruder [5]. Informally, the intruder can read, block, intercept, or forward any message sent by an honest agent. Verification consists of encoding the structure of a protocol in an appropriate formalism, and then finding attacks that a Dolev-Yao intruder is able to perform. Many different formalisms have been explored, including epistemic logics [4], multi-agent systems [6], strand spaces [13,7], multi-set re-writing formalisms [2] and logic programs under the stable model semantics [1].

Most existing work in protocol verification relies on the same semi-formal specification of protocols that we used in the introduction. Consider the Challenge-Response protocol and the attack described previously. Several points may be noted about the protocol specification. First, while the intent of the protocol and the attack are intuitively clear, the meaning of the exchanges in the protocol are ambiguous. Consider the first line of the Challenge-Response protocol: it cannot mean that A sends a message to B, since this may not be the case, as the attack illustrates. Nor can it mean that A *intends* to send a message to B, because in the attack it certainly isn't A's intention to send the message to the intruder! As well, in the first line of the protocol there is more than one action taking place, since in some fashion A is involved in sending a message and B is involved in the receipt of a message. Hence, the specification language is imprecise and ambiguous; notions of agent communication should be made explicit.

The specification also leaves crucial notions unstated, including: the goal of the protocol, the fact that N_A is a freshly generated nonce, and the capabilities of the intruder. Moreover, the specification does not take into account the broader context in which the protocol is to be executed. This context might contain other agents, interleaved protocol runs, and even constraints on appropriate behaviour for honest agents. For example, it is quite possible that a protocol could fail via what might be called a "stupidity attack", in which A simply sends the unencrypted nonce to the intruder. Certainly this would not be an expected outcome, but it is a logically possible compromise of the protocol. It may well be that there are other "untoward happenings" significantly more subtle than the stupidity attack; consequently, it is desirable to have a framework for specifying protocols in a general enough fashion in which such possibilities may be explicitly taken into account.

2.2   A Declarative Approach

We argue that in order to provide a robust demonstration of the security and correctness of a protocol, all aspects of a protocol exchange need to be specified. We suggest that an explicit, logical formalisation in the SitCalc provides a suitable framework. Our primary aim is to clearly formalize exactly what is going on in a cryptographic protocol in a declarative action formalism; such a formalization will provide a more nuanced and flexible model of agent communication.

We proceed as follows. The first step is to develop a suitable formal model of the message passing environment. In past work, we have used a simple transition system representation [9]. While this is sufficient for the high-level analysis of a protocol in terms of the beliefs of the agents, it is not expressive enough to precisely encode all aspects of a message passing system. In §3, we present a SitCalc model of the message passing environment in which agents, messages, and keys are all first-class objects that are composed, manipulated and exchanged through a small set of actions.

Our goal is to translate the standard "arrows and colons" representation of a protocol directly into an executable GoLog program that corresponds to our SitCalc theory of message passing. As such, in §4, we demonstrate how a simple syntactic variation on a standard protocol specification can be translated into a GoLog procedure that precisely encodes the messages that are sent and received in a protocol specification. Roughly, the idea is to iterate through all components of each message and translate each message exchange into a detailed procedure for composing, sending, receiving, and decomposing each message.

In §5, we present a high-level version of our verification algorithm. We use an iterative deepening approach to consider progressively longer sequences of message exchanges that encode a complete protocol run. We assume that every message is first received and decomposed by the intruder, who then has all components of the message for use in an attack. Although our model is highly expressive and the space of possible actions for the intruder grows quickly, we find that attacks are still found in a reasonable amount of time for many standard test protocols.

3 The Message Passing Domain

While our approach is to translate protocol specifications directly into GoLog, underlying our approach is a formal SitCalc model of the message passing domain. In this section, we briefly outline the details of our model using the Challenge-Response protocol as an example. We assume familiarity with the situation calculus [10].

3.1 Vocabulary

In our model, there are four primitive sorts of objects (beyond actions and situations): *agents, nonces, keys,* and *encrypted data.*

- **Agents**: Agent predicates include **Agent(a)**, **Intruder(a)** and **Principal(a)**. By default, the domain only has the lone intruder *intr* and two principals: *Alice*, and *Bob*.
- **Nonces:** For nonces, there is the candidacy predicate **Nonce(n)** and the functional fluent **fresh(s)** that generates a new unique nonce.
- **Keys: Key(k)** is true for any key, and there are predicates to distinguish symmetric keys, public keys, and private keys: **SymKey(k)**, **PubKey(k)**, and **PrivKey(k)**. There are also functional fluents to encode the relationship between key pairs.
- **Encrypted Data** is text encoded with a specific key, fulfilling **Encrypted(enc)**. The *decKey* function can be applied as follows **decKey($\{Bob\}_{K_{Alice}}$)** to garner the correct key: KP_{Alice}. **encKey** is similar, except it extracts the key used for encryption rather than decryption. Another extraction fluent is **dec(encrypted,k)**, which returns internal contents of *encrypted*, assuming that k is the correct key. To convert raw data and a key into the correct encoded object, use the function **enc(text,k)**.

Any list of the above objects is called *text*, designated by the corresponding candidacy predicate **Text(t)**. There are several functions for working with text, such as the length function **Length(text)** and the extraction functions such as **First(lst)**, **Second(lst)**, and **Third(lst)**.

A *message* is composed of text contents, a sender address, and a recipient address. For message fluents, there is the basic candidacy predicate **Msg(m)** along with the extraction functions **SenderAddr(m)**, **RecipAddr(m)**, and **Contents(m)**. To determine whether or not a message has been posted for access, we use the fluent **Sent(msg, s)**. If true, then it is possible for another agent to receive this message. To check whether a message was received by an agent, we evaluate **Recd(a, msg, s)**. This fluent is true only if the agent has received that message at least once. In the domain, we assume that any sent message can only be received once. Implicitly, this means that if a message *msg* is sent n times, that message can be received at most n times.

The notion of *having* a message or a key is fundamental in protocol verification. An agent's ownership of data fundamentally changes what text agents can send or encrypt, and this is represented in our model through the ownership fluent **Has(agent,object,s)**. The **Has** fluent also has an expression for defining the ownership of lists. For example, if $[l_1, l_2, \ldots]$ is some ordered list, then:

$$Has(agent, [l_1, l_2, \ldots], s) \leftrightarrow Has(agent, l_1, s) \wedge Has(agent, l_2, s) \wedge \ldots$$

Thus, if an agent has a list of objects then it owns each object in the list; conversely an agent owns a list of objects if it has each of the individual objects in the list.

3.2   The Initial Situation

In our model, it is necessary to explicitly specify which keys and text strings are owned by each agent at the outset. Consider a domain with three agents: the intruder *intr*, and two principals *Alice* and *Bob*. We can model a situation where all agents have a shared key as follows:

Agent	Owned Agent IDs	Owned Shared Keys
Alice	Alice, Bob, intr	$K_{Alice/Bob}, K_{Alice/intr}$
Bob	Alice, Bob, intr	$K_{Alice/Bob}, K_{Bob/intr}$
intr	Alice, Bob, intr	$K_{Alice/intr}, K_{Bob/intr}$

This table indicates that each agent is able to compose messages that involve the names of the other agents, and all pairs of agents share a key. As agents compose and exchange messages, each agent will acquire more information. It is also possible to encode non-standard initial situations; for example, it is straightforward to specify that a key has been compromised before a protocol run begins. This flexibility in changing the initial ownership of information is an advantage of our approach, as many attacks on protocols arise because an intruder obtains a key they are not expected to have.

3.3   Message Passing Actions

Using the constructed initial situation and the language of logical fluents, we can then define both the preconditions and postconditions of performing nonce generation, encryption, composition, message sending and message receiving. For simplicity, general type predicates, such as **Agent(a)**, are always part of the preconditions but are not listed in the following definitions:

1. **makeNonce(a,n,s)**
   PRECONDITIONS: IsFresh(n,s)
   EFFECT: Has(n,s)
2. **encrypt(a,text,k,s)**
   PRECONDITIONS: Has(a,text,s) $\wedge$ ($\exists$ k Pubkey(k,s) $\vee$ Has(a,k,s))
   EFFECT: Has(a, enc(text,k),s)
3. **decrypt(a, enc-text, k,s)**
   PRECONDITIONS:
   Has(a,enc-text,s) $\wedge$ Has(a,k,s)
   $\wedge$ ((SymKey(k, s) $\wedge$ (k = encKey(enc-text))) $\vee$ (AsymKey(k, encKey(enc-text),s)))
   EFFECT: Has(a,dec(enc-text,k),s)

4. **compose(a,m,se,re,t)**

This action allows for an agent to package text into a message with a **sender** and **recipient**.

PRECONDITIONS: Has(a,t,s)

EFFECT: Has(make-message(se,re,t),s)

5. **send(a,m)**

PRECONDITIONS:

Has(a,m,s) $\land$ ((a = intr) $\lor$ ((a = SenderAddr(m)) $\land$ (a $\neq$ RecipAddr(m)))

The precondition ensures that principals can only send messages where they are the *Sender*.

EFFECT: Sent(m,s)

6. **receive(a,m)**

PRECONDITIONS:

Sent(m,s) $\land$ ((a = intr) $\lor$ (a = RecipAddr(m))))

Principals can only receive a message if it is addressed to them.

EFFECTS: Has(a,Contents(m),s), Recd(a,m), a Sent(msg,s) is cancelled.

These simple definitions remove many ambiguities associated with a message passing domain, such as whether the intruder can intercept messages (it can) or if encryption can be broken (it cannot). The strength of this model is that a list of six actions and their preconditions/effects can define a large class of agent interactions.

4   Golog Encoding

To find attacks, we encode the SitCalc representation of a protocol as a Golog program. Golog is a high-level programming lanugage suitable for the implementation of SitCalc action theories, introduced in [11]. In this section, we sketch the details of our encoding.

4.1   Protocol Representations

The standard "arrows and colons" representation of a protocol is typically used in the protocol verification literature. As such, we use a similar representation; however, our representation differs in that it is based on a formal grammar that can easily be given as input to be parsed by our verification system. In the interest of space, we do not include the grammar here. However, the following table illustrates how it can be used to encode the Challenge Response Protocol:

Standard Representation	Grammar Representation
$A \to B : \{N_a\}_{K_{A/B}}$	A->B: {Na}K-A/B
$B \to A : N_a$	B->A: Na

Our representation of a protocol also includes a specification of initial key ownership, based on the following grammar:

```
"KEYS:"
{key-line}
key-line = role ":" key-list
key-list = key
key-list = key "," key-list
```

The key-list after each role indicates which keys an agent owns in the initial situation. In order to facilitate the representation of standard protocols, if no key ownership properties are set then the default is to assign a shared key to each pair of agents, as well as appropriate public-private key pairs for asymetric cryptography. Adding key-ownership details, the Challenge Response Protocol is encoded as follows:

```
A->B: {Na}K-A/B
B->A: Na
KEYS:
A: K-A/B
B: K-A/B
```

For any "arrows and colons" representation of a protocol, it is easy to produce this grammatical specification as input to the GoProVe verifier.

4.2   Compiling to GoLog: Motivation

Actions in a protocol must be compiled into multiple-step procedures in GoLog. For example, in the standard specification, one line encodes the act of sending a fresh nonce. A simple GoLog procedure for sending a fresh nonce is the following:

```
Agent-Nonce-Send(A,B,s):
    let Na = fresh(s)
    act(makeNonce(A,Na))
    let send-text = list(Na)
    let send-message = make-message(A, B, send-text)
    compose(A, B, send-text, send-message)
    send(A,send-message)
```

This procedure can be initiated by an agent A to generate a random number, compose a message including this number, and finally send the message to B. When B responds, A will need to perform conditional checks to see if the received message has the proper format and data. The main purpose of the compiler is to produce a series of procedures that send valid protocol messages, and verify that received messages match the transaction format. The GoLog representation of a protocol must also include constraints about protocol advancement, so that honest agents only continue a protocol if they believe that all previous sends/receives have been legitimate. In the next section, we describe the actions and checks that are required.

4.3   The Compilation Algorithm

Each step in a protocol involves one agent sending a message to another. The compilation algorithm translates each protocol step to a pair of procedures: one procedure to encode the sending action, and one procedure to encode the receiving action. In this section, we focus on the sending action.

Consider a message-send action of the following form:

$$A \to B : \{M_1, \ldots, M_{j_1}\}_{K_1}, \ldots, \{M_1, \ldots, M_{j_n}\}_{K_n}.$$

The algorithm for translating this step into a GoLog procedure is as follows:

1. Write GoLog commands to declare the procedure with a suitable name.
2. Initialize a list *enc* of encrypted components.
3. For each $\{M_1, \ldots, M_{j_i}\}_{K_i}$ with $1 \leq i \leq n$:
   (a) Initialize a list $data_i$ of (unencrypted) data components.
   (b) For each M_k with $1 \leq k \leq j_i$:
      i. Write GoLog commands to get the data M_k.
      ii. Append M_k to the list $data_i$.
   (c) Write GoLog commands to encrypt the list *data* with the key K_i.
   (d) Append $data_i$ to the list *enc*.
4. Write GoLog commands to compose and send message constraining items in *enc*.
5. Write GoLog commands to advance to the next protocol step.

To illustrate the process, we show how the first message of the Challenge-Response Protocol is encoded as a GoLog program. In this case, there is just one encrypted component $\{N_a\}_{K_{A/B}}$ and one data component N_a. The algorithm therefore simplifies to a five step procedure:

I. Write GoLog commands to declare procedure.
II. Write GoLog commands to get data corresponding to N_a.
III. Write GoLog commands to encrypt N_a with $K_{A/B}$.
IV. Write GoLog commands to compose and send the message.
V. Write GoLog commands to advance to the next protocol step.

For step (I), we give the procedure a name and check that the participants are distinct.

```
CR-A-1-proc(A, B, protocol-states, s):
    test(neq(A, B))
```

For step (II), we use the functional fluent *fresh* to generate a new nonce and the *makeNonce* action to specify it is created by A:

```
let Na = fresh(s)
act(makeNonce(A, Na))
```

In this particular case, there is only one item of text data to send. It is then passed to step (III), where we specify the key to be used for encryption and apply it to the list of text items using the *encrypt* action:

```
let K-A/B = get-shared-key(A, B)
let text-to-encrypt = list(Na)
let enc-text = enc(text-to-encrypt, K-A/B)
act(encrypt(A, text-to-encrypt, K-A/B))
```

Note that this step is essentially identical if multiple items of text are encrypted with the same key. In the general case, the output of this step is also a list, where each item of the list is a text component encrypted with a different key. For step (IV), we compose a message that consists of all encrypted components

```
let send-text = list(enc-text)
let send-message = make-message(A, B, send-text)
compose(A, B, send-text, send-message)
send(A, send-message)
```

Finally, for step (V.), we advance the state of the protocol:

```
new-protocol-state(A, send-message, (A, B), CR-A-1)
```

This completes the specification of the GoLog procedure for sending the first message of the Challenge Response Protocol. The procedure generated to receive this message is similar. The main difference is that we need to consider previously sent messages. This is simply a matter of additional checks on variable bindings.

In the interest of space, we do not include the complete encoding of the Challenge-Response protocol here. We simply remark that each message exchanged in the protocol is encoded through a send procedure and a receive procedure, as illustrated above.

5   The GoProVe Verifier

In this section, we present the approach used by GoProVe, our prototype verification software. Our software is able to simulate protocol runs with no intruder to check if a protocol is well formed, and it is also able to check for attacks on protocols in the presence of an intruder.

5.1   Protocol Simulation

Protocol simulation is the process of testing protocol runs where all agents, including the intruder, are honest. The process of protocol simulation involves the following sub-procedures:

1. **Initiate Protocol.** Select a pair of agents, and assign them the roles in the protocol.
2. **Simulation Control.** Check if a message has been sent. If a message was sent, then run a procedure to receive and react to that message. If no message has been sent, then check if there is a message available to be sent. If there is no message to send, the session is complete.
3. **Output.** Determine if a successful run has been completed for some agent pair.

This simple algorithm detects if a protocol run between any pair of agents can complete successfully. This is important for verification, because we need to be able to distinguish between a badly formed protocol with no successful runs, and an insecure protocol that is vulnerable to attack.

5.2   Protocol Verfification

For protocol verification, the simulation algorithm is extended by introducing an intruder. The intruder has special recieve/send procedures that allow the protocol to advance even if messages are intercepted or forged. Since a protocol run does not have a fixed (or unbounded) length in this context, the search for a run must use iterative deepening. The GoProVe verifier uses the following procedure to find attacks.

Initialize: Set message ownership for all agents.
Start: Choose a principal A to initiate protocol run with B. Set depth=0.
While(true)
> **Execute protocol action:** Send or receive message.
> Set depth = depth + 1.
> If (message was sent)
> > Intruder receives message, decrypts, composes all possible messages.
>
> If(attack occurred)
> > Return attack.
>
> If(depth $\geq$ max depth)
> > Backtrack.
>
> **Advance protocol**: Do one of the following:
> > Principal receives a sent message.
> > Intruder sends message.

There are many ways that the search space can be expanded indefinitely, but we have limited the behaviour of the intruder in a way that makes this less of a problem. For example, the set of nonces held by the intruder is fixed, so it is not possible for the intruder to expand the search space by continually generating nonces. In order to optimize performance, we also use a *Greedy Intruder Channel Model*. In this model, the intruder decrypts, parses and analyzes every received message immediately after it is received. While this might appear to slow down run times due to unecessary processing, it actually protects the verifier from combinatorial explosion. Since the intruder does all processing immediately, the algorithm above can actually be understood to progress rather than simply stall as an intruder loops through internal operations.

On finding a compromised goal, GoProVe outputs a description of any attacks, as well as information about agent beliefs regarding sessions, in the following format:

```
Protocol-ID: CR
      Status: completed
      Viewer: Alice
      Role B: Bob
      Role A: Alice

Messages:
      (Alice,Bob,[{N2}K-Alice/Bob])
      (Bob, Alice, [N2])
```

In the case of the Challenge-Response protocol, two different runs are output: one where Alice started a session with Bob and another where she thought that Bob was challenging her. This is an indication of an attack, because it indicates that Alice may have incorrect beliefs about the messages that have been exchanged in a protocol run.

5.3 Performance

The table below gives a list of the GoProVe verifier's runtimes for the Challenge Response protocol. Run times are also included for some other protocols: a variation

Protocol	Max-Depth	attack?	time
CR	3	Y	78 ms
CR-Server	7	Y	9 min
NS	4	Y	2 s
NS	4	Y	5 s
NSL	5	N	7 min

of Challenge Response with servers, Needham-Schroeder, and Needham-Schroeder-Lowe [12].

These results are promising, as protocol verification is a challenging task even for simple protocols. Moreover, since the verification task can be done offline, all of these run times would be appropriate for practical use in evaluating a new protocol.

6   Related Work

Most logic-based approaches to protocol verification are influenced to some degree by the pioneering BAN logic of [4]. This approach has been highly influential because it reduces protocol verification to reasoning about knowledge in a formal logic. However, BAN logic itself consists of an ad hoc set of rules of inference with no formal semantics. In this respect, our approach differs from the BAN tradition. Rather than defining a new protocol-specific logic, we encode protocols in a general purpose action formalism.

Hernández and Pinto propose an approach similar to ours, notably due to the fact that they also use the SitCalc [8]. However, they focus on producing proofs of correctness based on the actions of honest agents. By contrast, we explicitly model the actions of an intruder, and we view protocol verification as the process of "planning an attack." Our treatment of communication is also different: while Hernández and Pinto define an unreliable broadcast channel, we define a direct channel that allows the intruder the first opportunity to receive a message. As such, our approach is best viewed as an alternative to the Hernández-Pinto approach, rather than a continuation.

7   Discussion

We have described a declarative approach for the representation and analysis of cryptographic protocols. In our model, the effects of actions and the properties of the environment are explicitly specified in a Golog program, based on an underlying SitCalc action theory. Our approach is distinguished by the explicit representation of all aspects of a protocol, including capabilities of the intruder that are often hard-coded. As a result, our approach is more flexible and more elaboration tolerant than alternative approaches.

Specific advantages of our declarative model can be seen by considering simple variations on the standard Dolev-Yao intruder. For example, we have already discussed the fact that it is possible to modify the key-ownership settings. Similarly, it would be possible to constrain procedures with respect to the topology of a particular network. In contrast, in extant logical approaches to protocol verification, it is not straightforward to modify the model for a specific application.

Many proofs of protocol correctness assume that honest agents do not perform actions that compromise secret information; however, it is not always clear which actions are likely to do so. In our framework, we can discover these undesirable actions and we can formally specify axioms that restrict honest agents from performing them. To the best of our knowledge, this problem has not been addressed in related formalisms.

In addition to the theoretical advantages gained by using the SitCalc for encoding protocols, we also gain the practical advantage that it is relatively easy to implement a prototype verification system in Golog. Our software uses a formal grammar to represent the structure of a protocol, and translates protocols directly into Golog programs that encode SitCalc action theories. In principle, users need not have detailed knowledge of our SitCalc encoding in order to analyse the security of a protocol.

There are several directions for future work. One direction would be to explore a wider range of protocols, such as those designed for non-repudiation and fair exchange. Another improvement would be to reduce the running time, in order to address longer and more complex protocols. However, as a proof of concept, the current run times demonstrate that the we can use a flexible model of protocol execution based on a rigorous action formalism, while still finding attacks in a reasonable time.

References

1. Aiello, L.C., Massacci, F.: Planning attacks to security protocols: Case studies in logic programming. In: Kakas, A.C., Sadri, F. (eds.) Computat. Logic (Kowalski Festschrift). LNCS (LNAI), vol. 2407, pp. 533–560. Springer, Heidelberg (2002)
2. Armando, A., Compagna, L., Lierler, Y.: Automatic Compilation of Protocol Insecurity Problems into Logic Programming. In: Proceedings of JELIA, pp. 617–627 (2004)
3. Brackin, S., Meadows, C., Millen, J.: CAPSL Interface for the NRL Protocol Analyzer. In: Proceedings of ASSET 1999. IEEE Press (1999)
4. Burrows, M., Abadi, M., Needham, R.: A logic of authentication. ACM Transactions on Computer Systems 8(1), 18–36 (1990)
5. Dolev, D., Yao, A.C.: On the Security of Public Key Protocols. IEEE Trans. on Inf. Theory 2(29), 198–208 (1983)
6. Fagin, R., Halpern, J.Y., Moses, Y., Vardi, M.Y.: Reasoning about Knowledge. The MIT Press (1995)
7. Halpern, J.Y., Pucella, R.: On the Relationship between Strand Spaces and Multi-Agent Systems, CoRR, cs.CR/0306107 (2003)
8. Hernández-Orallo, J., Pinto, J.: Especificación formal de protocolos criptográficos en Cálculo de Situaciones. Novatica 143, 57–63 (2000)
9. Hunter, A., Delgrande, J.: Belief Change and Cryptographic Protocol Verification. In: Proceedings of AAAI (2007)
10. Levesque, H.J., Pirri, F., Reiter, R.: Foundations for the Situation Calculus. Linköping Electronic Articles in Computer and Information Science 3(18) (1998)
11. Levesque, H.J., Raymond, R., Lesperance, Y., Lin, F., Scherl, R.: GOLOG: A Logic Programming Language for Dynamic Domains. Journal of Logic Programming (31), 1–3 (1997)
12. Lowe, G.: Breaking and Fixing the Needham-Schroeder Public-Key Protocol Using FDR. In: Margaria, T., Steffen, B. (eds.) TACAS 1996. LNCS, vol. 1055, pp. 147–166. Springer, Heidelberg (1996)
13. Thayer, J., Herzog, J., Guttman, J.: Strand Spaces: Proving Security Protocols Correct. Journal of Computer Security 7(2-3), 191–230 (1999)

Identifying Explicit Discourse Connectives in Text

Syeed Ibn Faiz and Robert E. Mercer

Department of Computer Science
The University of Western Ontario
London, ON, Canada
{mibnfaiz,mercer}@csd.uwo.ca

Abstract. Explicit discourse relations in text are signalled by discourse connectives like *since, because, however*, etc. Identifying discourse connectives is a part of the bigger task called discourse parsing in which discourse coherence relations are extracted from text. In this paper we report improvements to the state-of-the-art for identifying explicit discourse connectives in the Penn Discourse Treebank and the Biomedical Discourse Relation Bank. These improvements have been achieved with maximum entropy (logistic regression) classifiers by combining machine learning features from previous approaches with new surface level features that capture information about a connective's surrounding phrases and new syntactic features that add more information from the path in the syntax tree connecting the root to the connective and from the clause following the connective by means of its syntactic head.

1 Introduction

Discourse connectives are those words or phrases that can signal discourse relations. For example, most common discourse connectives include *and, but, however, when*, etc. Identifying discourse connectives is important for discourse relation parsing which has useful application in many natural language processing tasks including question answering, semantic interpretation of natural language, textual entailment, anaphora resolution, and in our work on higher order relations in biomedical text [7]. Identification of discourse connectives is not straightforward because of ambiguity. Non-discourse usage can be found for most connective strings. For example, the following two sentences show discourse and non-discourse usage of *and*.

(1) He believes in what he plays, *and* he plays superbly. (discourse usage)

(2) That went over the permissible line for warm *and* fuzzy feelings.
    (non-discourse usage)

With the advent of the Penn Discourse Treebank, researchers have undertaken data-driven/machine learning approaches to the problem of identifying discourse connectives. In this paper we report improvements over such existing works by combining their strengths as well as introducing some novel features.

O. Zaïane and S. Zilles (Eds.): Canadian AI 2013, LNAI 7884, pp. 64–76, 2013.
© Springer-Verlag Berlin Heidelberg 2013

2   Related Works

Emily Pitler and Ani Nenkova have considered the problem of identifying explicit discourse connectives as a disambiguation task [13]. They have shown that syntactic features can be used to disambiguate discourse connectives. Besides the connective itself as a feature they have experimented with four main syntactic features: *self category*—the highest node in the syntax tree that dominates only the words in the connective, *parent category*—the immediate parent of the self category, the *left sibling* of the self category, and the *right sibling* of the self category. They apply two binary features which check whether the right sibling contains a VP and/or a Trace[1]. They also use pairwise interaction features between the connective and each of the syntactic features (e.g., connective and self category) and interaction features between each pair of syntactic features (e.g., self category and parent category).

Ben Wellner has reported two approaches for identifying connectives based on their syntactic context, one based on the syntactic constituency structure and another on the dependency structure [24]. In the first approach he considers features derived from the constituent parse tree. These features include connective features, the path from the connective head word to the syntactic root, some syntactic context features (e.g. whether an NP appears before a VP in this path) and some conjunctive features. In the dependency structure-based approach he uses features derived from the dependency tree. These features include the connective feature, contextual features (previous and next words and their parts-of-speech), features involving the parent and sibling of the connective head in the dependency tree and clause detection features (whether the parent and/or any sibling has a syntactic subject).

The most recent related work is that of Ziheng Lin *et al.* [10]. They improve on Pitler and Nenkova's work by introducing some new features. These new features include the previous and next words and their parts-of-speech and their interactions with the connective and its part-of-speech. They also included as a feature the path from the connective to the syntactic root and a compressed version of this path in which adjacent identical constituents are collapsed into one.

Ramesh and Yu [16] considers the problem of identifying discourse connectives from biomedical text. For that they use the Biomedical Discourse Relation Bank [15]. With a set of mostly orthographic features they have trained classifiers that are based on linear-chain first-order Conditional Random Field (CRF) models. A CRF is a probabilistic undirected graphical model used to label sequence data [22]. They train their models using ABNER [21], a biomedical named entity recognizer, with its default feature set, which includes standard bag-of-words, orthographic and n-gram features.

---

[1] In linguistic theories that include movement rules, *trace* is an empty category that indicates the position in the syntactic tree from which a phrase or clause has been moved. The PTB marks these traces. The automatic parser used in our work does not.

3   Corpora and Features

3.1   The Penn Discourse Treebank (PDTB)

The Penn Discourse Treebank (PDTB) [14] is the largest annotated discourse corpus. It annotates the same Wall Street Journal articles as the Penn Treebank (PTB) and the PropBank, effectively aligning three different types of annotation (syntactic, semantic and discourse). PDTB takes a lexically grounded and theory neutral approach. Instead of annotating all possible discourse coherence relations it annotates discourse relations at a lower level. At this level a discourse relation holds between two abstract objects, i.e. events, facts or propositions. The PDTB annotates both explicit and implicit discourse relations. An explicit discourse relation is signalled by an explicit connective like *because, since, therefore*, etc. There are 18,459 explicit discourse connective **tokens**[2] in the PDTB. Discourse connectives are often modified by adverbs such as *only, even, at least*, etc. The modified connectives like *only because, just when*, etc. are considered to belong to the same **type**[3] as that of their head (*because, when*, etc.). There are 100 unique connective types and 273 distinct discourse connectives in the PDTB.

3.2   The Biomedical Discourse Relation Bank (BioDRB)

The Biomedical Discourse Relation Bank (BioDRB) [15] annotates a subset of the full-text biomedical articles of the Genia corpus with discourse relations. It adopts the annotation guidelines of the PDTB, differing in the annotation process itself, and annotates both explicit and implicit relations. There are several discourse connectives in the BioDRB which are not annotated as discourse connectives in the PDTB. It has been reported that only 44% of the discourse connectives in the BioDRB also occur in the PDTB [16]. The remaining 56%, 178 unique discourse connectives in all, include many frequent cue phrases in the biomedical domain, such as *by, due to, in order to*, indicating possible domain-specific characteristics of the BioDRB. The designers of the BioDRB were inclined to explore such domain specific characteristics, which is why the annotators were strongly encouraged to identify additional connectives that might not appear as discourse connectives in the PDTB [15]. This contrast between the two corpora shows that the set of connective phrases which are more likely to signal discourse relations can differ across various domains and genres of text.

3.3   Features

As mentioned above, Pitler and Nenkova have used a set of syntactic features and a single surface level feature, namely the connective identity feature (the connective string itself). Henceforth, we refer to these features as P&N. Wellner

---

[2] A token is an occurrence of an explicit discourse connective.

[3] A connective type is an unmodified discourse connective such as *because* or *when*, i.e., a discourse connective in its base form.

has experimented with both syntactic and dependency features. Lin *et al.* have used all of the P&N features and have added some surface level features and a pair of syntactic path features. In our work we have analyzed all the features used in these earlier works and propose some new ones. We have grouped these features into two classes: **surface level features** and **syntactic features**.

Surface Level Features. Properties of the neighbouring words of a discourse connective can sometimes signal its presence. Both Wellner and Lin *et al.* have found such surface level features useful for connective identification. The features that they have considered for a connective, *C*, with previous word, *prev*, and next word, *next*, include: **prev**, **next**, part-of-speech (**POS**) of **prev**, **POS** of **next**, **C** + **prev**, **C** + **next**, **C** + **POS** of **next**, **C** + **POS** of **prev**, **POS** of **C** + **POS** of **prev**, **POS** of **C** + **POS** of **next**. In our experiment we have found that **prev** and **next** are good features especially when combined with the connective. For example, if a candidate connective string is followed by *of* or *to* (e.g. *as a result of, in addition to*) then it is not likely to be a discourse connective. Similarly when a candidate connective string *when* is preceded by either a year, month or week then *when* does not function as a discourse connective, rather it acts as a temporal modifier. In our experiment we have observed that the features derived from chunk tags (phrase level tags, such as, NP, VP, etc.) are more useful than the POS features. Using the conjunction of the connectives and the chunk tags of the neighbouring words instead of the POS tags gave us a better result on the development set. For example, whenever *as* is immediately followed by a VP it is unlikely to be a discourse connective. However, *when* immediately followed by a VP may often be a discourse connective. We approximated the chunk tag for a word by taking the label of its grand-parent in the constituent parse tree.

Syntactic Features. The path from the syntactic root to the connective in the constituent parse tree contains some of the most important syntactic information for connective identification. Wellner derives constituent features solely from this path. He uses the last few constituents of this path in different combinations and a collapsed version of the complete path in which adjacent constituents with identical labels are compressed into one. Pitler and Nenkova, on the other hand, do not use the complete path to derive features. They, instead, use syntactic context beyond this path through their left sibling and right sibling features. In our experiment on the development set we have found that augmenting the P&N features with information about the whole path is beneficial. However, it appears that using all of the constituents of that path individually is better than using the whole path as a single feature. We therefore use each individual constituent which is predecessor to the parent category constituent combined with its distance from the connective as a feature.

Pitler and Nenkova have shown that including more features about the right sibling can improve the result. In our experiment on the development set we achieve an improved result by using the part-of-speech of the syntactic head of the right sibling as a feature. To find the syntactic head we use the head finding rules from Collins' PhD thesis [3].

Knott has done an extensive study of discourse connectives and their properties [9]. He classifies discourse connectives into the following categories based on their syntactic types: subordinating conjunctions, coordinating conjunctions, discourse adverbials, prepositional phrases and phrases taking sentence complements. Following Wellner, we use the category of a connective as a feature. Moreover, we use this category in conjunction with the syntactic head of the right sibling as another feature.

4   Results

As mentioned above, there are 18,459 annotated tokens of the 100 explicit discourse connective types in the PDTB. The sentences containing these tokens constitute the set of positive training examples. The sentences containing occurrences of the connective strings in the text of the PDTB which are not annotated as discourse connectives are treated as negative training examples. Pitler and Nenkova and Lin *et al.* report their results which were obtained using a 10-fold cross validation over the PDTB sections 2-22. They used section 0 and 1 as the development dataset. Wellner, however, used sections 2-21 for training, sections 23-24 for testing and section 22 for developing his features. In this work, we follow the first two groups of researchers to prepare the development and training/validation datasets. We have observed that the PDTB section 24 contains many anomalous annotations. As Wellner mentioned in his thesis, coordinating connectives within VP-coordination appear frequently as discourse connectives in section 24, whereas according to the PDTB annotation guidelines such connectives should not be annotated as positive instances. In section 24, 183 instances of *and* are annotated as discourse connectives. In 77 of these instances, *and* is immediately inside a VP. In comparison, among the 2442 positive instances of *and* in sections 2-22, only 26 are immediately dominated by a VP. As we want to compare our feature set with that of all the previous works on a common setting, a 10-fold cross validation on sections 2-22 seems to be the better choice to us. An advantage of using 10-fold cross-validation is that we repeatedly use 90% of the data for training and the remaining 10% for testing. The resulting average accuracy, in most cases, is a reliable estimation of the true accuracy when the model is trained on all data and tested on unseen data [18]. Moreover, extensive testing on numerous datasets, with different learning techniques, has shown that 10 is about the right number of folds to get the best estimate of error, and there is also some theoretical evidence to back this up [26].

We have trained maximum entropy (logistic regression) classifiers using the features discussed above with MALLET[4], an open-source machine learning toolkit. Logistic regression is a discriminative probabilistic classification model. A logistic regression classifier is also called a maximum entropy classifier because it is obtained by following the maximum entropy principle which states that the best probability model for the data is the one which maximizes entropy over the set of probability distributions that are consistent with the evidence [17].

---

[4] http://mallet.cs.umass.edu

By maximizing entropy, in other words maximizing uniformity, this model avoids any unnecessary assumptions and becomes the least committed model. Binary logistic regression models the conditional distribution of the output given the observation as follows:

$$p(y = 1|x) = \frac{1}{1 + exp(-\sum^k \lambda_i f_i(x, y))}$$

where x is an observation, y is the output (class), $f_i(x, y)$ is a binary-valued feature function, the λ_is are the feature weights and k is the total number of features. The parameters of this model are computed by maximum likelihood estimations, i.e., finding the best fit to the training data. We have also experimented with a Naïve Bayes (NB) classifier and the AdaBoost ensemble method with decision tree classifiers. However, the logistic regression classifier gives us the best result. Although both logistic regression and NB classifiers consider the same hypothesis space, NB tries to find each feature's weight independently. So, for a feature set comprising many correlated features, logistic regression often outperforms NB. AdaBoost is a simple meta-algorithm that uses a set of weak classifiers and takes their votes to make a decision. This simple strategy often performs remarkably well for some problems. One advantage of AdaBoost is that it is often less prone to overfitting than other learning algorithms.

Most of the features that we have discussed above are in fact feature templates and can generate a large number of binary features. To minimize data sparsity and to reduce the risk of overfitting we applied feature pruning to remove the features which occurred fewer than two times.

Using the gold standard parses provided in the PTB, Pitler and Nenkova have achieved an F-score of 94.19%. We have replicated their work using their feature set using both gold standard parses and automatic parses generated by an automatic parser, namely the BLLIP reranking parser[5]. Using gold standard parses we get an F-score of 95.34%. F-score drops to 93.58% when we use the automatic parses. Interestingly, our replicated result is significantly better than that of the original work. The reasons behind this may include improvements in MALLET and/or differences in implementation. Lin *et al.* also replicated the Pitler and Nenkova work and achieved an F-score of 92.75% and 91.00% with gold standard parses and automatic parses, respectively.

Adding the new surface level and syntactic features to the P&N features improves the performance of the classifier. With the gold standard parses the F-score increases to 96.22%. As discussed above, the P&N features is a set of syntactic features that try to capture the syntactic context around a connective. However, the local context surrounding the connective is often indicative of discourse usage. The surface level features that we add to the P&N features serve this purpose by giving more information about the local context in the form of next or previous words and phrase types. Inclusion of these additional features help to identify the discourse usage of several subordinate conjunctions

---

[5] https://github.com/BLLIP/bllip-parser

and discourse adverbial better. The specific gains that we obtain are presented in the next section.

The statistical significance of this improvement has been measured by using the Wilcoxon signed-rank test [25]. The test was applied to the differences between the error rates for all 10 folds. The null hypothesis for the test is that the differences are distributed symmetrically around zero, i.e., the error rates are drawn from the same distribution. The test shows that the improvements we obtained are statistically significant at significance level $\alpha = 0.001$ ($p = 0.0009766$).

Our results are better than those reported by Lin *et al.* for their enhanced feature set. They achieved an F-score of 95.76% with gold standard parses. As mentioned earlier, Wellner has reported an evaluation of his classifier on the PDTB sections 23-24. To compare with our results we have trained a classifier with his feature set (both constituent and dependency features) and have evaluated it by doing a 10-fold cross validation over sections 2-22. Using gold standard parses and Wellner's feature set we get an F-score of 95.85%. Using a Wilcoxon signed-rank test, we found that our feature set gave us a statistically significant improvement over this replicated result at $\alpha = 0.001$ ($p = 0.0009766$).

Table 1 summarizes the results we have obtained with both gold standard parses (GS) and automatic parses (AUTO) using three different feature sets, namely P&N, Wellner's feature set (WN) and our feature set, F&M, composed of P&N together with our new features.

Table 1. Comparison of results on the PDTB with different feature sets. The **P&N** and **WN** columns present the results obtained by reimplementing the work of Pitler and Nenkova [13] and Wellner [24], respectively. The column **F&M** shows the results we obtained using our new set of features. GS and AUTO indicate that the results were obtained using gold-standard parses and automatic parses, respectively. These results were obtained by doing a 10-fold cross-validation over the PDTB sections 2-22. Bold numbers indicate the best results and italics indicates statistical significance.

	P&N		**WN**		**F&M**	
	GS	AUTO	GS	AUTO	GS	AUTO
Accuracy	97.24	96.11	97.57	97.34	**97.78**	**97.53**
Precision	94.08	90.37	95.54	95.10	**95.82**	**95.11**
Recall	**96.64**	**97.02**	96.18	95.85	**96.64**	96.52
F-Score	95.34	93.58	95.86	95.47	*96.22*	*95.81*

We have also evaluated our feature set using the BioDRB. This corpus differs from the PDTB in some important ways. Unlike the PDTB, where discourse annotations are aligned with both raw text and parse trees (the PTB), the BioDRB annotations are only aligned with raw text. This means that for each attribute (e.g., the connective phrase itself, the arguments of the connective) of a discourse relation, its annotation contains the offset address of the span of text in the full text article that corresponds to that attribute. To conduct our experiments on the BioDRB in the same way that we did on the PDTB, we need to preprocess it. The sentences in the 24 full text articles are segmented using an open-source natural language processing library named OpenNLP [1]. Parse trees for the segmented sentences are then produced by using the BLLIP re-ranking parser with the self-trained biomedical parsing model [11]. Then, we automatically align the discourse annotations with the parse trees and produce new annotations in the PDTB annotation format. We have made this preprocessed version of the BioDRB publicly available online[6].

One of our syntactic features, namely the connective category features, depends on the syntactic category of the connective. As discussed above, the category of a connective is given in the classification provided by Knott [9]. However, we have found many connectives (e.g., *following, by, to*) in BioDRB which are not classified in [9]. Therefore, we have manually analyzed those connectives and have categorized them into three classes: subordinating conjunctions, coordinating conjunctions and discourse adverbials. Our complete classification is provided in [7].

We have evaluated our feature set on the BioDRB in three different ways. Results of each of these tests are shown in Table 2. The column titled **PDTB-BioDRB** shows the results of using the binary maximum entropy classifier trained on the PDTB sections 2-22 and tested on the BioDRB. Since not all connectives in the BioDRB are also present in the PDTB, for this test we only considered the connectives which are common to both the PDTB and the Bio-DRB. Second, we have done a 10-fold cross-validation on the BioDRB considering only the common connectives. The column **BioDRB-CC** presents the results of this test. Finally, we have done a 12-fold cross-validation on the BioDRB considering all of the 179 connective types in it. The results we obtained are shown in column **BioDRB-12F-CV**.

Using our feature set we have achieved a significantly better result than what has been reported by Ramesh and Yu [16]. By doing a 12-fold cross-validation on the BioDRB, they obtain precision, recall and F-score of 79%, 63% and 69% respectively. The reasons why we achieve better results include the fact that they use the feature set of a biomedical named entity recognizer, named ABNER[7] [21]. The features that ABNER uses are mainly orthographic features, i.e., surface level features [20]. We have found that, for discourse connective identification, syntactic features are more powerful than surface level features.

With our set of features mentioned above, we obtain inferior results for the Bio-DRB in comparison to the results obtained with the PDTB. There are a few reasons behind this. First, the number of discourse connective types annotated in the BioDRB is greater than that in the PDTB (179 versus 100). Second, there are a few connective types in the BioDRB (e.g., *to, by*) which are very frequently used as non-discourse function words. Since they very rarely act as discourse connectives, they are difficult to identify when they do so. In addition, because of the presence of such connectives, the number of negative training examples in the BioDRB

---

[6] https://github.com/syeedibnfaiz/BioDRB
[7] http://pages.cs.wisc.edu/~bsettles/abner/

Table 2. Results of the evaluation of the feature set on the BioDRB. The **PDTB-BioDRB** column presents the evaluation of the classifier which was trained on the PDTB and tested on BioDRB. The results of doing a 12-fold cross-validation over the BioDRB is presented in the **BioDRB-12F-CV** column. The second column, **BioDRB-CC**, shows the results obtained by doing a 10-fold cross-validation over the BioDRB considering only the connectives which are common to both the PDTB and BioDRB. Bold numbers indicate the best results.

	PDTB-BioDRB	BioDRB-CC	BioDRB-12F-CV
Accuracy	91.43	93.73	**94.34**
Precision	86.16	**87.34**	85.17
Recall	75.00	**85.36**	79.80
F-Score	80.19	**86.28**	82.36

is five times greater than the number of positive training examples. Third, since there isn't a gold standard parsed treebank for the BioDRB, we had to depend on an automatic parser which may produce incorrect parses resulting in incorrect syntactic features for both the training and testing phases. The BLLIP reranking parser has an F-measure of 91.02% on section 23 of the PTB (equivalently, the PDTB) [2]. With this in mind, it should also be noted that a typical sentence in the BioDRB is structurally more complex than one in the PDTB. The average sentence length in the PDTB is 23 words, whereas in the BioDRB it is 29. To compare the syntactic complexity, we use the Yngve syntactic complexity measure, which relies on the right branching nature of English syntax trees with the assumption that deviating from that introduces complexity in the language [19]. According to this measure, we found the syntactic complexity of the sentences in the PDTB and BioDRB to be 3.31 and 3.55, respectively. Therefore, a parser is more likely to make errors when parsing a sentence from the BioDRB than when parsing a sentence from the PDTB.

5   Discussion

The previous section has focussed on demonstrating the improvements that we have achieved in the identification of explicit discourse connectives in two standard corpora. What follows discusses other findings drawn from our analysis.

From Table 1 it can be observed that our proposed feature set is more robust than the P&N features. Using the F&M features, the F-score achieved with gold standard parses and automatic parses differs by only 0.4 percentage points, whereas using the P&N features the difference is about 1.7 percentage points, showing that our feature set is less dependent on the quality of the parse.

Connective or connective category specific ML models have proven to be useful for discourse argument identification [6]. Following that path, we have experimented with connective category specific classifiers for discourse connective identification. However, we obtained results inferior to what we get using

Table 3. Connective type-wise results on the PDTB. The **P&N** columns are the results of the connective type-wise evaluation using our implementation using the feature set proposed in [13]. The **F&M** columns are the results of the connective type-wise evaluation using our proposed feature set. Bold numbers indicate the best results.

	Coordinating Conjunction		Subordinating Conjunction		Discourse Adverbial	
	P&N	**F&M**	**P&N**	**F&M**	**P&N**	**F&M**
Accuracy	98.81	**98.89**	96.69	**97.62**	93.92	**95.08**
Precision	96.95	**97.39**	91.86	**94.67**	93.69	**95.15**
Recall	**98.05**	97.94	96.17	**96.53**	95.09	**95.73**
F-Score	97.49	**97.66**	93.95	**95.59**	94.38	**95.44**

a general classifier. We have also evaluated the feature sets on each connective category. Table 3 shows the connective category-wise results of 10-fold cross-validation over the PDTB sections 2–22 using the P&N and F&M feature sets. Using the new syntactic and surface features we have been able to increase the F-score on the subordinating conjunction and discourse adverbial categories by 1.64 percentage point and 1 percentage point, respectively.

By analyzing classifier errors, we found that it is sometimes not possible to identify connectives based only on syntactic and surface level context, and it seems that some form of semantic understanding is required. For example, one problem with subordinating conjunctions like *when, after* and *before* is that they often occur as temporal modifiers and the syntactic context we consider does not indicate that. The following sentence shows one such scenario in which *when* is used as a temporal modifier.

(3) Notably, one of Mr. Krenz's few official visits overseas came a few months ago, *when* he visited China after the massacre in Beijing.

To overcome this problem we have experimented with a feature created just for subordinating conjunctions. This feature is computed by first taking the parent of the SBAR[8] that dominates the connective and then checking whether it dominates a terminal node having a temporal sense. Words including the names of months, dates, etc. are considered to have a temporal sense. Earlier, we have seen how such words combined with an immediately following connective like *when* are good surface level features. Here, we suggest examining a larger context than just the previous word. Inclusion of this feature slightly improved the performance of the classifier on connectives which are subordinating conjunctions. The F-score on this category improved from 95.59% to 95.77%. We believe that including such semantic features will further improve the classifier's performance. The F&M results in Table 1 do not include this feature.

We have found that only a handful of the connectives cause most of the errors. More precisely, 62% of the errors are due to misclassification of the connectives

---

[8] Indicates a clause introduced by a (possibly empty) subordinating conjunction.

and, as, when, after, also and *if.* We believe that some of the misclassifications of *and* are due to inconsistencies in the annotation of conjunctions in a coordinated verb phrase. Although the PDTB annotation guideline states such conjunctions are not annotated as discourse connectives, we found what we believe to be exceptions to this rule.

Besides using the feature-based ML method discussed above we have also experimented with kernel methods. The advantage of using kernel methods over feature-based methods is that we can use a rich structured representation of instances without having to explicitly explore a very large feature space [5,23]. This is possible because kernel functions, which are at the heart of kernel methods, directly compute a similarity score between a pair of instances without having to transform them into very high dimensional feature vectors. For the problem of discourse connective identification, we have shown how the syntactic context around the discourse connectives can be used to identify them. In a feature-based method we represent the context as a set of features observed in that syntactic context. When we use a kernel method we can represent the syntactic context directly as a tree and define a kernel function that can compute the similarity between a pair of such trees. In this experiment we have used a simple kernel known as the subset tree kernel (SST) [4]. This kernel computes the similarity between two trees based on the number of substructures that are in common between them. We represent a connective instance by a subtree that includes the connective, its parent, its left and right siblings and their immediate children. We used the SST kernel implementation of [12] which is built on top of a well known implementation of Support Vector Machines [8]. Using a 10-fold cross validation over the PDTB sections 2-22 we obtain precision, recall and F-score of 91.03%, 96.87% and 93.86%, respectively. For such a simple setting, these results seem quite promising to us. In this work, we have shown that considering surface level context beside the syntactic context helps to achieve a better result. Therefore, composite kernels that take both of these two types of context into account to compute similarity are likely to offer improved results. The simple SST kernel does not distinguish between subtrees appearing in the subtrees rooted at the left or right sibling of a connective. Having kernels that compute more intelligently the similarity between the syntactic contexts surrounding connectives may lead to a better result.

6   Conclusions and Future Work

In this paper we have considered the problem of identifying explicit discourse connectives in text. We have applied machine learning with a feature set that includes features proposed in existing works along with some new surface level and syntactic features to build maximum entropy (logistic regression) classifiers that achieve better performance than what has been obtained with previously proposed feature sets. The new surface level features capture information about a connective's surrounding phrases. The new syntactic features gather more information from the path in the syntax tree connecting the root of the tree and the connective and use the part of speech of the syntactic head of the clause following the connective. We have evaluated our feature set using two publicly available discourse annotated corpora, namely the Penn Discourse TreeBank and the Biomedical Discourse Relation Bank and achieved statistically significant improvement over existing works. We have also applied a kernel-based method to this problem and have found reasonably good results with a simple kernel function.

Identifying the explicit discourse connectives is the first step of a bigger problem, known as discourse parsing. The remaining steps include identifying the sense of the connectives and their arguments. We have approached these problems and the results we have obtained will be reported in another paper. With regards to the results of identifying explicit discourse connectives with a kernel-based method, they seem promising to us and we will continue to investigate more in this direction.

Acknowledgments. This work was partially funded through a Natural Sciences and Engineering Research Council of Canada (NSERC) Discovery Grant to Robert E. Mercer. Fruitful interactions with Emily Pitler and Ben Wellner and suggestions from the reviewers have improved the interpretation and presentation of our results.

References

1. Apache Software Foundation: Apache OpenNLP (2012)
2. Charniak, E., Johnson, M.: Coarse-to-fine n-best parsing and MaxEnt discriminative reranking. In: Proceedings of the 43rd Annual Meeting of the ACL, pp. 173–180 (2005)
3. Collins, M.: Head-driven statistical models for natural language parsing. PhD thesis, University of Pennsylvania (1999)
4. Collins, M., Duffy, N.: Convolution kernels for natural language. In: Advances in Neural Information Processing Systems 14, pp. 625–632. MIT Press (2001)
5. Cristianini, N., Shawe-Taylor, J.: An Introduction to Support Vector Machines and Other Kernel-based Learning Methods. Cambridge University Press (2000)
6. Elwell, R., Baldridge, J.: Discourse connective argument identification with connective specific rankers. In: Proceedings of the 2nd IEEE International Conference on Semantic Computing, pp. 198–205 (2008)
7. Ibn Faiz, M.S.: Discovering higher order relations from biomedical text. Master's thesis, The University of Western Ontario, London, Ontario, Canada (2012)
8. Joachims, T.: Making large-scale support vector machine learning practical. In: Schölkopf, B., Burges, C.J.C., Smola, A.J. (eds.) Advances in Kernel Methods: Support Vector Learning, pp. 169–184. MIT Press (1999)
9. Knott, A.: A Data-Driven Methodology for Motivating a Set of Coherence Relations. PhD thesis, University of Edinburgh, Edinburgh (1996)
10. Lin, Z., Ng, H.T., Kan, M.Y.: A PDTB-styled end-to-end discourse parser. CoRR. Volume arXiv:1011.0835 (2010)
11. McClosky, D., Charniak, E., Johnson, M.: Automatic domain adaptation for parsing. In: Human Language Technologies: The 2010 Annual Conference of the North American Chapter of the ACL (HLT 2010), pp. 28–36 (2010)

12. Moschitti, A.: Making tree kernels practical for natural language learning. In: Proceedings of the 11th Conference of the European Chapter of the ACL (EACL 2006), pp. 113–120 (2006)
13. Pitler, E., Nenkova, A.: Using syntax to disambiguate explicit discourse connectives in text. In: Proceedings of the ACL-IJCNLP 2009 Conference Short Papers (ACLShort 2009), pp. 13–16 (2009)
14. Prasad, R., Dinesh, N., Lee, A., Miltsakaki, E., Robaldo, L., Joshi, A.K., Webber, B.L.: The Penn Discourse TreeBank 2.0. In: Proceedings of the 6th International Conference on Language Resources and Evaluation, LREC 2008 (2008)
15. Prasad, R., McRoy, S., Frid, N., Joshi, A., Yu, H.: The biomedical discourse relation bank. BMC Bioinformatics 12(1), 188–205 (2011)
16. Ramesh, B.P.P., Yu, H.: Identifying discourse connectives in biomedical text. In: Proceedings of the American Medical Informatics Association Fall Symposium (AIMA 2010), pp. 657–661 (2010)
17. Ratnaparkhi, A.: Maximum entropy models for natural language ambiguity resolution. PhD thesis, University of Pennsylvania (1998)
18. Refaeilzadeh, P., Tang, L., Liu, H.: Cross-Validation. In: Liu, L., Özsu, M.T. (eds.) Encyclopedia of Database Systems, pp. 532–538. Springer (2009)
19. Roark, B., Mitchell, M., Hollingshead, K.: Syntactic complexity measures for detecting Mild Cognitive Impairment. In: Biological, Translational, and Clinical Language Processing, pp. 1–8. Association for Computational Linguistics (2007)
20. Settles, B.: Biomedical named entity recognition using conditional random fields and rich feature sets. In: Proceedings of the International Joint Workshop on Natural Language Processing in Biomedicine and its Applications (JNLPBA 2004), pp. 104–107 (2004)
21. Settles, B.: ABNER: An open source tool for automatically tagging genes, proteins, and other entity names in text. Bioinformatics 21(14), 3191–3192 (2005)
22. Sutton, C., McCallum, A.: An introduction to conditional random fields for relational learning. In: Getoor, L., Taskar, B. (eds.) Introduction to Statistical Relational Learning. MIT Press (2007)
23. Vapnik, V.N.: Statistical learning theory. Wiley (1998)
24. Wellner, B.: Sequence models and ranking methods for discourse parsing. PhD thesis, Brandeis University, Waltham, MA, USA (2009)
25. Wilcoxon, F.: Individual Comparisons by Ranking Methods. Biometrics Bulletin 1(6), 80–83 (1945)
26. Witten, I.H., Frank, E.: Data Mining: Practical Machine Learning Tools and Techniques with Java Implementations, 1st edn. Morgan Kaufmann (1999)

Unsupervised Extraction of Diagnosis Codes from EMRs Using Knowledge-Based and Extractive Text Summarization Techniques

Ramakanth Kavuluru[1,2,*], Sifei Han[2], and Daniel Harris[2]

[1] Division of Biomedical Informatics, Department of Biostatistics
[2] Department of Computer Science
University of Kentucky, Lexington, KY
{ramakanth.kavuluru,eric.s.han,daniel.harris}@uky.edu

Abstract. Diagnosis codes are extracted from medical records for billing and reimbursement and for secondary uses such as quality control and cohort identification. In the US, these codes come from the standard terminology ICD-9-CM derived from the international classification of diseases (ICD). ICD-9 codes are generally extracted by trained human coders by reading all artifacts available in a patient's medical record following specific coding guidelines. To assist coders in this manual process, this paper proposes an unsupervised ensemble approach to automatically extract ICD-9 diagnosis codes from textual narratives included in electronic medical records (EMRs). Earlier attempts on automatic extraction focused on individual documents such as radiology reports and discharge summaries. Here we use a more realistic dataset and extract ICD-9 codes from EMRs of 1000 inpatient visits at the University of Kentucky Medical Center. Using named entity recognition (NER), graph-based concept-mapping of medical concepts, and extractive text summarization techniques, we achieve an example based average recall of 0.42 with average precision 0.47; compared with a baseline of using only NER, we notice a 12% improvement in recall with the graph-based approach and a 7% improvement in precision using the extractive text summarization approach. Although diagnosis codes are complex concepts often expressed in text with significant long range non-local dependencies, our present work shows the potential of unsupervised methods in extracting a portion of codes. As such, our findings are especially relevant for code extraction tasks where obtaining large amounts of training data is difficult.

1 Introduction

Extracting codes from standard terminologies is a regular and indispensable task often encountered in medical and healthcare fields. Diagnosis codes, procedure codes, cancer site and morphology codes are all manually extracted from patient records by trained human coders. The extracted codes serve multiple purposes

---

* Corresponding author.

O. Zaïane and S. Zilles (Eds.): Canadian AI 2013, LNAI 7884, pp. 77–88, 2013.

including billing and reimbursement, quality control, epidemiological studies, and cohort identification for clinical trials. In this paper we focus on extracting international classification of diseases, clinical modification, 9th revision (ICD-9-CM) diagnosis codes from electronic medical records (EMRs), although our methods are general and also apply to other medical code extraction tasks.

Diagnosis codes are the primary means to systematically encode patient conditions treated in healthcare facilities both for billing purposes and for secondary data usage. In the US, ICD-9-CM (just ICD-9 henceforth) is the coding scheme still used by many healthcare providers while they are required to comply with ICD-10-CM, the next and latest revision, by October 1, 2014. Regardless of the coding scheme used, both ICD code sets are very large, with ICD-9 having a total of 13,000 diagnoses while ICD-10 has 68,000 diagnosis codes [1] and as will be made clear in the rest of the paper, our methods will also apply to ICD-10 extraction tasks. ICD-9 codes contain 3 to 5 digits and are organized hierarchically: they take the form abc.xy where the first three character part before the period abc is the main disease category, while the x and y components represents subdivisions of the abc category. For example, the code 530.12 is for the condition *reflux esophagitis* and its parent code 530.1 is for the broader condition of *esophagitis* and the three character code 530 subsumes all *diseases of esophagus*. Any allowed code assignment should at least assign codes at the category level (that is, the first three digits). At the category levels there are nearly 1300 different ICD-9 codes. In our current work, we only work on predicting the category level codes. That is, if the actual code is abc.xy, our methods will only be able to generate abc as the correct category code.

The process of assigning diagnosis codes is carried out by trained human coders who look at the entire EMR for a patient visit to assign codes. Majority of the artifacts in an EMR are textual documents such as discharge summaries, operative reports, and progress notes authored by physicians, nurses, or social workers who attended the patient. The codes are assigned based on a set of guidelines [2] established by the National Center for Health Statistics and the Centers for Medicare and Medicaid Services. The guidelines contain rules that state how coding should be done in specific cases. For example, the signs and symptoms (780-799) codes are often not coded if the underlying causal condition is determined and coded.

In this paper we propose an unsupervised ensemble approach to extract ICD-9 codes and test it on a realistic dataset curated from the University of Kentucky Medical Center. Our approach is based on named entity recognition (NER), knowledge-based graph mining, and extractive text summarization methods. We emphasize that automatic medical coding systems, including our current attempt, are generally not intended to replace trained coders but are mainly motivated to expedite the coding process and increase the productivity of medical record coding and management. Hence we take a recall oriented approach with a lesser emphasis on precision. In the rest of the paper, we first discuss related work and the context of our paper in Section 2. We describe our dataset in Section 3 and elaborate our methods in Section 4. We provide an overview of the evaluation measures in Section 5 and present our results in Section 6.

2   Related Work

Several attempts have been made to extract ICD-9 codes from clinical documents since the 1990s. Advances in natural language and semantic processing techniques contributed to a recent surge in automatic extraction. de Lima et al. [3] use a hierarchical approach utilizing the alphabetical index provided with the ICD-9-CM resource. Although completely unsupervised, this approach is limited by the index not being able to capture all synonymous occurrences and also the inability to code both specific exclusions and other condition specific guidelines. Gunderson et al. [4] extracted ICD-9 codes from short free text diagnosis statements that were generated at the time of patient admission using a Bayesian network to encode semantic information. However, in the recent past, concept extraction from longer documents such as discharge summaries has gained interest. Especially for ICD-9 code extraction, recent results are mostly based on the systems and dataset developed for the BioNLP workshop shared task on multi-label classification of clinical texts [5] in 2007.

An important issue in clinical document analysis is the absence of datasets that are free to use by other researchers due to patient privacy concerns and regulations. The BioNLP shared task [5] takes an important first step in providing such a dataset which consists of 1954 radiology reports arising from outpatient chest x-ray and renal procedures and are observed to cover a substantial portion of pediatric radiology activity. The radiology reports were also formatted in XML with explicit tags for *history* and *impression* fields. Finally, there were a total of 45 unique codes and 94 distinct combinations of these codes in the dataset. The dataset was split into training and testing sets of nearly equal size where example reports for all possible codes and combinations occur in both sets. This means that all possible combinations that will be encountered in the test set are known ahead of time. The top system obtained a micro-average F-score of 0.89 and 21 of the 44 participating systems scored between 0.8 and 0.9. Next we list some notable results that fall in this range obtained by various participants and others who used the dataset later. The techniques used range from completely handcrafted rules to fully automated machine learning approaches. Aronson et al. [6] adapted a hybrid MeSH term indexing program MTI that is in use at the National Library of Medicine (NLM) and included it with SVM and k nearest neighbor classifiers for a hybrid *stacked* model. Goldstein et al. [7] applied three different classification approaches - traditional information retrieval using the search engine library Apache Lucene, Boosting, and rule-based approaches. Crammer et al. [8] use an online learning approach in combination with a rule-based system. Farkas and Szarvas [9] use an interesting approach to induce new rules and acquire synonyms using decision trees.

We believe that the coverage of pediatric radiology activity, the small number of codes and their combinations where code combinations are known ahead of time, and the clear demarcation of history and impression fields do not provide a realistic representation of EMRs, especially for in-patient visits. It is well known that ICD-9 codes are extracted from the full EMR [10] for each in-patient visit where the EMR includes documents such as emergency department notes, discharge summaries, radiology reports, pathology reports, operative reports, progress notes, and multiple flow sheets. Aronson et al. [6] also discuss the narrow focus on cough/fever/pneumonia and urinary/kidney problems and the relatively error-free clinical text present in the BioNLP radiology report dataset as a possible limitation for the extensibility of techniques to generalized EMRs.

3   In-Patient EMR Dataset

As a first step to study automatic diagnosis coding at the EMR level, we curated a dataset of 1000 clinical document sets corresponding to a randomly chosen set of 1000 in-patient visits to the University of Kentucky (UKY) Medical Center in the month of February, 2012. We also collected the ICD-9-CM codes for these EMRs assigned by trained coders at the UKY medical records office. Aggregating all billing data, this dataset has a total of 7480 diagnoses leading to 1811 unique ICD-9 codes that map to 633 top level codes (three character categories). Using the (code, label, count) representation the top 5 most frequent codes are (401, *essential hypertension*, 325), (276, *Disorders of fluid electrolyte and acid-base balance*, 239), (305, *nondependent abuse of drugs*, 236), (272, *disorders of lipoid metabolism*, 188), and (530, *diseases of esophagus*, 169). The average number of codes is 7.5 per EMR with a median of 6 codes. There are EMRs with only one code, while the maximum number assigned to an EMR is 49 codes. For each in-patient visit, the original EMR consisted of several documents, some of which are not conventional text files but are stored in the RTF format. Some documents, like care flowsheets, vital signs sheets, ventilator records were not considered for this analysis. We have a total of 5583 documents for all 1000 EMRs. While our correct codes arise from billing data where different trained coders code each EMR (one per coder), the BioNLP shared task dataset is a high quality dataset coded by three different companies with a final correct code set generated by consolidation of the three sets of codes.

Before we proceed to our methods, we note that after a discussion with our medical records officer, we learned that the coders do not necessarily code conditions purely from a billing perspective. On the contrary, they are trained to extract all codes following the coding guidelines even if the patient may not be billed for them eventually. However, as explained towards the end of Section 1, the coding guidelines might not allow certain codes even though they are discussed in the EMRs because of some specific restrictions on how coding should be done. In our unsupervised methods we do not model the logic behind the coding guidelines and hence our approach is primarily a recall oriented one. However, we use text summarization techniques to weed out codes (so, to improve precision) that are extracted from noun phrases (in EMR narratives) using certain statistical measures (more on this later).

4   Our Approach

To extract diagnosis codes, we used a combination of three methods: NER, knowledge-based graph mining, and extractive text summarization. In this section we elaborate on the specifics of each of these methods. Before we proceed, we first discuss the Unified Medical Language System (UMLS), a biomedical knowledge base used in our NER and graph mining methods.

4.1   Unified Medical Language System

The UMLS[1] is a large domain expert driven aggregation of over 160 biomedical terminologies and standards. It functions as a comprehensive knowledge base and facilitates interoperability between information systems that deal with biomedical terms. It has has three main components: Metathesaurus, Semantic Network, and SPECIALIST lexicon. The Metathesaurus has terms and codes, henceforth called *concepts*, from different terminologies. Biomedical terms from different vocabularies that are deemed synonymous by domain experts are mapped to the same Concept Unique Identifier (CUI) in the Metathesaurus. The semantic network acts as a typing system that is organized as a hierarchy with 133 *semantic types* such as *disease or syndrome, pharmacologic substance,* or *diagnostic procedure.* It also captures 54 important relationships (or relation types) between biomedical entities in the form of a relationship hierarchy with relationships such as *treats, causes,* and *indicates.* The Metathesaurus currently has about 2.8 million concepts with more than 12 million relations connecting these concepts. Although relations in the Metathesaurus have relation types that are beyond the 54 available through the semantic network, here we would like to limit ourselves to high level relation types such as *parent, child, rel_narrow,* and *rel_broad.* The high level relations can be represented as $C1 \rightarrow\ < rel - type > \rightarrow\ C2$ where $C1$ and $C2$ are concepts in the UMLS and $< rel - type > \in \{parent, child, rel\_narrow, rel\_broad\}$. The semantic interpretation of these relations (or triples) is that the $C1$ is related to $C2$ via the relation type $< rel - type >$. The *child* (resp. *parent*) relationship means that concept $C1$ has $C2$ as a child (resp. parent). The *rel_broad* (resp. *rel_narrow*) type means that $C1$ represents a broader (resp. narrower) concept than $C2$. For example, the concept *hypertensive disease* is a broader concept compared to *systolic hypertension.* These broad and narrow relationships are created by experts to capture those relationships that cannot be captured by the more rigid parent/child relationships in different source vocabularies. The SPECIALIST lexicon is useful for lexical processing and variant generation of different biomedical terms.

4.2   Named Entity Recognition: MetaMap

NER is a well known application of natural language processing (NLP) techniques where different entities of interest such as people, locations, and institutions are automatically recognized from mentions in free text (see [11] for a

---

[1]   UMLS Reference Manual: http://www.ncbi.nlm.nih.gov/books/NBK9676/

survey). Named entity recognition in biomedical text is difficult because linguistic features that are normally useful (e.g., upper case first letter, prepositions before an entity) in identifying generic named entities are not useful when identifying biomedical named entities, several of which are not proper nouns. Hence, NER systems in biomedicine rely on expert curated lexicons and thesauri. In this work, we use MetaMap [12], a biomedical NER system developed by researchers at the NLM. MetaMap uses a dictionary based approach (using the UMLS concept names as the dictionary) in combination with heuristics for partial mapping (based on lexical information in the SPECIALIST lexicon) to extract UMLS concepts. MetaMap can process a textual document as a whole but can also generate UMLS concepts from individual noun phrases that are passed as input to it. The latter option is more helpful to identify more specific concepts from longer phrases. Since more specific diagnosis codes are more valuable than generic codes (systolic hypertension vs hypertension), we used the latter approach called "term processing" in MetaMap's manual. MetaMap also identifies negations of concepts and hence we only used non-negated disorders when extracting codes. So as the first step in our code extraction pipeline, we extract biomedical named entities by running MetaMap on noun phrases that satisfy the following regular expressions based on those used in the paper [13].

1. `Noun* Noun`.
2. `(Adj|Noun)+ Noun`, and
3. `((Adj|Noun)+ | ((Adj|Noun)* (NounPrep)?)(Adj|Noun)*)Noun`

Here `Adj` stands for adjective and `NounPrep` stands for a noun followed by a preposition. Note that we allow the presence of a single preposition to capture phases like "malignant neoplasm of colon". We also allow single token noun phrases that just consist of one noun. For instance, both "hypertension" and "systolic hypertension" will be processed by MetaMap for concept extraction when the latter phrase occurs in text. However, we have a way of assigning more weight to specific codes using key phrase extraction covered in Section 4.4. Once we obtain non-negated UMLS concepts using MetaMap from these phrases, we convert these concepts to ICD-9 diagnosis codes when possible as explained next.

ICD-9-CM is one of the over 160 source vocabularies integrated into the UMLS Metathesaurus. As such, concepts in ICD-9-CM also have a concept unique identifier (CUI) in the Metathesaurus. As part of its output, for each concept, MetaMap also gives the source vocabulary. The concepts from MetaMap with source vocabulary ICD-9-CM finally become the set of extracted codes for each EMR document set. However, this code set may not be complete because of missing relationships between UMLS concepts. That is, in our experience, although MetaMap identifies a disorder concept, it might not always map it to a CUI associated with an ICD-9 code; it might map it to some other terminology different from ICD-9, in which case we miss a potential ICD-9 code because the UMLS mapping is incomplete. We deal with this problem and explore a graph based approach in the next section.

4.3   UMLS Knowledge-Based Graph Mining

As discussed in Section 4.2, the NER approach might result in poor recall because of lack of completeness in capturing synonymy in the UMLS. However, using the UMLS graph with concepts (or equivalently CUIs) as nodes and the inter-concept relationships connected by relation types *parent* and *rel_broad* as edges, we can map a original CUI without an associated ICD-9 code to a CUI with an associated diagnosis code. We adapt the approach originally proposed by Bodenreider et al. [14] for this purpose. The mapping algorithm starts with a CUI c output by MetaMap that is not associated with an ICD-9 code and tries to map it to an ICD-9 codes as follows.

1. Recursively, construct a subgraph G_c (of the UMLS graph) consisting of ancestors of the input non-ICD-9 CUI c, using the *parent* and *rel_broad* edges. Build a set I_c of all the ICD-9 concepts associated with nodes added to G_c along the way in the process of building G_c. Note that many nodes added to G_c may not have associated ICD-9 codes.
2. Delete any concept c_1 from I_c if there exists another concept c_2 such that
   - c_1 is an ancestor of c_2, and
   - The length of the shortest path from c to c_2 is less than the length of the shortest path from c to c_1.
3. Return the ICD-9 codes of remaining concepts in I_c and the corresponding shortest distances from c.

Note that the algorithm essentially captures ancestors of the input concept and tries to find ICD-9 concepts in them. We also see that instead of returning a single code, the algorithm returns a set of ICD-9 codes (possibly singleton or empty). If the set has more than one code, all resulting ICD-9 codes are included in the extracted code set for performance evaluation purposes.

4.4   Extractive Text Summarization: C-Value Method

Extractive text summarization is an approach where short summaries of a collection of documents are generated by selecting a few sentences or phrases from those documents that represent the gist of the collection in some way. Key phrase extraction algorithms including the C-value method [13] and TextRank [15] belong to a category of summarization algorithms that extract top phrases that capture a summary of a collection of documents. In this paper, we apply the C-value method to rank the noun phrases that were used to extract ICD-9 codes in Section 4.2. The ranking on the noun phrases automatically imposes a ranking on the codes extracted from them using the approaches outlined in Sections 4.2 and 4.3. The C-value of a noun phrase is computed based on its frequency and the frequencies of longer phrases that contain it in the given set of documents. We use all the documents in an EMR as the corpus to extract candidate noun phrases for code extraction. Hence, we also use the same set of documents to compute the frequencies of all phrases required to compute the C-value of a given phrase. The C-value formula can be written as

$$C(p) = \begin{cases} \log_2(len(p)) \cdot f(p) & \text{if } p \text{ is not nested} \\ \log_2(len(p)) \cdot \left(f(p) - \frac{1}{|T_p|} \sum_{q \in T_p} f(q) \right) & \text{if } p \text{ is nested} \end{cases}$$

where $C(p)$ is the C-value of phrase p, $len(p)$ is number of words in p, and T_p is the set of the longer noun phrases that contain p, and $f(p)$ is the frequency of p in all documents from an EMR. If p is not nested, it implies that it does not appear in longer phrases. When it is nested, we discount its C-value based on the number of its occurrences in longer phrases (the $\sum_{q \in T_p} f(q)$ part) and dampen this discount based on the number of unique longer phrases that contain it (the $\frac{1}{|T_p|}$ part). We chose to include all codes arising from phrases whose C-value is ≥ 1. Although we had phrases that had C-value as high as 20, including only codes whose phrases had very high C-values resulted in many missed codes. Hence we chose those codes arising from phrases with C-values ≥ 1 based on the needs of our recall oriented task. Before applying the C-value filter to eliminate nonsignificant codes extracted using MetaMap and graph based methods, we also applied a different filter that eliminated codes that are extracted from a set of very frequent phrases (mostly single nouns) that result in common symptoms like cold. This is akin to a stop word list used in information retrieval and text classification research.

5   Evaluation Measures

Before we discuss our findings, we establish notation to be used for evaluation measures. Let M be the set of all EMR records; here $|M| = 1000$ since we have 1000 EMRs. Let E_i and B_i, $i = 1, \ldots, 1000$, be the set of extracted codes using our methods from EMR documents and the corresponding set of billing codes respectively for the EMR of the i-th in-patient visit. Since the task of assigning multiple codes to an EMR is the multi-label classification problem, there are multiple complementary methods [16] for evaluating automatic approaches for this task. Here we use EMR-based precision and recall and code label based micro and macro precision, recall, and F-score. First we discuss the EMR-based measures. The average EMR-based precision P_{emr} and recall R_{emr} are

$$P_{emr} = \frac{1}{|M|} \cdot \sum_{i=1}^{|M|} \frac{|E_i \cap B_i|}{|E_i|} \quad \text{and} \quad R_{emr} = \frac{1}{|M|} \cdot \sum_{i=1}^{|M|} \frac{|E_i \cap B_i|}{|B_i|}.$$

On the other hand, considering each code as a label, we define the code-based measures. For each code C_j in the billing code set C of the dataset, we have code-based precision $P(C_j)$, recall $R(C_j)$, and F-score $F(C_j)$ defined as

$$P(C_j) = \frac{TP_j}{TP_j + FP_j}, \quad R(C_j) = \frac{TP_j}{TP_j + FN_j}, \quad \text{and} \quad F(C_j) = \frac{2P(C_j)R(C_j)}{P(C_j) + R(C_j)},$$

where TP_j, FP_j, and FN_j are true positives, false positives, and false negatives, respectively of code C_j. Now code-based macro average precision, recall, and F-score are defined as

$$P_c^{macro} = \frac{\sum_{j=1}^{|C|} P(C_j)}{|C|}, \quad R_c^{macro} = \frac{\sum_{j=1}^{|C|} R(C_j)}{|C|}, \quad \text{and } F_c^{macro} = \frac{\sum_{j=1}^{|C|} F(C_j)}{|C|},$$

respectively. Finally, the code-based micro precision, recall, and F-score are defined as

$$P_c^{micro} = \frac{\sum_{j=1}^{|C|} TP_j}{\sum_{j=1}^{|C|}(TP_j + FP_j)}, \quad R_c^{micro} = \frac{\sum_{j=1}^{|C|} TP_j}{\sum_{j=1}^{|C|}(TP_j + FN_j)},$$

$$\text{and } F_c^{micro} = \frac{2P_c^{micro} R_c^{micro}}{P_c^{micro} + R_c^{micro}},$$

respectively. While the macro measures consider all codes as equally important, micro measures tend to give more importance to codes that are more frequent.

6 Results and Discussion

First we present our EMR based average precision and recall results in Table 1. In our experiments, ICD-9 codes that are associated with concepts at a distance greater than 1 from the input concept in the graph mining approach (Section 4.3) improved recall by only 1% with a 2% decrease in precision. Hence here we only report results when the shortest distance between the input concept and the ICD-9 ancestors is ≤ 1. Distance zero codes are those that are directly obtained from MetaMap output without having to use the graph mining method. The "No C-value" column in all the tables in this section means that C-value restriction is not applied to the noun phrases used for code extraction.

Using the graph mining approach we see an improvement of 12% in recall from 0.3 to 0.42. Without any recall loss, the C-value method improves precision by 7% when using the graph mining approach and by 6% when not using it. Thus we see a clear advantages of the key phrase scoring approach in increasing precision and the knowledge based graph mining approach in increasing recall. The 99% confidence interval ranges when using C-value ≥ 1 without graph mining are $0.28 \leq R_{emr} \leq 0.32$ and $0.50 \leq P_{emr} \leq 0.56$; the same ranges using both C-value and graph mining are $0.38 \leq R_{emr} \leq 0.44$ and $0.45 \leq P_{emr} \leq 0.49$.

Next we present our macro averaged recall, precision, and F-scores in Table 2. These results provide a contrast to the observations made in the EMR based measures. Although there is an 8% increase in recall using the graph mining

Table 1. EMR-Based Average Precision and Recall

	without graph-mining		graph distance ≤ 1	
	No C-value	C-value ≥ 1	No C-value	C-value ≥ 1
R_{emr}	0.30	0.30	0.42	0.42
P_{emr}	0.47	0.53	0.40	0.47

Table 2. Code-Based Macro Precision and Recall

	without graph-mining		graph distance ≤ 1	
	No C-value	C-value ≥ 1	No C-value	C-value ≥ 1
R_c^{macro}	0.58	0.53	0.66	0.62
P_c^{macro}	0.74	0.79	0.64	0.69
F_c^{macro}	0.57	0.56	0.57	0.58

approach, we notice that the recall gain comes at an expense of 10% loss in precision. However, we believe this is a still a reasonable although not ideal situation especially considering that our goal is a recall oriented approach to expedite the coding process. The C-value method increases precision by an amount equal to the loss in recall. However, this is not ideal as recall is more important to us. But we note that macro measures give equal importance to all codes. That is codes that occur very infrequently are also scored the same way frequent scores are scored.

Before we discuss micro measures, we note that several codes that were in billing were never extracted using our methods (more on this in the Discussion section). The F-scores for all those codes will be zero. Thus, we chose to compute micro measures over subsets of the set of all billing codes C. We choose two particular subsets. The first set is the set of all codes in billing that were retrieved at least once using a particular configuration of our methods; so these are the set of codes C_j for which the F-score $F(C_j) > 0$, whose results are presented in Table 3. Since the F-score changes with the method, we also show the number of codes that satisfy $F(C_j) > 0$ for each technique as the last row.

We also computed micro measures over the set of codes that satisfy $F(C_j) > 0.5$ whose results are shown in Table 4. We realize that showing micro measures for codes that were extracted at least once may overestimate the performance of methods, which is not our intention. Our only purpose of showing these results is to demonstrate how our unsupervised approach works on a subset of the codes and to quantify the difference between different components of our approach.

Table 3. Code-Based Micro Measures with $F(C_j) > 0$

	without graph-mining		graph distance ≤ 1	
	No C-value	C-value ≥ 1	No C-value	C-value ≥ 1
R_c^{micro}	0.48	0.40	0.57	0.48
P_c^{micro}	0.53	0.64	0.44	0.53
F_c^{micro}	0.50	0.49	0.50	0.50
$|\{j : F(C_j) > 0)\}|$	277	271	370	365

Table 4. Code-Based Micro Measures with $F(C_j) > 0.5$

	without graph-mining		graph distance ≤ 1	
	No C-value	C-value ≥ 1	No C-value	C-value ≥ 1
R_c^{micro}	0.64	0.61	0.69	0.62
P_c^{micro}	0.70	0.75	0.66	0.71
F_c^{micro}	0.67	0.67	0.68	0.66
$|\{j : F(C_j) > 0.5)\}|$	170	168	237	240

Out of 633 total possible codes, the NER approach extracted 277 (43% of 633) and using the graph mining approach we have 370 (58% of 633). In both tables, the pattern we see in the macro measures repeats where the increase in recall due to the graph mining approach is offset by a decrease in precision.

To understand the nature of our errors, we went back and looked at some of the codes that were causing high recall and precision errors. For example, from Table 3, we can see that even with the graph based approach only 58% of the codes were extracted. We found out that this is because there are a set of codes that never get extracted due to the complex ways in which they manifest in free text. A main class of such codes is the set of E (external cause of injury or poison) and V (encounters with circumstances other than disease or injury) codes. These codes do not generally manifest in free text as noun phrases and have evidence spread throughout the document with non-local dependencies. Out of a total of 92 E and V codes in our dataset, 85 were never extracted; the micro average recall over all E and V codes is 0.01. Similar to E and V codes, there are other classes of codes that rely on non-local dependencies in textual documents. One of them is the set of codes that deal with pregnancy and childbirth (codes 630-670). In our dataset, over a total of 28 codes that belong to this class, the average recall was 0.25. When it comes to precision errors, most of the errors were caused by common symptoms such as *pain* and *bruising* that are generic and are not coded in many cases based on coding guidelines.

7 Conclusion

In this paper we presented an unsupervised ensemble approach to extract diagnosis codes from EMRs. We used a biomedical NER system MetaMap for the basic recognition of biomedical concepts in EMRs and mapped them to ICD-9 codes using the UMLS Metathesaurus . We then used graph mining to exploit the UMLS relationship graph to extract candidate ICD-9 codes for those disorder concepts output by MetaMap that did not have an associated ICD-9 code. We show a 12% improvement in EMR based average recall with this approach. Next, we used key phrase extraction using the C-value method to improve EMR based average precision by 7%. To our knowledge, our results are the first to report on EMR level extraction of diagnosis codes using a large set of codes. Although machine learning approaches are important, we believe that unsupervised knowledge-based approaches are essential especially given that large amounts of biomedical data might not be available owing to privacy issues involving patient data. As future work, we plan to extract full ICD-9 codes instead of top level category codes and are working on combining our unsupervised methods with machine learning approaches to build a hybrid extraction system.

Acknowledgements. This publication was supported by the National Center for Research Resources and the National Center for Advancing Translational Sciences, US National Institutes of Health (NIH), through Grant UL1TR000117. The content is solely the responsibility of the authors and does not necessarily represent the official views of the NIH.

References

1. American Medical Association: Preparing for the icd-10 code set (2010),
   http://www.ama-assn.org/ama1/pub/upload/mm/
   399/icd10-icd9-differences-fact-sheet.pdf
2. National Center for Health Statistics and the Centers for Medicare and Medicaid
   Services (2011), http://www.cdc.gov/nchs/icd/icd9cm.htm
3. de Lima, L.R.S., Laender, A.H.F., Ribeiro-Neto, B.A.: A hierarchical approach
   to the automatic categorization of medical documents. In: Proceedings of the 7th
   Iintl. Conf. on Inf. & Knowledge Mgmt., CIKM 1998, pp. 132–139 (1998)
4. Gundersen, M.L., Haug, P.J., Pryor, T.A., van Bree, R., Koehler, S., Bauer, K.,
   Clemons, B.: Development and evaluation of a computerized admission diagnosis
   encoding system. Comput. Biomed. Res. 29(5), 351–372 (1996)
5. Pestian, J.P., Brew, C., Matykiewicz, P., Hovermale, D.J., Johnson, N., Cohen,
   K.B., Duch, W.: A shared task involving multi-label classification of clinical free
   text. In: Proceedings of the Workshop on BioNLP 2007, pp. 97–104 (2007)
6. Aronson, A.R., Bodenreider, O., Demner-Fushman, D., Fung, K.W., Lee, V.K.,
   Mork, J.G., Neveol, A., Peters, L., Rogers, W.J.: From indexing the biomedical literature to coding clinical text: experience with mti and machine learning approaches. In: Biological, Translational, and Clinical Language Processing,
   pp. 105–112. Assc. for Comp. Ling. (2007)
7. Goldstein, I., Arzumtsyan, A., Uzuner, O.: Three approaches to automatic assignment of icd-9-cm codes to radiology reports. In: Proceedings of AMIA Symposium,
   pp. 279–283 (2007)
8. Crammer, K., Dredze, M., Ganchev, K., Pratim Talukdar, P., Carroll, S.: Automatic code assignment to medical text. In: Biological, Translational, and Clinical
   Language Processing, pp. 129–136. Assc. for Comp. Ling. (2007)
9. Farkas, R., Szarvas, G.: Automatic construction of rule-based icd-9-cm coding systems. BMC Bioinformatics 9(S-3) (2008)
10. Pakhomov, S.V.S., Buntrock, J.D., Chute, C.G.: Automating the assignment of
    diagnosis codes to patient encounters using example-based and machine learning
    techniques. J. American Medical Informatics Assoc. 13(5), 516–525 (2006)
11. Nadeau, D., Sekine, S.: A survey of named entity recognition and classification.
    Lingvisticae Investigationes 30(1), 3–26 (2007)
12. Aronson, A.R., Lang, F.M.: An overview of metamap: historical perspective and
    recent advances. J. American Medical Informatics Assoc. 17(3), 229–236 (2010)
13. Frantzi, K.T., Ananiadou, S., Tsujii, J.: The $C - value/NC - value$ Method of
    Automatic Recognition for Multi-word Terms. In: Nikolaou, C., Stephanidis, C.
    (eds.) ECDL 1998. LNCS, vol. 1513, pp. 585–604. Springer, Heidelberg (1998)
14. Bodenreider, O., Nelson, S., Hole, W., Chang, H.: Beyond synonymy: exploiting
    the umls semantics in mapping vocabularies. In: Proceedings of AMIA Symposium,
    pp. 815–819 (1998)
15. Mihalcea, R., Tarau, P.: Textrank: Bringing order into text. In: Proceedings of
    EMNLP, pp. 404–411 (2004)
16. Tsoumakas, G., Katakis, I., Vlahavas, I.P.: Mining multi-label data. In: Data Min-
    ing and Knowledge Discovery Handbook, pp. 667–685 (2010)

Maintaining Preference Networks
That Adapt to Changing Preferences

Ki Hyang Lee[1], Scott Buffett[2], and Michael W. Fleming[1]

[1] University of New Brunswick, Fredericton, NB, Canada, E3B 5A3
{kihyang.lee,mwf}@unb.ca
[2] National Research Council Canada, Fredericton, NB, Canada, E3B 9W4
Scott.Buffett@nrc.gc.ca

Abstract. Decision making can be more difficult with an enormous amount of information, not only for humans but also for automated decision making processes. Although most user preference elicitation models have been developed based on the assumption that user preferences are stable, user preferences may change in the long term and may evolve with experience, resulting in dynamic preferences. Therefore, in this paper, we describe a model called the dynamic preference network (DPN) that is maintained using an approach that does not require the entire preference graph to be rebuilt when a previously-learned preference is changed, with efficient algorithms to add new preferences and to delete existing preferences. DPNs are shown to outperform existing algorithms for insertion, especially for large numbers of attributes and for dense graphs. They do have some shortcomings in the case of deletion, but only when there is a small number of attributes or when the graph is particularly dense.

Keywords: Preference networks, User modeling, Transitive reduction.

1   Introduction

Decision making can be incredibly difficult when there is an enormous amount of information, not only for humans but also for automated decision making processes. For example, in amazon.com, there are millions of items for sale at a time, and it seems almost impossible for a buyer to navigate all the items by hand to find the books that he/she wants to buy. Even when product search tools are applied, without a proper user preference model, search results are often inaccurate, and thus may result in many pages of output for users to examine. Therefore, in this paper, we propose a dynamic preference model that starts with a small amount of information about a user's preferences and that can be modified as those preferences change.

Preference elicitation has been studied to help autonomous decision agents by extracting knowledge of a user's preferences in order to maximize utility over the possible outcomes in a particular decision problem [1]. Such decision problems may be simple, such as choosing the best product at a satisfactory price for a consumer, or very complex, such as participating in a number of multi-issue negotiations to settle a contract. To date, many preference elicitation approaches separate the preference

O. Zaïane and S. Zilles (Eds.): Canadian AI 2013, LNAI 7884, pp. 89–99, 2013.

elicitation and decision making stages, by attempting to gain as much information as possible on the user's preferences before commencing the decision-making stage [6, 9]. While some recent approaches do recognize that preference elicitation must be done quickly and in real time [11], and others recognize that it sometimes must be carried out before the set of features for which preferences are being elicited is fully known [10], to the best of our knowledge, researchers in the field have yet to recognize that the preference elicitation process should never truly end. This is because of the fact that preferences are always changing, especially in the long term, affected by human genetic evolution, technological change, the evolution of social systems, and the changing availability of environmental and other natural resources [3]. Preferences may evolve with experience, resulting in flexible future preferences [4]. Also, a user's preferences are likely to change as a result of the elicitation process itself [5]. Preferences can also change often in the short term, as evidenced by the observation that most people do not order the same meal every time they enter a particular restaurant. Thus we argue that a model for user preferences must be easily and quickly updated at any time, even during the decision process, to allow a decision agent to react properly in the face of changing preferences.

To address the problem of evolving preferences, we propose the use of a dynamic preference network to model the preferences of a particular user. The network is maintained in such a way as to allow changes to be made quickly and without the need to ever reconstruct the model from scratch. Thus the network has very desirable anytime properties, allowing an agent to access the most up-to-date preference model during the decision-making process. To facilitate this preference modeling, while allowing for fast preference retrieval, we maintain only a transitive reduction of the preference network, and propose an algorithm for transitively reduced graph maintenance that vastly outperforms existing algorithms when used for our purpose.

In Section 2, we briefly review related work. In Section 3, we introduce our dynamic preference network and discuss a new methodology for maintaining these networks, introducing new insertion and deletion algorithms. In Section 4, we discuss test results of the new methodology and how it compares to the current methodology. Finally, conclusions are presented in Section 5.

2   Related Work

2.1   Preference Models

Preference modeling can use numerical representations, weighted and conditional logics or graphical representations [12]. The Conditional Preference Network (CP-net), proposed by Boutilier et al. [6], is a well known graphical representation of conditional preference relations. Conditional preferences are those where tastes are dependent on the outcome of one or more particular attributes. CP-nets are a relatively compact, intuitive, and structured specification of preference relations. While CP-nets are used to reason about conditional preferences over values within a subset of attributes, a Conditional Outcome Preference Network (COP-net), proposed by Chen et al. [2], has extended conditional preference representation for representing user preferences and for predicting preferences over all feasible outcomes for any type of conditional preference. A COP-network is a directed acyclic graph, with n vertices and m arcs. Each vertex represents an outcome and each directed arc (v_i, v_j) represents a preference relation between vertex v_i and vertex v_j meaning that a user prefers v_i to v_j. Therefore, n vertices represent a set of n outcomes. Chen et al.[2] provided reduced COP-nets, corresponding to a transitive reduction of initial COP-nets, to reduce the redundancy of the initial network. This transitive reduction of the COP-net makes it easy to calculate utility estimates using an approach called the longest path method.

2.2   Transitive Reduction Maintenance

Little research has been done with fully dynamic maintenance algorithms for transitive reduction [8]. J.A. La Poutré and Jan Van Leeuwen, in 1988, presented an algorithm for the problem of efficiently updating the transitive closure and transitive reduction of a directed graph [7]. The total time complexity of an edge insertion is $O(n^2 e_{new})$ and the total time complexity of an edge deletion is $O(n^2 e_{old})$, where n is the number of vertices, e_{new} is the number of edges in the resulting graph after an edge insertion, and e_{old} is the number of edges in the original graph before an edge deletion. King and Sagert, in 2002, presented a fully dynamic algorithm for maintaining transitive closure using an adjacency matrix [8]. In this paper, the authors provided an algorithm with $O(n^2)$ time for each update and $O(1)$ time for each query, in directed acyclic graphs. The key idea of King's algorithm is to track the number of distinct directed paths from each vertex i to each vertex j in the graph, and for each insertion or deletion, add or subtract the number of directed paths between affected vertices. Since King's algorithm keeps track of the number of distinct directed paths between two vertices in the graph, it requires updating each affected cell in the path matrix.

3   Preference Network Maintenance

3.1   Dynamic Preference Networks (DPN)

In this paper, we present a new model that extends the previous model proposed by Chen et al. [2] in order to represent changing preferences. We refer to this new model as a dynamic preference network (DPN). A DPN = (V, A) is a directed graph where each vertex in V corresponds to a unique outcome of interest, and an arc (v_1, v_2) in A indicates that the outcome represented by v_1 is preferred over the outcome represented by v_2. The transitive property also holds, meaning that for any v_1 that is a proper ancestor of v_2, v_1 is preferred over v_2. If there is no directed path between two vertices, then the preference over the two corresponding outcomes is not known with certainty. Note that such a preference may exist, and there are algorithms [2] for estimating the relative utilities of the two outcomes by analyzing the overall structure of the graph. The DPN is reduced if, for any arc (v_1, v_2), there does not exist a directed path from v_1 to v_2 through v_3, where v_3 is not equal to v_1 or v_2.

3.2   Algorithm for Maintaining Transitive Reduction

In order to provide efficient maintenance of the transitive reduction, we consider the fact that there are 4 basic conditions that can occur between two vertices i and j. If there is no path and no arc from i to j, then it is in condition 1. If there is an arc (i, j) but no other directed path, then it is in condition 2. If there is no arc but a directed path from i to j, then it is in condition 3. If there is an arc (i, j) and a longer directed path from i to j, then it is in condition 4.

In our method, we maintain two $n \times n$ binary matrices, M_1 and M_2, where n is the number of vertices in a DPN, to represent these 4 conditions. Matrix M_1 stores the information about arcs representing a set of preferences elicited from a user. Therefore, if there is an arc (i, j) in a DPN, then $M_1(i, j) = 1$. Matrix M_2 is used for a set of distinct directed paths, and so if there is a directed path (of length greater than 1) from i to j in a DPN, then $M_2(i, j) = 1$. Therefore, condition 1 can be represented with $M_1(i, j) = 0$ and $M_2(i, j) = 0$. Condition 2 can be represented with $M_1(i, j) = 1$ and $M_2(i, j) = 0$. Condition 3 can be represented with $M_1(i, j) = 0$ and $M_2(i, j) = 1$. Condition 4 can be represented with $M_1(i, j) = 1$ and $M_2(i, j) = 1$. Let G=<V, E> be a directed acyclic graph representing a DPN. The binary matrices M_1 and M_2 of G are |V| x |V| matrices with values {0, 1}.   Building initial matrices M_1, M_2 is done as follows:

1. $M_1 (i, j), M_2 (i, j) = 0$ for every $i, j \in V$.
2. Update $M_1 (i, j) = 1$, if $(i, j) \in E$ for $i, j \in V$.
3. Add $M_2 (i, j) = 1$, if there is a directed path of length > 1 from i to j for $i, j \in V$.

Fig.1 shows an example of a DPN with a set of vertices V={v_1, v_2, v_3, v_4} and a set of arcs E={(v_1, v_2), (v_1, v_3), (v_2, v_3)} and the 2 corresponding binary matrices M_1 and M_2.

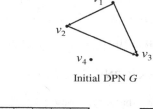

Initial DPN *G*

	v_1	v_2	v_3	v_4
v_1	0	1	<u>1</u>	0
v_2	0	0	1	0
v_3	0	0	0	0
v_4	0	0	0	0

Matrix 1 (M₁)

	v_1	v_2	v_3	v_4
v_1	0	0	<u>1</u>	0
v_2	0	0	0	0
v_3	0	0	0	0
v_4	0	0	0	0

Matrix 2 (M₂)

Fig. 1. DPN example 1

Finally, a reduced DPN can be derived from Matrix 1 by removing redundant arcs, which can be done by converting $M_1(i, j)$ to 0, if $M_1(i, j)=1$ and $M_2(i, j)=1$. We define this operation as '-'. In Fig. 2, the underlined values for $M_1 (v_1, v_3)$ and $M_2 (v_1, v_3)$ are both equal to 1, and so the value of the reduced matrix M_1-M_2 (v_1, v_3) is 0. Then the reduced DPN G^R has a set of arcs $E^R = \{(v_1, v_2), (v_2, v_3)\}$, and thus the arc (v_1, v_3) has been removed. Fig. 2 shows an initial DPN and a reduced DPN.

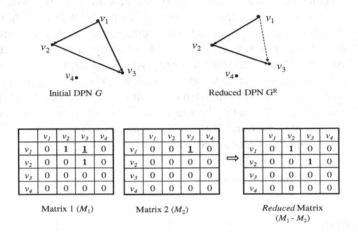

Initial DPN G Reduced DPN G^R

	v_1	v_2	v_3	v_4
v_1	0	1	1	0
v_2	0	0	1	0
v_3	0	0	0	0
v_4	0	0	0	0

Matrix 1 (M_1)

	v_1	v_2	v_3	v_4
v_1	0	0	1	0
v_2	0	0	0	0
v_3	0	0	0	0
v_4	0	0	0	0

Matrix 2 (M_2)

$\Rightarrow$

	v_1	v_2	v_3	v_4
v_1	0	1	0	0
v_2	0	0	1	0
v_3	0	0	0	0
v_4	0	0	0	0

Reduced Matrix
$(M_1 - M_2)$

Fig. 2. DPN and reduced DPN

3.3   New Preference Insertion

Based on the 4 basic conditions, we develop the following three observations, which are also illustrated in Figure 3.

- Observation 1 If there is any vertex v_i and paths $p_1 = (v_1, v_2, ..., v_i)$ and $p_2 = (v_i, v_{i+1}, ..., v_n)$, then there is a path p from v_1 to v_n, $p = (v_1, v_2, ..., v_i, ..., v_n)$.
- Observation 2 If there is an arc (v_i, v_j) and paths $p_1 = (v_1, v_2, ..., v_i)$ and $p_2 = (v_j, v_{j+1}, ..., v_n)$, then there is a path p from v_1 to v_n, $p = (v_1, v_2, ..., v_i, v_j, ..., v_n)$.
- Observation 3 If there is an arc (v_i, v_j) and a path $p_1 = (v_j, v_{j+1}, ..., v_n)$ then there is a path p from v_1 to v_n, $p = (v_i, v_j, ..., v_n)$.

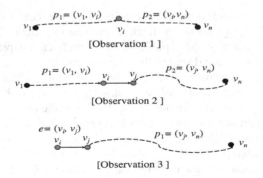

Fig. 3. Observation 1, Observation 2 and Observation 3

From these three observations, we can develop our algorithm for a new preference insertion. Since a preference is represented by an arc in a DPN, we consider inserting a new arc (v_i, v_j) to the initial DPN when there is a new preference added such that a user prefers an outcome v_i to an outcome v_j. Since a new arc (v_i, v_j) is inserted, for all vertices r that have a path to v_i and all vertices s that have a path from v_j, then, by Observation 2 with new arc (v_i, v_j), there is now a directed path from r to s. Therefore, $M_2(r, s)=1$. When inserting a new arc(v_i, v_j) in a directed acyclic graph, there are three steps. The first step is: for all $r \in V$ such that $M_1(r, v_i) =1$ or $M_2(r, v_i)=1$, if $M_2(r, v_j)=0$, then $M_2(r, v_j)=1$. The second step is: for all $s \in V$ such that $M_1(v_j, s)=1$ or $M_2(v_j, s)=1$, if $M_2(v_i, s)=0$, then $M_2(v_i, s)=1$. The third step is: for all $r, s \in V$ from step 1 and step 2, if $M_2(r, s)=0$, then $M_2(r, s)=1$. Finally, the fourth step is to find the reduced matrix M_1-M_2: for all $i, j \in V$, if $M_1(i, j)=1$ and $M_2(i, j)=0$, then $M_1-M_2(i, j)=1$; otherwise $M_1-M_2(i, j)=0$. Fig. 4 shows the new insertion algorithm for a DPN.

Algorithm Insert $((v_i, v_j), G)$

   -to insert *a single edge* (v_i, v_j) to G.

$M_1(v_i, v_j) \leftarrow 1$

for all $r \in V$ such that $M_1(r, v_i) =1$ or $M_2(r, v_i)=1$ do   *-Step 1*
   if $M_2(r, v_j)=0$, *then* $M_2(r, v_j)=1$.

for all $s \in V$ such that $M_1(v_j, s)=1$ or $M_2(v_j, s)=1$ do   *-Step 2*
   if $M_2(v_i, s)=0$, *then* $M_2(v_i, s)=1$

for all $r, s \in V$ from *Step* 1 and *Step* 2   *-Step 3*
   if $M_2(r, s)=0$, *then* $M_2(r, s)=1$

Return $M_1 - M_2$ *-Step 4*

Fig. 4. Insertion algorithm for DPNs

3.4   Preference Deletion

In our DPN, we also provide a method for maintaining a reduced DPN when deleting an arc. A user may change his/her preferences because of a change in the preference itself, because of a change in the variables used for preference expression or because of a change in the preference domain [10]. For example, consider a user who simply changes his/her preference over a car. The user now prefers a black color for a car whereas the user preferred white before. In this case, we make this change by first deleting the user's previous preference: a user prefers white over black, and second, by adding the new preference: a user prefers black over white.

To maintain a reduced DPN with an arc(v_i, v_j) deleted, we have to answer the question "Is there a directed path from a vertex v_l to a vertex v_n not containing the arc(v_i, v_j) ?". We call this question the sensitivity query. To answer the sensitivity query, we develop Observations 4 and 5 based on the previous Observations 1, 2 and 3.

- **Observation 4** Let there be two directed paths $p_1 = (v_1, v_2, \ldots, v_k, \ldots, v_{i-1}, v_i)$ and $p_2 = (v_j, v_{j+1}, \ldots, v_{j+m}, \ldots, v_{n-1}, v_n)$. If there is an arc (v_k, v_{j+m}), then there is a path p from v_1 to v_n, including an arc (v_k, v_{j+m}), namely $p = (v_1, v_2, \ldots, v_k, v_{j+m}, \ldots, v_n)$.
- **Observation 5** Let there be two directed paths $p_1 = (v_1, v_2, \ldots, v_k, \ldots, v_{i-1}, v_i)$ and $p_2 = (v_j, v_{j+1}, \ldots, v_{j+m}, \ldots, v_{n-1}, v_n)$. If there is a *directed path* from a vertex v_k in p_1 to a vertex v_{j+m} in p_2, then there is path p from v_1 to v_n, including two vertices v_k and v_{j+m}, namely $p = (v_1, v_2, \ldots, v_k, \ldots, v_{j+m}, \ldots, v_n)$.

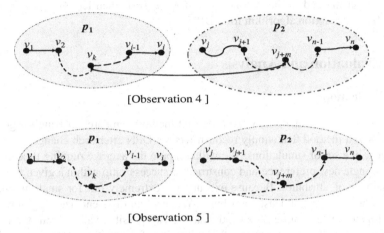

[Observation 4]

[Observation 5]

Fig. 5. Observation 4 and Observation 5

Algorithm Delete $((v_i, v_j), G)$

– to delete a single edge (v_i, v_j) from G.

If $M_1(v_i, v_j) = 1$, *then* $M_1(v_i, v_j) = 0$

If $M_2(v_i, v_j) = 0$, *then* do

Step 1. Define the set of vertices $I = \{i \in V$ such that $M_1(i, v_i) = 1$ or $M_2(i, v_i) = 1$ or $i = v_i\}$ and $J = \{j \in V$ such that $M_1(v_j, j) = 1$ or $M_2(v_j, j) = 1$ or $j = v_j\}$

Step 2. Define the sets $I_1 = \{i_1 \in I \mid \exists\, k \in V \setminus I$ such that $M_1(i, k) = 1\}$ and $K = \{k \in V \setminus I \mid \exists\, i \in I$ such that $M_1(i, k) = 1\}$.

For each $i_1 \in I_1$:

Step 2-1 Define the set $I_2 = \{i_2 \in I \mid \exists i_1 \in I_1$ such that $M_1(i_2, i_1) = 1$ or $M_2(i_2, i_1) = 1\}$ and $K_1 = \{k \in K \mid$ such that $M_1(i_1, k) = 1\}$. If $k \in K_1 \cap J$ and $M_2(i_2, k)$ is unmarked, then mark $M_2(i_2, k)$.

Step 2-2 Define the set $J_2 = \{j_2 \in J \mid \exists k \in K$ such that $M_1(k, i_2) = 1$ or $M_2(k, i_2) = 1\}$. For all $i_1 \in I_1$, $j_2 \in J_2$ such that $M_2(i_1, j_2)$ is unmarked, then mark $M_2(i_1, j_2)$.

Step 2-3, for all $i_2 \in I_2$ and $j_2 \in J_2$ from step 2-1 and step 2-2, if $M_2(i_2, j_2) = 1$ and unmarked, then mark $M_2(i_2, j_2)$.

Step 3. for all $i \in I$, $j \in J$ which are not marked from the step 2, if $M_2(i, j) = 1$, then $M_2(i, j) = 0$.

Step 4. return $M_1 - M_2$

Fig. 6. Deletion algorithm for DPNs

We develop our deletion algorithm based on the 4 basic conditions. For the deletion of an arc(v_i, v_j) in a DPN, the first step is to find all vertices in the reduced DPN that will be affected by removing the arc(v_i, v_j): Define the set of vertices I to include v_i and all ancestors of v_i. Define the set J to include v_j and all descendants of v_j. Second, mark the pairs of vertices that are involved in any directed path from I to J not containing the arc(v_i, v_j), by applying Observation 4 and 5 to find if there are paths from vertices in I to vertices in J not containing the arc(v_i, v_j). Third, for all $i \in I$ and $j \in J$ which are not marked in the second step, if $M_2(i, j)=1$, then let $M_2(i, j) = 0$. Fig. 6 shows the new deletion algorithm for a DPN.

4   Evaluation and Analysis

4.1   Evaluation

In this section, we evaluate how the proposed method compares to King's algorithm, as well as to a method that simply reconstructs the DPN after each change, in terms of efficiency, in a set of simulations. We will compare the average running time taken to update a single new preference and construction success ratio within a given time for a large number of attributes. Because execution is extremely fast for small numbers of attributes, we start with at least 5 attributes. We test four different sizes of graphs (n=5, 6, 7 and 8 attributes), with each attribute having two possible values. Thus the number of outcomes is 2^n. In order to test different densities for each graph, we choose different values for the number of initial preferences generated. Also, if the number of attributes is greater than 8, the running time is significantly increased with King's algorithm and so we will examine the percentage of the time in which construction can be completed within a given time. For each set of test conditions, we randomly generate 100 different initial sets of preferences, with any redundant and inconsistent preferences removed, and a set of 100 new preferences (updates) for each of these graphs. This provides us with a total of 10,000 test cases for each set of conditions.

4.1.1   Proposed vs. Current Methods
In this experiment, we examine how the efficiency of the proposed method compares to that of King's algorithm and a method involving DPN reconstruction. We first compare the average update time (insert/delete) of the three methods and then we compare the construction success ratios within a given time.

4.1.1.1   Average Update Time. In this set of experiments, we compare the average time required for inserting and deleting preferences, using three different approaches: the proposed method, King's method, and a technique that involves reconstructing the graph from scratch after the insertion/deletion of a preference. Fig. 7. a) shows the average insertion time with different density graphs using 8 attributes. Fig. 7. b) shows the average insertion time for different numbers of attributes. The results demonstrate that the proposed method outperforms King's method. The proposed insertion algorithm decreases its insertion time as the graph becomes denser and does not increase its insertion time as the number of attributes increases.

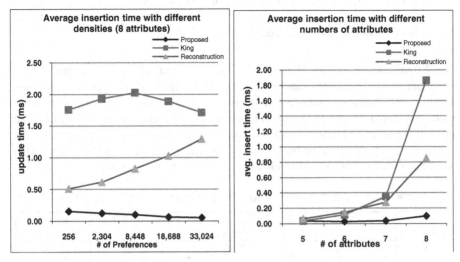

Fig. 7. a) Average insertion time with different densities b) Average insertion time with different numbers of attributes

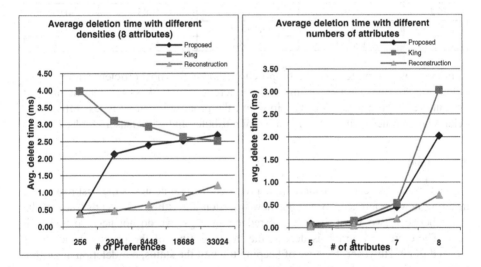

Fig. 8. a) Average deletion time with different densities b) Average deletion time with different numbers of attributes

Fig. 8. a) shows the average deletion time with different density graphs using 8 attributes. Fig. 8. b) shows the average deletion time for different numbers of attributes. Each result shows that the reconstruction time is less than maintaining a reduced DPN using either King's deletion algorithm or the proposed deletion algorithm. However, the reconstruction method still requires that the consistency of the graph be checked, since in order to find the longest path to estimate utilities we have to ensure that the graph is acyclic. With the consistency condition, the proposed method is more efficient with large numbers of attributes except in the case of dense graphs. For dense graphs or a small number of attributes, King's method is always more efficient than the proposed method.

4.1.1.2 Construction Success Rate Within a Given Time Limit. Since the running time with more than 8 attributes can be unreasonably long, in this experiment, we will examine how often the initial construction of a graph with a large number of attributes can be completed in a given limited amount of time (500ms). In this experiment, we compare the proposed method to King's method. Table 1 shows the results of this experiment. The proposed method is able to complete the initial construction within the 500ms time limit at a significantly higher rate than King's method.

Table 1. Construction success rate with different numbers of attributes in 500ms

# of Attributes	# of Preferences	Initial graph (# of edges)	Proposed Method (%)	King's Method (%)
9	512	508.00	100.00	6.09
10	1,024	1,021.00	66.61	0.77
11	2,048	2,041.00	17.02	0.10
12	4,096	4,095.00	3.85	0.02
13	8,192	8,187.00	0.96	0.00
14	16,384	16,378.00	0.19	0.00

Finally, we can notice that the proposed method is the most efficient with building initial graphs and with inserting new preferences regardless of the number of attributes and the graph's density. However, the proposed method is less efficient in deleting preferences with a small number of attributes and with denser graphs.

4.2   Analysis

Of the three approaches considered, the proposed method achieves the best results for inserting new preferences. Since our new insertion algorithm only keeps track of the status between two vertices, once we reduce an edge, we do not need to update it later. Therefore, if a graph is denser, the insertion time of the proposed method is reduced more. However, because of keeping track of the status, our deletion algorithm takes more time to update the status after deleting a preference. In practice, changes in user preferences are more likely to involve adding new preferences than deleting preferences. Therefore, reducing insertion time is a more important part of the entire maintenance process. The proposed method boasts the smallest insertion time, compared to the existing methods.

Also, our proposed method handles larger numbers of attributes than the existing method. Since the existing method is based on keeping the number of paths between two vertices, it requires an object matrix, whereas the proposed method uses two bit matrices. As a result, the proposed method reduces its space and its running time.

5   Conclusions

In this paper, we provide a dynamic preference model that supports preference addition and preference deletion after an initial preference model has been built. Our results clearly show that the proposed method is efficient in maintaining a model of preferences in a DPN. Although it sacrifices some efficiency in deleting preferences from denser graphs, it shows significant savings in insertion time and in space. Since, in practice, deleting preferences is caused mostly by direct conflicts after adding new preferences, reducing insertion time can lead to overall improvement. Our main contribution is that we apply the concept of transitive reduction maintenance to the problem of maintaining an evolving preference model, which includes the specification of new insertion and deletion algorithms. Another contribution is that the space and average time needed to perform this maintenance is significantly reduced in many cases by using a binary representation of the dynamic preference model.

References

1. Buffett, S., Fleming, M.: Persistently Effective Query Selection in Preference Elicitation. In: The 2007 IEEE/WIC/ACM International Conference on Intelligent Agent Technology (IAT 2007), Fremont, California, USA, pp. 491–497 (2007)
2. Chen, S., Buffett, S., Fleming, M.W.: Reasoning with conditional preferences across attributes. In: Kobti, Z., Wu, D. (eds.) Canadian AI 2007. LNCS (LNAI), vol. 4509, pp. 369–380. Springer, Heidelberg (2007)
3. Miguel, F., Ryan, M., Scott, A.: Are preferences stable? The case of health care. Journal of Economic Behavior & Organization 48, 1–14 (2002)
4. Koopmans, T.: On flexibility of future preference. In: Cowles Foundation Discussion Papers (1962)
5. Pu, P., Faltings, B., Torrens, M.: Effective interaction principles for on-line product search environments. In Proceedings of the 3rd ACM/IEEE International Conference on Web intelligence. IEEE Press (2004)
6. Boutilier, C., Brafman, R., Domshlak, C., Hoos, H., Poole, D.: CP-nets: A Tool for Representing and Reasoning with Conditional Ceteris Paribus Preference Statements. Journal of Artificial Intelligence Research 21, 135–191 (2004)
7. La Poutré, J.A., Leeuwen, J.V.: Maintenance of transitive closures and transitive reductions of graphs (1988)
8. King, V., Sagert, G.: A Fully Dynamic Algorithm for Maintaining the Transitive Closure. Journal of Computer and System Sciences 65, 150–167 (2002)
9. Chajewska, U., Koller, D., Parr, R.: Making rational decisions using adaptive utility elicitation. In: AAAI 2000, Austin, Texas, USA, pp. 363–369 (2000)
10. Boutilier, C., Regan, K., Viappiani, P.: Simultaneous Elicitation of Preference Features and Utility. In: Proceedings of the Twenty-fourth AAAI Conference on Artificial Intelligence (AAAI 2010), Atlanta, GA, pp. 1160–1167 (2010)
11. Guo, S., Sanner, S.: Real-time Multiattribute Bayesian Preference Elicitation with Pairwise Comparison Queries. Appearing in Proceedings of the 13th International Conference on Artificial Intelligence and Statistics (AISTATS), Chia Laguna Resort, Sardinia, Italy. JMLR: W&CP, vol. 9 (2010)
12. Kaci, S.: Working with Preferences: Less Is More. Cognitive Technologies, pp. 11–193. Springer (2011)

Fast Grid-Based Path Finding for Video Games

William Lee and Ramon Lawrence

University of British Columbia

Abstract. Grid-based path finding is required in many video games and virtual worlds to move agents. With both map sizes and the number of agents increasing, it is important to develop path finding algorithms that are efficient in memory and time. In this work, we present an algorithm called DBA* that uses a database of pre-computed paths to reduce the time to solve search problems. When evaluated using benchmark maps from Dragon Age[TM], DBA* requires less memory and time for search, and performs less pre-computation than comparable real-time search algorithms. Further, its suboptimality is less than 3%, which is better than the PRA* implementation used in Dragon Age[TM].

1 Introduction

As games have evolved, their size and complexity has increased. The virtual worlds and maps have become larger as have the number of agents interacting. It is not uncommon for hundreds of agents to be path finding simultaneously, yet the processing time dedicated to path finding has not substantially increased. Consequently, game developers are often forced to make compromises on path finding algorithms and spend considerable time tuning and validating algorithm implementations.

Variations of A* and PRA* are commonly used in video games [9]. The limitation of these algorithms is that they must plan a complete (but possibly abstract) path before the agent can move. Real-time algorithms such as kNN LRTA* [3] and HCDPS [8] guarantee a constant bound on planning time, but these algorithms often require a considerable amount of pre-computation time and space.

In this work, we propose a grid-based path finding algorithm called DBA* that combines the real-time constant bound on planning time enabled by using a precomputed database as in HCDPS [8] with abstraction using sectors as used in PRA* [9]. The result is a path finding algorithm that uses less space than previous real-time algorithms while providing better paths than PRA* [9]. DBA* was evaluated on Dragon Age[TM]maps from [10] with average suboptimality less than 3% and requiring on average less than 200 KB of memory and between 1 and 10 seconds for pre-computation.

2 Background

Grid-based path finding is an instance of a heuristic search problem. The algorithms studied in this paper, although potentially adaptable to general heuristic search, are specifically designed and optimized for 2D grid-based path finding. States are vacant square grid cells. Each cell is connected to four cardinally (i.e., N, E, W, S) and four

O. Zaïane and S. Zilles (Eds.): Canadian AI 2013, LNAI 7884, pp. 100–111, 2013.

diagonally neighboring cells. Out-edges of a vertex are moves available in the cell. The edge costs are 1 for cardinal moves and 1.4 for diagonal moves. Standard octile distance is used for the heuristic.

Algorithms are evaluated based on the quality of the path produced and the amount of time and memory resources consumed. *Suboptimality* is defined as the ratio of the path cost found by the agent to the optimal solution cost minus one and times 100%. Suboptimality of 0% indicates an optimal path and suboptimality of 50% indicates a path 1.5 times as costly as the optimal path.

An algorithm is also characterized by its *response time*, which is the maximum planning time per move. The *overall time* is the total amount of time to construct the full solution, and the *average move time* is the overall time divided by the number of moves made. Memory is consumed for node expansions (e.g. open and closed lists in A*), for storing abstraction information (e.g. regions), and for storing computed paths. *Per agent memory* is memory consumed for each search agent. *Fixed memory* is memory consumed that is shared across multiple, concurrent path finding agents.

2.1 A*

A* [5] is a common algorithm used for video game path finding as it has good performance characteristics and is straightforward to implement. A* paths are always optimal as long as the heuristic function is admissible. However its overall time depends on the problem size and complexity, resulting in highly variable times. The biggest drawback is that its response time is the same as its overall time as a complete solution is required before the agent moves. A*'s memory use is variable and may be high depending on the size of its open and closed lists and the heuristic function used.

2.2 PRA*

The PRA* variant implemented in Dragon Age [9] was designed to improve on A* performance for path finding. Response time and per agent memory are reduced by abstracting the search space into sectors and first computing a complete solution in the abstract space. The abstraction reduces the size of the problem to be solved by A*. The abstraction requires a small amount of fixed memory. It also results in solutions that are suboptimal (between 5 to 15%). The small trade-off in space and suboptimality is beneficial for faster response time.

2.3 Real-Time Algorithms

There have been several real-time algorithms proposed that guarantee a constant bound on planning time per action (response time) including TBA* [1], D LRTA* [4], kNN LRTA* [3], and HCDPS [8]. TBA* is a time-sliced version of A* that exhibits the same properties as A* with the added ability to control response time and per move planning time. All of the other algorithms rely on some form of pre-computation to speed up online search.

D LRTA* [4] performs clique abstraction and generates next hop information between regions allowing an agent to know the next region to traverse to. The size of the abstraction and its pre-computation time were significant.

kNN LRTA* [3] created a database of compressed problems and online used the closest problem in the database as a solution template. Database construction time is an issue, and there is no guarantee of complete coverage. The major weakness with these two algorithms is that they fall back on LRTA* when no information is in the database, which may result in very suboptimal solutions.

HCDPS [8] performs offline abstraction by defining regions where all states are bi-directionally hill-climbable with the region representative. A compressed path database was then constructed defining paths between all region representatives. This allows all searches to be done using hill-climbing, which results in minimal per agent memory use. HCDPS is faster than PRA* with improved path suboptimality, but its abstraction consumes more memory.

2.4 All-Pairs Shortest Path Algorithms

LRTA* with subgoals [6] pre-computes a subgoal tree from each goal state where a subgoal is the next target state to exit a heuristic depression. Online, LRTA* will be able to use the subgoal tree to escape a heuristic depression.

It is also possible to compute a solution to the all-pairs shortest path problem and store it in a compressed form. Algorithms such as [2] store for each pair of states the next direction or state to visit along an optimal path. Depending on the compression, it is possible to achieve perfect solutions at run-time very quickly with the compromise of considerable pre-computation time and space to store the compressed databases.

2.5 Summary

All of these algorithm implementations balance search time versus memory used. The "best" algorithm depends on the video game path finding environment and its requirements. There are three properties that algorithms should have to be useful in practice:

- Solution consistency - The quality of the solutions (suboptimality) should not vary dramatically between problems.
- Adequate response time - The hard-limits on the response time dictated by the game must be met.
- Memory efficiency - The amount of per agent memory and fixed memory should be minimized.

The goal is to minimize suboptimality, response time, and memory usage.

3 Approach Overview

This work combines and extends the best features of two previous algorithms to improve these metrics. Specifically, the memory-efficient sector abstraction developed for [9] is integrated with the path database used by HCDPS [8] to improve suboptimality, memory usage, and response time. The DBA* algorithm performs offline pre-computation before online path finding. The offline stage abstracts the grid into sectors and regions and pre-computes a database of paths to navigate between adjacent regions. A sector is an n x n grid of cells (e.g. 16 x 16). The sector number for a cell (r, c) is calculated by

$$\lfloor \frac{r}{n} \rfloor * \lceil \frac{w}{n} \rceil + \lfloor \frac{c}{n} \rfloor \tag{1}$$

where w is the width of the grid map. A region is a set of cells in a sector that are mutually reachable without leaving the sector. Regions are produced by performing one or more breadth-first searches until all cells in a sector are assigned to a region. A sector may have multiple regions, and a region is always in only one sector. A region center or representative state is selected for each region. The definition and construction of regions follows that in [9].

DBA* then proceeds to construct a database of optimal paths between the representatives of adjacent regions using A*. Each path found is stored in compressed format by storing a sequence of subgoals, each of which can be reached from the previous subgoal via hill-climbing. Hill-climbing compressible paths are described in [8].

To navigate between non-adjacent regions, an R x R matrix (where R is the number of regions) is constructed where cell (i, j) contains the next region to visit on a path from R_i to R_j, the cost of that path, and the path itself (if the two regions are adjacent). The matrix is initialized with the optimal paths between adjacent regions, and dynamic programming is performed to determine the costs and next region to visit for all other matrix cells.

Online searches use the pre-computed database to reduce search time. Given a start cell s and goal cell g, the sector for the start, S_s, and for the goal, S_g, are calculated using Equation 1. If a sector only has one region, then the region is known immediately. Otherwise, a BFS bounded within the sector is performed from s until it encounters some region representative, R_s. This is also performed for the goal state as well to find the goal region representative, R_g.

Given the start and goal region representatives, the path matrix is used to build a path between R_s and R_g. This path may be directly stored in the database if the regions are adjacent, or is the concatenation of paths by navigating through adjacent regions from R_s to R_g. The complete path consists of navigating from s to the region representative R_s, then following the subgoals to region representative R_g, then navigating to g.

The response time is almost immediate as the agent can start navigating from s to the start region representative R_s as soon as the BFS is completed. The complete path between regions can be done iteratively with the agent following the subgoals in a path from one region to the next without having to construct the entire path. The overall time and number of states expanded are reduced as the only search performed is the BFS to identify the start and goal regions if a sector contains multiple regions.

4   Implementation Example

We describe the implementation using a running example.

4.1 Offline Abstraction

The first step in offline pre-computation is abstracting the search space into sectors as in [9]. Depending on the size of the map, the map is divided into fixed-sized sectors (e.g. 16 x 16 or 32 x 32). Note that there is no restriction that the sector sizes be square or a power of 2. Each sector is divided into one or more regions using BFS. The sector size serves as an upper bound for region size and limits the expansion of BFS during online path finding. If the sector size is larger, BFS within the sector will take longer. If the sector size is small, the database will be larger. In Figure 1 is a 6 sector subset of a map. The sector size is 16 x 16. Region representatives are shown labeled with letters. Sector 5 (middle of bottom row) has two regions E and F.

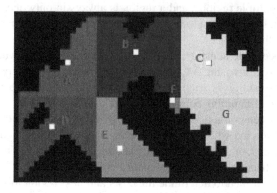

Fig. 1. Regions and Sector Abstraction Example

Region representatives are computed by summing the row (col) of each open state in the region and then dividing by the number of open states. If this technique results in a state that is a wall, adjacent cells are examined until an open state is found. DBA* and PRA* use this same technique for region representative selection in the experiments. Other methods such as proposed in [9] could also be used.

Unlike abstraction using cliques or hill-climbable regions, sector-based regions are built in $O(n)$ time where n is the number of grid cells. Each state is expanded only once by a single BFS. In comparison, the abstraction algorithm in HCDPS may expand a given state multiple times.

The second major advantage is that the mapping between abstract state and base state (i.e. what abstract region a given base state is in) does not need to be stored. Without compression, this mapping consumes the same space as the map. Compression can reduce the abstraction to about 10% of the map size [8].

Instead, to determine the region (and its region representative) for a given cell, first the sector is calculated. If the sector has only one region, then no search is required. Otherwise, a BFS is performed from the cell until it encounters one of the region representative states listed for the sector. This BFS is bounded by the size of the sector, and thus allows for a hard guarantee on the response time.

4.2 Offline Database Generation

After abstraction, a database of paths between adjacent region representatives are computed using A* and stored in a path database. These paths are used to populate a R^2 path matrix where R is the total number of regions. Dynamic programming is performed on the matrix to compute the cost and next hop region for all pairs of regions. Note that a complete path between all pairs of regions is not stored. The path matrix only stores the cost and the next region to traverse to. Paths are only stored between adjacent regions. This is very similar to how a network routing table works where paths are outgoing links and a routing table stores the address and cost of the next hop to route a message towards a given destination.

Algorithm 1. Offline Database Generation

```
// Compute optimal paths between adjacent regions using A*
for i = 0 to numRegions do
    for j = 0 to numNeighbors of region[i] do
        path = astar.computePath(region[i].center, region[j].center)
        matrix[i][j].cost = cost of path
        matrix[i][j].path = compress(path)
        matrix[i][j].next = j
    end for
end for
// Update matrix with dynamic programming
changed = true
while (changed) do
    for i = 0 to numRegions do
        for j = 0 to numNeighbors of region[i] do
            for k = 0 to numRegions do
                if (matrix[i][k].cost > matrix[i][j].cost+matrix[j][k].cost) then
                    matrix[i][k].cost = matrix[i][j].cost+matrix[j][k].cost
                    matrix[i][k].next = j
                    changed = true
                end if
            end for
        end for
    end for
end while
```

As an example, in Figure 2 is the path matrix for the 6 sector map in Figure 1. Entries in the matrix generated by dynamic programming are in italics. For example, the cost of a path from A to C goes through B with a cost of 27.6.

4.3 Online Path Finding

Given a problem from start state s to goal state g, DBA* first determines the start region and goal representatives R_s and R_g, using BFS if multiple regions are in the sector.

	A	B	C	D	E	F	G
A	-	**B** 13.8	B 27.6	**D** 13.2	**E** 20.0	B 15.6	B 37.4
B	**A** 13.8	-	**C** 14.8	**D** 21.6	**E** 20.0	**F** 11.8	**G** 23.6
C	B 27.6	**B** 14.8	-	B 36.4	B 34.8	**F** 9.8	**G** 13.6
D	**A** 13.2	**B** 21.6	B 36.4	-	**E** 14.6	B 36.4	B 45.2
E	**A** 20.0	**B** 20.0	B 34.8	**D** 14.6	-	B 31.8	B 43.6
F	B 15.6	**B** 11.8	**C** 9.8	B 36.4	B 31.8	-	**G** 13.0
G	B 37.4	**B** 23.6	**C** 13.6	B 45.2	B 43.6	**F** 13.0	-

Fig. 2. Path Matrix for Example Map

In Figure 3, the start state s is in region E, and the goal state g is in region G. Since sector 5 contains two regions, a BFS was required to identify region E. Region G was determined by direct lookup as there was only one region in the sector.

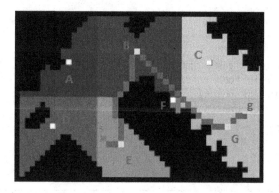

Fig. 3. Online Path Finding Example

The minimal response time possible before the algorithm can make its first move is the time for the BFS to identify the start region representative. At that point, the agent can navigate from s to R_s using the path found during BFS.

The algorithm then looks in the path matrix to find the next hop to navigate from R_s to R_g. In this case, that is to region B. As these are neighbor regions, a compressed path is stored in the database, and the agent navigates using hill-climbing following its subgoals. Once arriving at the representative for region B, it then goes to R_g (representative of region G). If a BFS path was computed to find R_g that path can be used. Otherwise, DBA* completes the path by performing A* bounded within the sector from R_g to g. The maximum move time is the first response time for BFS. The maximum per agent memory is the number of states in a sector (as may need to search entire sector with BFS).

4.4 Optimizations

Several optimizations were implemented to improve algorithm performance.

Path Trimming. The technique of path trimming was first described in PRA*[9] (remove last 10% of states in each concatenated path and then plan from the end of the path to the next subgoal) and extended in HCDPS (start and end optimizations). The general goal is to reduce the suboptimality when concatenating smaller paths to produce a larger path. Combining several smaller paths may produce longer paths with "bumps" compared to an optimal path. These techniques smooth the paths to make them look more appealing and have lower suboptimality.

Consider the path in Figure 3. It is visibly apparent that the agent could have navigated a better path by not navigating from s to the region representative of E then to the region representative of B. The subgoals on the path are shown, which in this case are just the region representatives themselves. Instead, the agent can perform a hill-climbing check from s to the first subgoal (which is R_B in this case) resulting in a much shorter path. A hill-climbing check if successful will shorten the path. If a hill-climbing check fails, then the agent would continue on its original planned path. Note that the agent does not actually move during the hill-climbing check. This optimization can be applied whenever there is a transition between paths put together using concatenation. The optimization is applied when navigating from s to R_s, between each path fragment in the path matrix, and from R_g to g. The idea is to try to go directly to the next subgoal in those cases with hill-climbing. Figure 4 shows the shorter path after applying this optimization which reduces suboptimality from 23% to 11%.

Increasing Neighborhood Depth. A second approach to reducing suboptimality is to increase the number of base paths computed. Instead of only pre-computing paths between adjacent regions, it is possible to increase the neighborhood depth to compute paths between regions up to L steps away. This has been used in HCDPS [8] to reduce the number of path concatenations which hurts suboptimality. The tradeoff is a larger database size and longer pre-computation time. In the running example, computing paths between regions up to 2 away results in a direct path between region E and region G and reduces suboptimality for the example problem to 0% (when used in combination with path trimming). Increasing neighborhood depth decreases suboptimality but does not guarantee an optimal path.

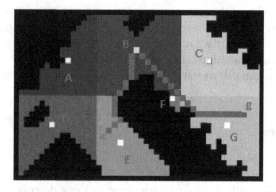

Fig. 4. Path Trimming Reduces Suboptimality When Combining Path Fragments

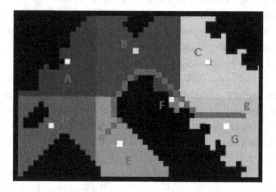

Fig. 5. Solution Path when Database Stores Paths Between Regions Up to 2 Regions Away

5   Experimental Results

Algorithms were evaluated on ten of the largest standard benchmark maps from Dragon Age: Origins™available at http://movingai.com and described in [10]. The 10 maps selected were hrt000d, orz100d, orz103d, orz300d, orz700d, irz702, orz900d, ost000a, ost000t, and ost100d. These maps have an average number of open states of 96,739 and total cells of 574,132. For each map, 100 of the longest sample problems were run from the problem set.

The algorithms compared included DBA*, PRA* as implemented in Dragon Age [9], and HCDPS. HCDPS was run for neighborhood depth $L = \{1, 2, 3, 4\}$. PRA* was run for sector sizes of 16 x 16, 32 x 32, 64 x 64, and 128 x 128 and uses the 10% path trimming and re-planning optimization. PRA* and DBA* both did not apply region center optimization, as region representatives were computed by averaging the rows and columns of all open states in the region. DBA* was run under all combinations of neighborhood level and grid size. LRTA* with subgoals [7] was also evaluated. Algorithms were tested using Java 6 under SUSE Linux 10 on an Intel Xeon E5620 2.4 GHz processor with 24 GB of memory.

In the charts, each point in the plot represents an algorithm with a different configuration. For PRA*, there are 4 points corresponding to 16, 32, 64, and 128 sector sizes. For HCDPS there are 4 points corresponding to levels 1, 2, 3, and 4. For DBA* that combines both sets of parameters, there are separate series for each sector size (16, 32, 64, 128), and each series consists of 4 points corresponding to levels 1, 2, 3, and 4.

5.1 Online Performance

Online performance consists of three factors: suboptimality of paths, total memory used, and average move time. Figure 6 displays suboptimality versus move time. DBA* variants (except for 128) are faster than HCDPS and PRA* and most variants have better suboptimality. DBA* has the same or better suboptimality than HCDPS for smaller sector sizes. Larger sector sizes for both DBA* and PRA* hurt suboptimality as the optimizations do not always counteract navigating through region representatives that may be off an optimal path. Larger sector sizes also dramatically increase the time for PRA* (PRA* 128 has 20 μs move time) as the amount of abstraction is reduced and the algorithm is solving a problem using A* that is not much smaller in the abstract space. Since DBA* uses its path database rather than solving using A* in the abstract space, its move time does not increase as much with larger sectors. The additional time is mostly related to the BFS required to find the region representatives in the start and goal sectors. LRTA* with subgoals (shown as sgLRTA* in the legend) has different performance characteristics than the other algorithms. sgLRTA* is almost optimal. Its move time is relatively high because the subgoal trees are large, and it takes the algorithm time to identify the first subgoal to use.

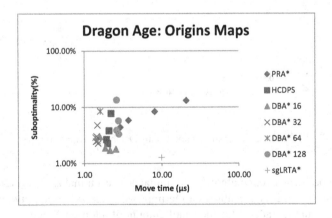

Fig. 6. Suboptimality versus Move Time

Figure 7 compares suboptimality versus total memory used. PRA* uses less memory as storing the regions and sectors requires minimal memory. The additional memory used by DBA* to store paths between regions amounts to 50 to 250 KB, but improves suboptimality from about 6% with PRA* to under 3%. DBA* dominates HCDPS in this metric. sgLRTA* is near perfect for solution quality, but consumes significantly more memory to the point that it is not practical in this domain.

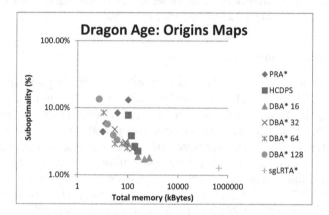

Fig. 7. Suboptimality versus Total Memory Used

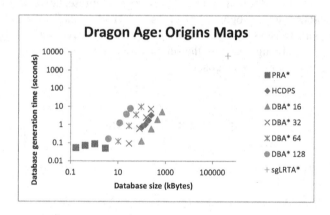

Fig. 8. Database Generation Time versus Database Generation Size

For online performance, no algorithm dominates as each makes a different tradeoff on time versus space. It is arguable that the improved path suboptimality and time of DBA* is a reasonable tradeoff for the small amount of additional memory consumed. Unlike PRA*, DBA* is optimized for static environments.

5.2   Pre-computation

Pre-computation, although done offline, must also be considered in terms of time and memory required. The results are in Figure 8. PRA* consumes the least amount of memory as it only generates sectors and does not generate paths between sectors. HCDPS and DBA* both perform abstraction and path generation, although DBA* in most configurations is faster with a smaller database size. sgLRTA* that computes and compresses all paths takes considerably longer and more space than all other algorithms.

6   Conclusions and Future Work

Grid-based path finding must minimize response time and memory usage. DBA* has lower suboptimality than other abstraction-based approaches with a faster response time. It represents a quality balance and integration of the best features of previous algorithms. Future work includes defining techniques for efficiently updating pre-computed information to reflect grid changes.

References

1. Björnsson, Y., Bulitko, V., Sturtevant, N.: TBA*: Time-bounded A*. In: Proceedings of the International Joint Conferences on Artificial Intelligence (IJCAI), pp. 431–436 (2009)
2. Botea, A.: Ultra-fast optimal pathfinding without runtime search. In: Proceedings of the Second Artificial Intelligence and Interactive Digital Entertainment Conference (AIIDE), pp. 122–127 (2011)
3. Bulitko, V., Björnsson, Y., Lawrence, R.: Case-based subgoaling in real-time heuristic search for video game pathfinding. Journal of Artificial Intelligence Research 39, 269–300 (2010)
4. Bulitko, V., Luštrek, M., Schaeffer, J., Björnsson, Y., Sigmundarson, S.: Dynamic control in real-time heuristic search. Journal of Artificial Intelligence Research 32, 419–452 (2008)
5. Hart, P., Nilsson, N., Raphael, B.: A formal basis for the heuristic determination of minimum cost paths. IEEE Transactions on Systems Science and Cybernetics 4(2), 100–107 (1968)
6. Hernández, C., Baier, J.A.: Fast subgoaling for pathfinding via real-time search. In: Bacchus, F., Domshlak, C., Edelkamp, S., Helmert, M. (eds.) Proceedings of the International Conference on Artificial Intelligence Planning Systems (ICAPS), pp. 327–330. AAAI (2011)
7. Hernández, C., Baier, J.A.: Real-time heuristic search with depression avoidance. In: Proceedings of the International Joint Conference on Artificial Intelligence (IJCAI), pp. 578–583 (2011)
8. Lawrence, R., Bulitko, V.: Database-driven real-time heuristic search in video-game pathfinding. IEEE Transactions on Computer Intelligence and AI in Games PP(99) (2013)
9. Sturtevant, N.: Memory-efficient abstractions for pathfinding. In: Proceedings of Artificial Intelligence and Interactive Digital Entertainment (AIIDE), pp. 31–36 (2007)
10. Sturtevant, N.R.: Benchmarks for grid-based pathfinding. IEEE Transactions on Computer Intelligence and AI in Games 4(2), 144–148 (2012)

Detecting Statistically Significant Temporal Associations from Multiple Event Sequences

Han Liang and Jörg Sander

Dept. of Computing Science, University of Alberta
Edmonton, Canada, T6G 2E8
{hliang2,joerg}@ualberta.ca

Abstract. In this paper, we aim to mine temporal associations in multiple event sequences. It is assumed that a set of event sequences has been collected from an application, where each event has an id and an occurrence time. Our work is motivated by the observation that in practice many associated events in multiple temporal sequences do not occur concurrently but sequentially. We proposed a two-phase method, called *Multivariate Association Miner* (MAM). In an empirical study, we apply MAM to two different application domains. Firstly, we use our method to detect multivariate motifs from multiple time series data. Existing approaches are all limited by assuming that the univariate elements of a multivariate motif occur completely or approximately synchronously. The experimental results on both synthetic and real data sets show that our method not only discovers synchronous motifs, but also finds non-synchronous multivariate motifs. Secondly, we apply MAM to mine frequent episodes from event streams. Current methods are all limited by requiring users to either provide possible lengths of frequent episodes or specify an inter-event time constraint for every pair of successive event types in an episode. The results on neuronal spike simulation data show that MAM automatically detects episodes with variable time delays.

1 Introduction

Nowadays, more and more temporal data in the form of event sequences is being generated. Each distinct event sequence consists of events of the same type, where each event has an id and an occurrence time. In practice associated events in different event sequences do often not occur concurrently but with a temporal lag. For example, in network monitoring, where people are interested in the analysis of packet and router logs, different types of events occurring sequentially can be recorded in a log file. The goal is to discover the temporal relations of these events, which indicate the performance of the network.

We propose a two-phase method, called *Multivariate Association Miner* (MAM), to detect temporal associations from multiple event sequences. First, we discover bivariate associations from pairs of event sequences by comparing the observed distribution of the temporal distances of their event occurrences with a theoretically derived null distribution. Second, we build a bivariate association graph and search each of its paths for a multivariate association.

O. Zaïane and S. Zilles (Eds.): Canadian AI 2013, LNAI 7884, pp. 112–125, 2013.

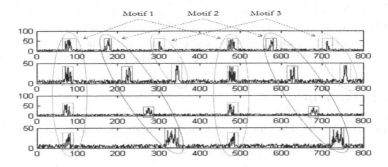

Fig. 1. Illustration of three multivariate motifs. An ellipse represents a multivariate motif occurrence, and a rectangle denotes a univariate element.

While our method can be applied in different application scenarios, in this paper, we focus on comparing its performance on two related tasks. Firstly, we apply MAM to detect multivariate motifs from multivariate time series data. In a univariate time series, a *motif* is a set of time series subsequences that exhibit high similarity and occur frequently in the whole time series [2]. The occurrence of a motif corresponds to some meaningful aspect of the data. In a d-dimensional multivariate time series containing d univariate time series with corresponding time points, a n-dimensional *multivariate motif* $(n \leq d)$ is a set of n-dimensional tuples of univariate elements, where the univariate elements from different dimensions have a temporal association, i.e., they occur concurrently as a *synchronous* multivariate motif (e.g., motif 1 in Figure 1) or sequentially as a *non-synchronous* multivariate motif (e.g., motif 2 and motif 3 in Figure 1). Existing methods of multivariate motif discovery are all limited by assuming that the univariate elements of a multivariate motif occur completely or approximately synchronously. Our experimental results confirm that our method successfully discovers both synchronous and non-synchronous multivariate motifs.

Secondly, we use our method to detect frequent episodes from event streams. Frequent episode discovery is a framework for detecting temporal patterns in symbolic temporal data [4]. The input data of this framework is a sequence of event occurrences with each characterized by an event type and an occurrence time. For example, an event sequence with three occurrences can be represented as follows: $< (A, 1.6), (B, 4.9), (C, 5.1) >$. The detected temporal patterns, referred to as *episodes*, are essentially small, temporally ordered sets of event types. Depending on different types of temporal orders over their event types, episodes are classified into two categories: serial episodes and parallel episodes. A *serial episode* requires its event types to occur sequentially. A *parallel episode* does not require any specific ordering of the event types. Current methods on frequent episode discovery are limited by requiring users to either provide possible lengths of frequent episodes or specify an inter-event time constraint for every pair of successive event types in an episode. Our empirical results show that our method is very effective in detecting episodes with variable lengths.

The reminder of the paper is organized as follows. Section 2 reviews the related work of temporal pattern discovery. Section 3 gives basic definitions. Sections 4 and 5 propose the principles of our method. Sections 6 and 7 describe experimental settings and results. Section 8 concludes the paper.

2 Related Work

Multivariate Motif Discovery. Current methods of multivariate motif discovery can be classified into three categories.

(1) Representing a multivariate time series as a set of multi-dimensional points. Methods in this group treat each univariate time series as a dimension and retrieve a set of d-dimensional points from d equal-length univariate time series. Minnen *et al.* proposed a method that represents data points symbolically based on a vector quantization and uses a suffix tree to locate motif seeds [7]. Their later work located multivariate motifs as regions of high density in the d-dimensional space [6]. Multivariate motifs discovered by these methods must span all dimensions and their univariate elements must be equally sized.

(2) Transforming a multivariate time series into a univariate time series. Tanaka *et al.* used *Principal Component Analysis* (PCA) to transform a multivariate time series into a univariate time series and applied a *Minimum Description Length* (MDL) principle on the projected time series to extract univariate motifs [13]. To handle multivariate time series data, other work in this category extend *Symbolic Aggregate Approximation* (SAX), a technique to reduce the dimensionality of univariate time series subsequences [2]. Minnen *et al.* developed a method that applies SAX on each of the univariate time series and concatenates SAX words from each dimension occurring together in a sliding window [8]. These methods all implicitly assume that the univariate elements in a multivariate motif must be completely synchronous.

(3) Combining a set of univariate motifs into a multivariate motif. Vahdatpour *et al.* constructs a coincidence graph based on the temporal relations of univariate motifs [14]. A graph is initially built, where a vertex represents a univariate motif and the weight of an edge between two vertexes indicates the frequency with which the occurrences of the two corresponding univariate motifs temporally overlap. Starting from the motif with the highest occurrences, a graph clustering algorithm iteratively detects multivariate motifs by comparing the weights of edges connected to this motif to a user-defined threshold. This method allows the univariate motifs to have different lengths and permits that multivariate motifs can span only a subset of dimensions.

Frequent Episode Discovery. The methods applying the framework of frequent episode discovery to neuronal spike data are classified into two categories.

(1) Mining serial and parallel episodes using an Apriori-style procedure. The methods detect serial episodes using a procedure similar to the Apriori algorithm [1]. Mannila *et al.* first presented the framework of frequent episode discovery [4]. The frequency of an episode is defined as the number of sliding windows in which

the episode occurs. Laxman *et al.* proposed their work based on a new frequency measurement, which counts the number of non-overlapped occurrences for an episode [3]. These methods all limit their searching scope by requiring users to provide the size of sliding windows or specify an inter-event time constraint for every pair of successive event types in an episode.

(2) Mining statistically significant episodes. The algorithms detect statistically significant serial episodes. Sastry *et al.* designed a statistical test to determine the significance level of discovered frequent episodes, based on the intuition that the interaction between two event sequences can be captured by the conditional probability of observing an event from one sequence after a time delay given that an event has occurred on the other sequence [11]. Patnaik *et al.* presented another approach by proposing what they so called *"excitatory dynamic networks"* (EDNs), where nodes denote event types and edges represent temporal associations among nodes [9]. The authors also defined their so-called *"fixed-delay episode"*, where the time-delay between every pair of event types in an episode is fixed. To obtain the marginal probabilities for an EDN, the occurrence-based frequencies of fixed-delay episodes are used to compute the probabilities for each node. These methods need users to specify an inter-event time-delay for every pair of successive event types in an episode.

3   Background and Definitions

An *event sequence* $\xi = < e_1, e_2, \ldots, e_m >$ is an ordered set of m events e_i. Each e_i in ξ denotes a tuple (e_id, t_i), where e_id represents the event id and t_i is the occurrence time of the event. All event occurrences in ξ are of the same type.

We introduce a *bivariate association* A_{ab}^d $(a \neq b)$, between two event sequences ξ_a and ξ_b, as a subset of the Cartesian product of ξ_a and ξ_b, as following:

Definition 1. *Let ξ_a and ξ_b be two event sequences. A set $A_{ab}^d \subseteq \xi_a \times \xi_b$ is called a bivariate association in (ξ_a, ξ_b) with mean temporal distance d if for all $(e, e') \in A_{ab}^d : t \leq t' \wedge t' - t \sim \Phi(\cdot) \wedge E(t' - t) = d$, and there is a one-to-one correspondence between the sets $\{e | \exists e' : (e, e') \in (A_{ab}^d)\}$ and $\{e' | \exists e : (e, e') \in (A_{ab}^d)\}$, where t (resp. t') is the occurrence time of event e (resp. e'), $\Phi(\cdot)$ denotes a a known distribution (e.g., uniform or Gaussian) that the temporal distance between two associated events follows, and $E(t' - t) = d$ is the expected temporal difference between associated events in A_{ab}^d.*

A *multivariate association* $MA_{1 \ldots k}^{d_1 \ldots d_{k-1}}$ between k event sequences $\xi_1, \ldots, \xi_k$ is defined as:

Definition 2. *Let $\xi_1, \ldots, \xi_k$ be k different event sequences. A set $MA_{1 \ldots k}^{d_1 \ldots d_{k-1}} \subseteq \xi_1 \times \ldots \times \xi_k$ is called a multivariate association in $(\xi_1, \ldots, \xi_k)$ if for all $(e^1, \ldots, e^k) \in MA_{1 \ldots k}^{d_1 \ldots d_{k-1}} : (e^i, e^{i+1})$ is an instance of a bivariate association in (ξ_i, ξ_{i+1}) with mean temporal distance d_i for all $1 \leq i \leq k - 1$.*

Note that bivariate and multivariate associations do not involve all event occurrences in the involved event sequences, but represent typically only small subsets, which are embedded in the event sequences.

In many real-world applications, especially in applications where the event sequences are the result of the superimposition of many low intensity, *arbitrary* point processes, the collected event sequences can be modeled as Poisson processes. For our approach, we assume that the event sequences collected from an application can be modeled as Poisson processes. The properties of a Poisson process that we will exploit in our approach are (i) the **inter-arrival times** T_i between consecutive event occurrences are independent and follow an exponential distribution with rate $\mu = 1/\lambda$, where λ denotes the intensity of the Poisson process, and (ii) the ith **arrival times** S_i, i.e., the times until the ith event occurrence from the starting point of the process, follow a Gamma distribution with shape parameter $\alpha = i$ and scale parameter $\beta = \lambda$.

4 Detecting Bivariate Associations

To determine whether two event sequences ξ_a and ξ_b are temporally associated, we analyze what we define as **forward distances**, which are the difference in time between the events $e \in \xi_a$ and the events $e' \in \xi_b$ occurring after e.

Definition 3. *The set of forward distances between event sequences ξ_a and ξ_b is given as $FD_{ij} = \{dist | \exists e \in \xi_a \exists e' \in \xi_b, t \leq t' \wedge dist = t' - t\}$, where t (resp. t') is the occurrence time of e (resp. e').*

To compute the forward distances from an event on a sequence ξ_a to events in a sequence ξ_b, we can think of projecting the event onto sequence ξ_b and denoting the projected position as h. The forward distance from h to its right nearest event on sequence ξ_b can be denoted as Z_1. Since we compute Z_1 for each event of sequence ξ_a, Z_1 can be treated as a random variable whose distribution can be derived from the so-called *Waiting Time Paradox* for Poisson processes [5]. The theorem states that (i) Z_1 follows the same exponential distribution as the inter-arrival times on sequence ξ_b, with mean $\mu = 1/\lambda_b$, and (ii) the forward distances Z_j from time h to the jth event after h, can be modeled as the arrival times of the jth event of a Poisson process starting at time h, following a Gamma distribution with shape parameter $\alpha = j$ and scale parameter $\beta = \lambda_b$.

Knowing the distribution of the individual forward distances, we can express the distribution of all forward distances x from all events on a finite sequence ξ_a to all events on a finite sequence ξ_b as a mixture of these individual distributions:

$$f_n(x) = \sum_{j=1}^{N} W_j \times g(x; j, \lambda_b), \tag{1}$$

where $f_n(x)$ is our expected null distribution of forward distances (i.e., when there is no temporal association), N is the number of individual distribution components, which equals the number of forward distances the first event on sequence ξ_a has. $g(x; j, \lambda_b)$ is the Gamma distribution that Z_j follows, and W_j represents the weight of the jth component. Figure 2 illustrates how we derive the weight for each component density from the properties of the involved Poisson

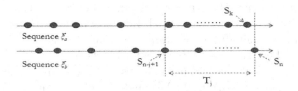

Fig. 2. Determining the Weights of Individual Gamma Distribution Components

processes. In the figure, S_n is the arrival time of the last event on sequence ξ_b and there are k events on sequence ξ_a that occur before S_n. S_{n-j+1} denotes the arrival time of the jth last event on sequence ξ_b. Every event on ξ_a that occurs before S_{n-j+1} will have all forward distances to events on ξ_b up to and including their jth right neighbor. However, each event on ξ_a after S_{n-j+1} will not have a distance to its jth right neighbor and will not contribute a distance to Z_j. If T_j denotes the time interval between S_{n-j+1} and S_n, its expected length $E(T_j)$ is $(j-1)/\lambda_b$. The expected number of events on sequence ξ_b that are in time interval T_j can be calculated by $(j-1)\lambda_a/\lambda_b$. Let N_j represent the number of distances in Z_j; its expected number $E(N_j)$ can be calculated by $k - [(j-1)\lambda_a/\lambda_b]$. Hence, we can compute each weight W_j by $E(N_j)/\sum_{i=1}^{N} E(N_i)$.

The intuition behind our method is: when two event sequences contain a bivariate association, the observed number of forward distances around the mean distance of the association will be larger than expected under the null distribution. To estimate the actual, observed distribution of forward distances, we generate a histogram of the actual forward distances, with a bin size determined by using Shimazaki's method [12]. We design a statistical test to determine the probability that a bin B contains the observed number $ON(B)$ of forward distances under the null hypothesis (i.e., forward distances are distributed according to $f_n(x)$). The probability P_B that a randomly chosen forward distance falls into B under the null hypothesis can be derived as following:

$$P_B = \int_l^u f_n(x)dx \tag{2}$$

where l and u denote the lower and upper bound of B, respectively. Given n observed forward distances, the distribution of the test statistic $ON(B)$ under the null hypothesis can be modeled by a Bernoulli experiment repeated independently n times with a success probability of P_B. Consequently, $ON(B)$ follows a binomial distribution with parameters n and P_B. Let α_0 be a significance level. Let α be the probability that we observe $ON(B)$ under the null hypothesis. B is statistically significant at significance level α_0 if $\alpha \leq \alpha_0$. To avoid false positives, we perform a Bonferroni adjustment by setting the significance level to α_0/m, where m denotes the number of tests.

Given a statistically significant bin B', we assume that there is a true bivariate association A_{ab}^d contained in the two corresponding sequences ξ_a and ξ_a. We estimate the mean temporal distance d of A_{ab}^d as the mean of the distances inside B'. Supposing that the temporal distances between associated events follow a distribution with a relatively small range around the mean (which is reasonable for many meaningful applications), we can use the bin size of B' as this range to extract the actual associated event occurrences. For an event occurrence on ξ_a, we only consider the event occurrences on ξ_b that are in the time interval $[mean - range/2, mean + range/2]$. Although there are some applications where multiple-to-one correspondence of associated events may exist, in this paper, we only focus on one-to-one correspondence of associated events. If there is more than one event occurrence on ξ_b in this range, we select the event occurrence on ξ_b that is closest to the mean temporal distance.

5   Detecting Multivariate Associations

To find multivariate associations, we construct a directed graph where a vertex denotes a bivariate association. We add a directed edge to the graph from a vertex v_i to a vertex v_j if the bivariate association of v_i ends in an event sequence that the bivariate association of v_j originates from. We use Algorithm 1 to discover multivariate associations. We first search the graph for "*start vertices*", i.e., vertices having no incoming edges, and store them in a set SV. In the case where all the vertices in a graph have incoming edges, we regard each vertex as a start vertex. Given a start vertex sv_i, we retrieve all of the paths beginning at sv_i and store them in a path set P_i. Starting from a randomly chosen path p in P_i, we search along p for its maximum sub-path sp_{max}, which is checked by Algorithm 2. If sp_{max} is a multivariate association, we store it in a temporary set $Temp$. We start our search again with another randomly chosen path in the remaining part of P_i and repeat Step 5-7 until all of the paths in P_i are processed. After iterating all of the start vertices in SV, we copy every detected multivariate association from $Temp$ to a result set R, and remove all the corresponding vertices and their associated edges from the graph. Finally, since it is possible that some multivariate associations are sub-paths of others, we remove these redundant associations from R. We repeat Step 1-11 until the graph is empty.

Algorithm 2 tests whether a multivariate association exists along a path, in the sense that the number of chains of connected associated pairs of events on this path is larger than expected when assuming that chains form by chance. We search for a multivariate association on a path $p = v_1 \ldots v_l$ as following. Let k be the number of chains of connected associated pairs of events from v_1 to v_{i-1} on p. Let x be the number of chains of connected associated pairs of events from v_1 to v_i on p. We use m and n to denote the number of associated pairs of events in v_{i-1} and v_i, respectively. If we randomly select an associated pair of events from v_{i-1}, the probability δ that this associated pair is on a chain from v_1 to v_i can be estimated by $\delta = k/m$. The distribution of observed number x of chains from v_1 until v_i can be modeled as a Bernoulli experiment, which is repeated independently n times with the success probability of δ. Hence, x follows a binomial distribution with parameters n and δ. Using this null distribution, we perform a statistical test to determine if the observed number of chains is

---

Algorithm 1. Multivariate Association Discovery (MAD)

Input: $G(V, E)$ - bivariate association graph
Output: R - a set of paths, each representing a multivariate association

1: **while** G is not empty **do**
2:   SV = vertices without incoming edges; if $SV = \emptyset$: $SV = V$
3:   **for** sv_i is in SV **do**
4:     Retrieve each path starting at sv_i from G and store them in a set P_i;
5:     **for** each path $p \in P_i$ **do**
6:       Starting from sv_i, search along p for its maximum sub-path sp_{max}, which is verified to be a multivariate association by Algorithm 2;
7:       Store sp_{max} in a temporary set $Temp$;
8:   **for** each path $p \in Temp$ **do**
9:     Remove all the vertices on p as well as their associated edges from G;
10:     Store p in a result set R and remove all the sub-paths of p from R;
11: **return** S

---

significant, where the significance level is adjusted by the number of tests, which equals the number of vertices on p. If the number of chains from v_1 to v_i is statistically significant, we continue with the next vertex v_{i+1} on p; otherwise, we return the sub-path from v_1 to v_{i-1} as a multivariate association.

6   Empirical Study on Multivariate Motif Discovery

We apply our method to detect multivariate motifs from multiple time series sequences and compare it with the work of Vahdatpour *et. al.* [14], which is one of the most effective methods for multivariate motif discovery. To use our method to discover multivariate motifs, we consider each univariate motif occurrence as an event and the set of motif occurrences retrieved from the same univariate time series can be transformed into an event sequence. The significance level of all statistical tests was set to 10^{-11}, and the threshold of determining the minimum correlation of two univariate motifs in Vahdatpour's method was set to 0.05 as in [14].

To evaluate MAM in terms of generality and robustness, we first conducted three groups of experiments with synthetic data. In each group, we generated a set of univariate time series with a length of 2×10^7 time units. We implanted varying numbers of occurrences of a multivariate motif and *"noise"* univariate motif occurrences (i.e., those that do not participate in the multivariate motif) into these time series, so that each univariate time series had a total of 1×10^4 occurrences. Both the length of noise univariate motif occurrences and the length of univariate elements in the multivariate motif equaled 20 time units. The performance of a method was evaluated using the F-measure, which is the weighted average of precision p and recall r.

Algorithm 2. Candidate Association Verification (CAV)

Input: $p = v_1 \ldots v_l$ - a path in G; α_0 - significance level
Output: sp_{max} - a multivariate association

1: $index = 1$;
2: **for** each vertex v_i $(2 \leq i \leq l)$ on p **do**
3:     $k \longleftarrow$ the number of chains of connected associated pairs of events starting from v_1 until v_{i-1} on p;
4:     $x \longleftarrow$ the number of chains of connected associated pairs of events starting from v_1 until v_i on p;
5:     $m \longleftarrow$ the number of associated pairs in v_{i-1};
6:     $n \longleftarrow$ the number of associated pairs in v_i;
7:     $\delta = k/m \longleftarrow$ the probability that an associated pair in v_{i-1} is on a chain starting from v_1 until v_{i-1};
8:     $\alpha = Binomial(x, n, \delta) \longleftarrow$ the probability of observing x chains starting from v_1 until v_i;
9:     **if** $\alpha \leq \alpha_0/l$ **then**
10:         $index = i$;
11:     **else**
12:         break;
13: **return** $sp_{max} = v_1 \ldots v_{index}$

In the first group of experiments, we created synthetic data sets containing five randomly generated univariate time series, where a 5-variate motif and some noise univariate motif occurrences were implanted. The 5-variate motif consisted of 4 bivariate motif components, each of which had a fixed standard deviation of temporal distances equal to 2.5 time units. We varied both the percentage of 5-variate motif occurrences from 10% to 100%, and the mean temporal distance between 10 and 5000 time units. Table 1(a) shows that MAM not only detected this multivariate motif when its univariate elements temporally overlap (i.e., the cases when the mean temporal distance equals 10 or 20 time units) but also found it as its univariate elements had varying temporal lags. Our method is also robust, especially in the situation where the number of noise univariate motif occurrences was 9 times larger than the one of multivariate motif occurrences. Table 1(b) shows that Vahdatpour's method detected nothing when the univariate elements of this implanted multivariate motif were non-synchronous.

In the second group, we varied both the percentage of 5-variate motif occurrences from 10% to 100%, and the standard deviation of the temporal distances of the bivariate motif components between 10 and 100 time units. Each bivariate motif component had now a fixed mean of temporal distances equal to 500 time units. Table 2 shows that our method detected the multivariate motif in most of the cases, and we confirmed via experiments (not shown here) with standard deviations from 40 to 125 time units that: the larger the standard deviation

Table 1. % of Multivariate Motif Occurrences vs. Mean Temporal Distance

(a) Experimental Results of MAM

Mean / Per.	10	20	200	1000	5000
10%	0.948	0.946	0.943	0.945	0.942
30%	0.953	0.954	0.952	0.954	0.957
50%	0.964	0.961	0.964	0.963	0.966
70%	0.976	0.975	0.973	0.975	0.976
90%	0.986	0.990	0.988	0.988	0.985

(b) Experimental Results of Vahdatpour's method

Mean / Per.	10	20	200	1000	5000
10%	0.889	0.496	0.0	0.0	0.0
30%	0.959	0.519	0.0	0.0	0.0
50%	0.974	0.531	0.0	0.0	0.0
70%	0.985	0.553	0.0	0.0	0.0
90%	0.990	0.564	0.0	0.0	0.0

becomes, the more difficult it is to detect the multivariate motif, because some instances of the bivariate components can no longer be detected. The results of Vahdatpour's method show that it never found this multivariate motif.

Table 2. % of Motif Occurrences vs. Standard Deviation of Temporal Distances

Vari. / Per.	2.5	5	10	20	30
10%	0.947	0.925	0.910	0.753	0.621
30%	0.953	0.946	0.937	0.887	0.743
50%	0.965	0.955	0.942	0.926	0.881
70%	0.974	0.962	0.956	0.943	0.901
90%	0.985	0.976	0.964	0.952	0.927

Finally, we generated a complex synthetic data set of ten randomly generated univariate time series with a length of 2×10^8 time units, where we implanted five multivariate motifs. Each multivariate motif had 1000 occurrences. We also added 5000 noise univariate motif occurrences to each dimension in the data set. Table 3 lists the properties of the implanted multivariate motifs. MAM obtained scores of 1.0 for the bivariate and the 3-variate motifs, 0.959 for the 5-variate motif, 0.946 for the 8-variate motif and 0.968 for the 10-variate motif. Vahdatpour's method performed well on bivariate and 3-variate motifs, but it detected none of the other non-synchronous multivariate motifs.

Table 3. The Properties of Implanted Multivariate Motifs

Motifs / Properties	bi-variate	3-variate	5-variate	8-variate	10-variate
Mean	2	20	800	3000	5000
Standard Deviation	0.5	2.5	10	10	5
Dimensions	1-2	3-5	1-5	1-8	1-10

To explore the utility of our method in real applications, we first used Smart-Cane data sets collected from a wearable system [15]. This system is developed as a device to monitor senior or impaired people in their assisted walking behavior. Three data sets, each of which has eight univariate time series, are generated by the sensors of the system. There exists a synchronous multivariate motif in these data sets, which corresponds to the normal use of the cane when walking

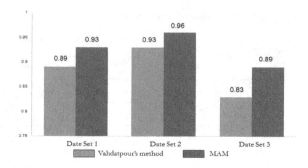

Fig. 3. The Accuracy of Vahdatpour's method and MAM for Three SmartCane Data Sets

(i.e., normal activity). Figure 3 shows the accuracy of normal activity discovery of both methods. Again, even for this task of synchronous motifs, MAM outperforms Vahdatpour's method.

We further evaluated MAM and Vahdatpour's method using a data set where non-synchronous multivariate motifs may exist. The data set consists of recordings of shovel operations provided by an industrial company. We attempt to detect a variety of patterns, such as dig-cycles. The power consumed by three motors (i.e., Crowd power, Hoist power, and Swing power) is recorded as a time series, and the power profiles of the motors can provide information about the shovel's activities. We ran our method and Vahdatpour's method using the same parameters as done for the SmartCane data. MAM detected several multivariate motifs, compared to Vahdatpour's method which detected only a multivariate motif that was a synchronous subset of a larger motif detected by MAM. We are currently in the process of characterizing and interpreting the usefulness of the temporal associations found in this data set.

7   Empirical Study on Frequent Episode Discovery

We also applied MAM to discover frequent episodes from neural spike train data and compared it with the work of Patnaik *et. al.* [9], which is one of most effective methods for detecting temporal patterns from spike train data. We ran MAM using the same parameters as done for multivariate motif discovery.

We evaluate our method and Patnaik's method on simulation data collected from a mathematical model of spiking neurons [10]. This model allows for temporal associations with variable time delays of associated spikes, which mimic the situation in conduction pathways of real neurons. Each generated spike train (a sequence of spikes made by a neuron) follows an inhomogeneous Poisson process. We use this model to assess the performance of a method in discovering several episodes implanted into several simulation data sets. Figure 4 illustrates these episodes, where nodes denote spike trains and directed arcs represent temporal orders of firing spikes among trains. For each episode the values above a

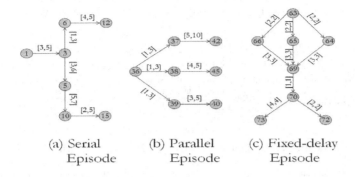

(a) Serial   (b) Parallel   (c) Fixed-delay
Episode   Episode   Episode

Fig. 4. Three Episodes Implanted into Synthetic Data Sets

Table 4. Spike Train Simulation Data Sets

(a)

Name	Length (ms)	Base Fir. Rate $\hat{\lambda}_0$	Act. Prob. ρ
A_1	60000	0.01	0.9
A_2	60000	0.015	0.9
A_3	60000	0.02	0.9
A_4	60000	0.025	0.9
B_5	60000	0.02	0.8
B_6	60000	0.02	0.85

(b)

Name	Length (ms)	Base Fir. Rate $\hat{\lambda}_0$	Act. Prob. ρ
B_7	60000	0.02	0.9
B_8	60000	0.02	0.95
C_9	60000	0.02	0.9
C_{10}	90000	0.02	0.9
C_{11}	120000	0.02	0.9

directed arc indicate the range of time delays between associated spikes. Each
data set contains 100 spike trains (the spike trains, which are not involved in
the implanted episodes, fire independently). Table 4 lists the properties of these
data sets. The second column shows the length of a data set (i.e., the number of
time slices in the data sequence), the third column shows the base firing rate of
neurons, and the fourth column presents the activation probability of a neuron
(i.e. the conditional probability that a neuron fires given its stimulus received in
the recent past). We arrange these data sets into three groups: A, B and C. We
measure the performance of a method by using again the F-measure.

First, we summarize the results of the two methods on the data sets of A-
group. We created these data sets by changing the base firing rate $\hat{\lambda}_0$ of neurons in the mathematical model. The larger value we assign to $\hat{\lambda}_0$, the more
spikes are generated on a train. Table 5(a) shows that MAM successfully detected all the implanted episodes by achieving high scores in all of the cases.

Table 5. Implanted Episodes vs. Base Firing Rate $\hat{\lambda}_0$

(a) Experimental Results of MAM

Episode Types \ λ_0	0.01	0.015	0.02	0.025
Serial Episode		0.998	0.996	0.995	0.993
Parallel Episode		0.998	0.997	0.996	0.995
Fixed-delay Episode	1.0	1.0	0.999	0.998

(b) Experimental Results of Patnaik's method

Episode Types \ λ_0	0.01	0.015	0.02	0.025
Serial Episode	0.0	0.0	0.0	0.0
Parallel Episode	0.0	0.0	0.0	0.0
Fixed-delay Episode	1.0	1.0	1.0	1.0

Table 6. Implanted Episodes vs. Activation Probability ρ

(a) Experimental Results of MAM

Episode Types ρ	0.8	0.85	0.9	0.95
Serial Episode	0.997	0.996	0.996	0.996
Parallel Episode	0.997	0.997	0.996	0.995
Fixed-delay Episode	1.0	1.0	0.999	0.999

(b) Experimental Results of Patnaik's method

Episode Types ρ	0.8	0.85	0.9	0.95
Serial Episode	0.0	0.0	0.0	0.0
Parallel Episode	0.0	0.0	0.0	0.0
Fixed-delay Episode	1.0	1.0	1.0	1.0

Table 7. Implanted Episodes vs. Data Set Length L

(a) Experimental Results of MAM

Episode Types Length(ms)	60000	90000	120000
Serial Episode	0.998	0.998	0.998
Parallel Episode	0.999	0.998	0.998
Fixed-delay Episode	0.1	0.1	0.999

(b) Experimental Results of Patnaik's method

Episode Types Length(ms)	60000	90000	120000
Serial Episode	0.0	0.0	0.0
Parallel Episode	0.0	0.0	0.0
Fixed-delay Episode	1.0	1.0	1.0

Table 5(b) presents that: although Patnaik's method was effective in finding the fixed-delay episode, it detected nothing when the time delay of two associated spikes varies. To reduce the computational complexity, Patnaik's method searches only for fixed-delay episodes to construct a dynamic Bayesian network to encode temporal associations among spike trains.

Second, we evaluate the competing methods using the data sets of B-group. This time we created the data sets by varying the activation probability ρ of a neuron. The larger value we set to ρ, the more occurrences of an episode are implanted into the data. The results in Table 6(a) show that our method successfully discovered these implanted episodes from the data. The results in Table 6(b) present that Patnaik's method found neither serial nor parallel episodes.

Finally, we evaluate the performances of these methods as the data set length was varied. Table 7(a) shows that our method worked constantly well by achieving high scores in all of cases, indicating that the performance of our method is not affected by the data set length. Table 7(b) shows that Patnaik's method still failed to detect either serial or parallel episodes.

8   Conclusion and Future Work

We presented a two-phase method that mines temporal associations in multiple event sequences. In an empirical study, we first used it to detect multivariate motifs. The results on both synthetic and real data sets showed that our method found both synchronous and non-synchronous multivariate motifs. We then applied our method to discover frequent episodes from event streams. The results on neuronal spike simulation data presented that our method effectively discovered episodes with variable lengths. In future work, we will investigate other application domains, such as network monitoring.

References

1. Agrawal, R., Srikant, R.: Fast algorithms for mining association rules in large databases. In: VLDB 1994, pp. 487–499 (1994)
2. Chiu, B., Keogh, E., Lonardi, S.: Probabilistic discovery of time series motifs. In: KDD 2003, pp. 493–498 (2003)
3. Laxman, S., Sastry, P., Unnikrishnan, K.: A fast algorithm for finding frequent episodes in event streams. In: KDD 2007, pp. 410–419 (2007)
4. Mannila, H., Toivonen, H., Verkamo, A.: Discovering frequent episodes in sequences. In: KDD 1995, pp. 210–215 (1995)
5. Meester, R.: A Natural Introduction to Probability Theory (2004)
6. Minnen, D., Isbell, C., Essa, I.: Discovering multivariate motifs using subsequence density estimation and greedy mixture learning. In: AAAI 2007, pp. 615–620 (2007)
7. Minnen, D., Starner, T., Essa, I., Isbell, C.: Discovering characteristic actions from on-body sensor data. In: ISWC 2006, pp. 11–18 (2006)
8. Minnen, D., Starner, T., Essa, I., Isbell, C.: Improving activity discovery with automatic neighborhood estimation. In: IJCAI 2007, pp. 2814–2819 (2007)
9. Patnaik, D., Laxman, S.: Discovering excitatory networks from discrete event streams with applications to neuronal spike train analysis. In: ICDM 2009, pp. 407–416 (2009)
10. Raajay, V.: Frequent episode mining and multi-neuronal spike train data analysis. Master's thesis, IISc, Bangalore (2009)
11. Sastry, P., Unnikrishnan, K.: Conditional probability-based significance tests for sequential patterns in multineuronal spike trains. Neural Computation (2010)
12. Shimazaki, H., Shinomoto, S.: A method for selecting the bin size of a time histogram. Neural Computation 19(6), 1503–1527 (2007)
13. Tanaka, Y., Iwamoto, K., Uehara, K.: Discovery of time-series motif from multidimensional data based on mdl principle. Machine Learning 58, 269–300 (2005)
14. Vahdatpour, A., Amini, N.: Toward unsupervised activity discovery using multidimensional motif detection in time series. In: IJCAI 2009, pp. 1261–1266 (2009)
15. Wu, W., Au, L., Jordan, B.: The smartcane system: an assistive device for geriatrics. In: BodyNets 2008, pp. 1–4 (2008)

Selective Retrieval for Categorization
of Semi-structured Web Resources

Marek Lipczak[1,2], Tomasz Niewiarowski[1], Vlado Keselj[1], and Evangelos Milios[1]

[1] Dalhousie University, Halifax, Canada, B3H 1W5
{lipczak,tomasz,vlado,eem}@cs.dal.ca
[2] 2nd Act Innovations Inc., Halifax, Canada

Abstract. A typical on-line content directory contains factual information about entities (e.g., address of a company) together with entity categories (e.g., company's industries). The categories are a salient element of the system as they allow users to browse for entities of a chosen type. Assigning categories manually can be a challenging task, considering that an entity can belong to few out of hundreds of categories (e.g., all possible industry types). Instead we suggest to augment this process with an automatic categorization system that suggests categories based on the entity's home page. To improve the accuracy of results, the system follows links extracted from the home page and uses retrieved content to expand an entity's term profile. The profile is later used by a multi-label classification system to assign categories to the entity. The key element of the system is a link ranking module, which uses home page features (e.g., position and anchor text of links) to select links that are most likely to improve the categorization results. Evaluation on a data set of nearly ten thousand company home pages confirmed that the link ranking approach allows the system to limit the retrieval and processing costs to allow real-time responses and still outperform the categorization results of baseline systems.

Keywords: web classification, multi-label classification, selective crawling, learning to rank, semi-structured web, Enterprise Content Management.

1   Introduction

The basic feature of on-line content directories is to store and organize information about specific entities in a structured form. For example, local directory services (e.g., Yelp[1]) store information about millions of local businesses, giving users access to factual information like address or hours of operation. An important feature of such systems is categorization of entities which allows users to browse the repository looking for entities of a specific type (e.g., Greek food restaurants). A similar model applies to a broad range of services including conference management systems which store contact information of potential

---

[1] http://www.yelp.com/about

O. Zaïane and S. Zilles (Eds.): Canadian AI 2013, LNAI 7884, pp. 126–137, 2013.

reviewers (factual information) and their areas of experience (categories that can be used to browse for suitable reviewers). The services that we specifically focus on are Enterprise Content Management (ECM) systems designed to store and provide access to company-related information including a company's business partners or competitors. Here the system stores the factual information about the company (e.g., address) and the industries the company belongs to as categories. In all presented examples the success of the service depends greatly on the quality and amount of information stored in the system. Typically the information is fetched to the system manually from an entity's web resource. In this paper we use the term *web resource* as the *home page* (the landing page for an entity's url) and *child pages* (all pages linked from the home page within the same domain). In comparison to the extraction of factual information, which is relatively easy to process even for an inexperienced user, the categorization task requires much more effort and experience. The content of an entity's web resource must be summarized and matched with potentially hundreds of available categories. This task is much better suited for a computer system, which can learn a model representing each category and match it against the profile of the entity extracted from its web resource. The categories can be automatically assigned to the entity or presented to a human categorizer for validation. This reduces the cognitive effort of a categorization task to a simple recognition of correct categories from a small recommendation set.

To extract the profiles of entities, the categorization system should retrieve the content of its home page. Using the home page, the system can follow the link structure within its web domain to retrieve more entity-related content and this way improve the accuracy of categorization process. However, each additional retrieval task increases the time and processing cost. Considering that a typical company home page contains on average fifty links, processing all of them is not feasible in real time. In addition, it is possible that newly retrieved information will be in fact harmful for the categorization results [5,14]. Let us consider the problem of finding industry categories for a typical multinational company. Its home page is likely to contain many links that lead to pages of the company's international offices or recent news about the company. These links are not only costly to retrieve, they are also likely to introduce noise in the categorization process. The key objective of our work was to create a system able to intelligently select a few links worth following. As we demonstrate, this *selective retrieval* approach can not only limit the number of retrieved pages in a practical range of 2 to 8 links, but also lead to superior categorization results. The proposed system extends the concept of *selective crawling*, which is an optimization technique that aims to reduce the cost and length of the crawling process by following only the relevant or important links [2]. Unlike the traditional selective crawling approaches, which are based on web-based characteristics of links (e.g., PageRank), our approach works locally, utilizing the fact that categorized entities come from the same domain (e.g., companies) therefore they share similarities in the link structure of their web resources. Based on that, the system is able to predict the usefulness of linked content and follow only links that are likely to contribute to the categorization process.

2   Related Work

The objective of our system is to categorize an entity based on its web resource, which can be considered as a web classification task [17]. A web page can be classified based on its content [14] or classes of its neighbours in the web graph [1]. Our system combines these two approaches by extending the content of the entity's home page by the content of selected child pages (i.e., pages linked from the home page). As there is no guarantee that child pages come from the same class, using their content can potentially decrease the quality of classification. In fact it is a typically observed outcome [5,14]. To overcome this problem Chakrabarti et al. [5] proposed to pre-classify the neighbour pages and use predicted classes as features for the classification of the home pages. Oh et al. [14] suggested to use only the content of the neighbours that are sufficiently similar to the home page. Both methods rely on the retrieval of a large number of neighbours which is time consuming and therefore impractical in our problem setting. Instead, we need a system that is able to predict the usefulness of child pages without the need of retrieving their content.

The proposed system, in its much smaller scale, tackles the same problem that is faced by modern web crawlers. The huge growth of the World Wide Web made it the main source of information, but also created issues with the quality, coverage and cost of retrieved information. These issues are addressed by selective crawling methods [2], which aim to extract the most useful information from the web understanding that exhaustive web crawling is cost prohibitive. The most frequently discussed type of selective crawling is *focused crawling* [6] – the retrieval of web pages from a specific category. In the context of focused crawling task we can say that our system tries to solve a reversed problem in which we look for web pages that would allow us to define the category of retrieved web resource. Nevertheless, we can see the overlap in attributes [15,16,21] (i.e., features extracted from anchor text and url content) that can be used to achieve both goals. Most of the proposed focused crawlers are based on binary classifications methods (follow, not-follow) [16], which is not applicable to our problem. Instead we propose the use of a ranking algorithm to prioritize the retrieval of most useful pages. The ranking algorithm is trained using similarities in the page link structure.

The semi-structured character of web information has been used in various information extraction systems [4,9,13]. The focus of these systems is to extract entity factual information from a single structured web page coming from an on-line catalog (e.g., extracting product attributes from an on-line store catalog [13]). The structured character of these pages comes from the fact that they are dynamically generated from a database. In comparison, our system operates on human-designed home pages of entities. Roth et al. [19] performed a study on the structure of web pages from three domains (online shop, news portal, company home page) showing that users have specific expectations and mental models for each of these. Matching these expectations is the key factor in the web page usability. Therefore, we can expect similar structure of web resources coming from the same domain.

3   Selective Retrieval for Web Categorization

The categorization of web-resources can be considered as a multi-label classification task with a relatively large fixed number of classes (potentially exceeding 100). Multi-label character is caused by the fact that represented entities are likely to have more than one category (e.g., company belonging to more than one industry). The objective of the system is to assign a set of categories from a limited vocabulary of categories to each classified entity. To make the system practically usable we have to consider not only the accuracy of classification, but also the efficiency of producing the results. Unlike most approaches which aim to improve the classification algorithm, we put the main focus on the process of efficient and effective retrieval of profiles that are used in the classification process. The system structure is presented in Fig. 1.

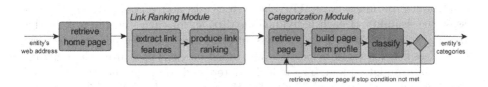

Fig. 1. Selective retrieval process. The link ranking module ranks links extracted from the home page to minimize the number of pages that have to be retrieved in the categorization module. The categorization module starts with the most promising links and proceeds retrieval and categorization iteratively until stop condition is met.

3.1   Link Ranking Module

The categorized entities come from the same domain, therefore the web resources representing them are likely to share some characteristics. Although there is no formal link structure of resources, we can expect similar links to be present on the home page of the entity. For example, a web page of a restaurant should contain links to its location and menu. The first link is not useful for categorization purposes and can in fact introduce noise to the process. On the other hand, the second link is likely to allow the system to come up with correct categories of the entity. The objective of the link ranking module is to examine the links present on the home page of the entity and rank them so links that are most likely to lead to such useful content are retrieved first. To produce the ranking the system can use only the information that is present on the home page. Considering this limitation we came up with the following set of link features divided into three groups:

- **Position features** represent the location of the link in the page structure. The features include the normalized distance from the top of the page, positions in different levels of menu, page header or footer, which are frequently occurring elements of entity web resources.
- **Anchor text features** consist of the most popular terms form the link anchors (excluding stop words). Examples of terms frequently occurring in company's pages are *contact, about, services, careers, products,* all of them leading to related structure elements of the page. In addition to text features, the system uses a special feature for links represented as a picture.
- **Link features** consist of a set of character 4-grams generated from link *href* attribute, as there are no clear word boundaries in urls. The purpose of these features is to represent similarities between urls that share the same element like "contact_us", "about" in link to a specific element of the page structure. An important ability of these features is to catch links to alternative language versions which are likely to contain phrases like "/fr/", "es/i".

To train the link ranking module the system requires a subset of categorized entities coming from the same domain (e.g., a set of companies with their industry types) with their web resources. In web resources, we use only the direct children of the home page, as with the increasing number of steps we are likely to introduce too much noise [17]. The system uses a cross-validation setting to assess the usability of each individual child page, for each web resource in the training set. To achieve that the system produces a classification model (using a classifier described in Section 3.3) based on each page from the training set in cross-validation fold. The model is later used to predict entity categories based on child page content. The higher the accuracy of a given child page the more likely it is to contribute to the final categorization result where information from all retrieved pages is combined. After running the process for all folds, the system has the accuracy based ranking of child pages for each training entity. The rankings are later used to train the model of a ranking algorithm. After experimenting with a range of ranking algorithms including RankNet [3], Rank-Boost [7], AdaRank [23] implemented in RankLib package[2] we decided to use *SVMrank* [11], a ranking method based on Support Vector Machines regression [10], as it was more effective and efficient than its competitors. Given the ranking model the system is able to automatically order child pages of newly observed web resources according to their predicted categorization accuracy.

3.2   Categorization Module

Each entity is represented by a profile of terms extracted from the retrieved pages. We use a standard bag-of-words approach counting the occurrences of each word in the page content. The categorization module uses the ranked list of links to iteratively extend the entity profile. The process is repeated until the stop condition is satisfied. The basic stop condition used in the system is based

---

[2] http://people.cs.umass.edu/~vdang/ranklib.html

on the total number of retrieved pages. In this case, the categorization results are produced based on the top N pages from the ordered list. The optimal number of retrieved pages is likely to vary depending on the web resource, therefore, we decided to evaluate two additional stop conditions. The *profile size* condition stops the process when the total number of features extracted from the retrieved pages reaches a certain threshold. We can assume that given a rich term profile the classifier should be able to make the correct decision. The *outcome stability* condition is based on the fact that with increasing number of pages, newly added pages are less likely to change the outcome of the classifier. As we discuss in the next section, the classifier used in the system produces confidence scores, the system accumulates the scores until the score for the top class reaches a certain threshold at which point the process is ended.

3.3 Classifier

The classifier builds its models based on training instances which are entities for which the category assignment is known. As each training entity is represented by its home page and all child pages there are two methods of building the train models. In the *cumulative* approach, the system concatenates the content of all pages and uses the extracted term profile as a single training instance. In *per page* approach each page is a separate training instance. The second approach results in larger number of more fine-grained training instances, but at the same time it is biased as entities with a large number of child pages contribute more training instances. Analogously, the same two approaches can be applied to test instances. In this case, the cumulative profile is expanded step-by-step with each retrieved page. In the per page setting each page is classified independently and a voting mechanism is used to combine classification scores. In the experiments, we used a bag-of-words approach to build the entity term profiles. The terms were selected from page content after the removal of HTML mark-up.

As an entity can be assigned to many categories, we need a multi-label classification method. We experimented with two multi-label classification approaches: a label powerset classifier [22] and a chain classifier [18]. Both methods rely on the correlation between the occurrence of various classes. For example, if a company belongs to *chemical* industry, it is more likely to belong to *pharmaceutical* industry than a *banking* industry. Surprisingly, in preliminary experiments neither method was able to outperform a baseline approach, which was an ensemble of binary Naive Bayes Multinomial classifiers [12], both being significantly slower. Therefore, in our experiments we decided to use the baseline approach. In the ensemble each classifier is a binary classifier trained for a single entity category. Naive Bayes classifiers produce a classification score in a range $[0, 1]$ (where results over 0.5 are considered as a positive classification) – the score is used as the confidence of a vote in the per page approach.

4   Evaluation

4.1   Dataset and Experimental Design

To obtain an evaluation data set we used Wikipedia, which contains a large number of company articles with company industry types and the links to company home pages. We selected companies with categories from a set of 100 most frequently occurring industry types. The industry types were extracted from Wikipedia's category field, using YAGO [20] as a preprocessing tool. We considered only companies for which an English web page could be found and disregarded web pages for which the textual content could not be extracted. For all companies we crawled the full content of their web resources (home page and all child pages). The result data set contained 9822 companies.

To test the performance of the system, we considered the scenario discussed in Section 1. The objective of the system is to propose to the user an exhaustive set of categories that are likely to be applicable to given entity. The quality of such a set is measured by *recall*, which is the number of correct categories proposed by the system divided by the number of all correct categories for the entity. As a large set of results would be hard to process for the user, we decided to consider only the top five results for classification result sets larger than five.

The dataset was randomly divided into a training set (80% of entities) and a test set (20% of entities). The training set was further divided into five folds which were used in a cross-validation setting to produce the accuracy scores for all child pages in the training set. These were later used to train the link ranking module. For each test entity we first produced the ranking of links from the home page (using the link ranking module) and then categorized the entity using classification models trained for all training instances.

4.2   Results

In our experiments we were interested in three aspects of system performance: (1) the comparison of cumulative and per page models for classification; (2) the usefulness of selective retrieval based on link ranking; (3) the comparison of stop conditions in the categorization module.

Model and Profile Preparation. To create a baseline for the link ranking method we have experimented with three profile types based on the home page only, the about page (when available), and the home page extended by all child pages. The second profile is based on the fact that company home pages often link to a special *about* page with a description of the company. Searching for phrase "about" in anchor text or url we found the about page for nearly 60% of companies in test data set. To compare the quality of profile extracted from about page and home page keeping the full test set, we modified the first profile type replacing the home page with about page for these 60% of cases.

As described in Section 3.3, the classification algorithm can be trained and evaluated with two different approaches. Given the set of pages belonging to the same entity the cumulative approach combines their content and treats them as a single training or test instance, whereas the per page approach uses each page separately as an individual training instance. While testing, the per page approach classifies each page separately and combines the results using a voting mechanism. We trained the classifier using both approaches on the full content of pages from the training set. Later we classified the test instances using only the content of the home page, about page or the home page extended by all child pages. In the last case testing was done using the same method for profile preparation as used in training.

For all profile types the per page approach outperformed the cumulative approach; however, the difference in the results was minimal (Table 1). We used the per page models and profiles in the rest of the experiments. Larger differences can be noticed comparing the profile types. Interestingly, the use of about page decreases the performance comparing to the home page. Although, the content of about page may seem as a good description of the company, it is usually too short and meant to be attractive rather than informative. The home and child pages profile produces the most accurate results. However, to obtain these results the system has to retrieve and process on average fifty additional pages, which makes the cost of this approach prohibitive. It is possible, however, that retrieving all pages to the full content profile is not necessary, even using basic ordering techniques. In the next experiment, we tested two of such approaches and compared them to the proposed selective retrieval approach.

Table 1. Results of two model building approaches for home page, about or home page (depending on availability) and home page extended by all child pages. The use of about pages decreases the performance, while adding the content of child pages improves the performance at the cost of retrieval of additional pages.

	cumulative	per page
home page profile	0.702	0.705
about or home page profile	0.682	0.687
home and all child pages profile	0.715	0.720

Page Selection in Retrieval Process. To observe how the link order generated in the link ranking module affects the categorization, we compared its results with two simple approaches based on the order in which the links can be found on the home page and on randomized link order. In all cases we allowed the system to retrieve at most N pages including the home page of the resource and recorded categorization accuracy for an increasing number of retrieved pages.

Surprisingly, the home page order approach, which in most cases is the default crawling method, results in decreased classifier accuracy with the first retrieved pages (Fig. 2). After retrieving six pages the system returns to its performance for the home page only. With increasing number of pages the results improve further; nevertheless, retrieval based on page order of links does not seem to be a good strategy. The use of random order of links allows us to avoid the initial decrease of performance. The accuracy of the full content profile is reached after the retrieval of 16 pages, which on average is 33% of all pages linked from the home page. Both basic approaches are clearly outperformed by the results of the link ranking system. By using a single top link from the ranking the classifier is able to achieve higher recall than full content profile. The system reaches top performance with the limit of 8 retrieved pages, which on average is 15% of the total number of links from the home page. Putting a hard limit on the number of retrieved pages allows the system to retrieve them in parallel which even further improves the processing time and allow real-time responses.

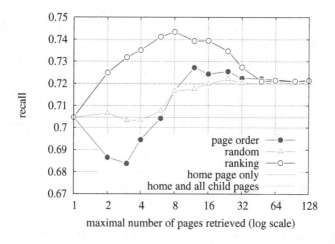

Fig. 2. Three link ranking approaches. Thanks to the link ranking module, retrieval of even a single child produces better results than processing of all child links. On the contrary, following the basic link order decreases the performance with few first links.

To understand why first links on the home page are likely to decrease classification accuracy we examined the weight vector generated during the training of the support vector machines algorithm used in the link ranking module. The weights represent features used in the ranking algorithm and its magnitude represents the importance of the feature (for that reason the weight vector can be a base of a feature selection mechanism [8]). The features with highest negative weights are anchor words like *contact* or *careers* which are likely to be be placed in the header of the page. Among link URL n-grams we can find phrases like */es/* which marks links to alternative language versions of the page. All of these links are likely to mislead the classification process and are successfully removed by the proposed link ranking approach. On the other hand, the most useful links can be found in the page menu and their anchor text contain terms like *about* or *products*. These links are likely to be found in the middle of the page. This is why the retrieval of 12 − 16 pages based on the home page link order allows the system to outperform the classifier based on full content profiles.

Choosing the Stop Condition. In the previous experiment we used the simplest stop condition – the maximal number of retrieved pages. We compared it to two other stop conditions described in Section 3.2. The profile size condition is based on the total number of features extracted from the retrieved pages; the output stability condition is based on the predicted ability of newly retrieved pages to change the results of the classifier. For each stop condition we selected a set of thresholds, which resulted in a similar average number of pages retrieved per entity. This way we were able to compare the performance of the system with similar work-load given different stop conditions.

Comparing all stop conditions we can observe that the number of pages and profile size have similar characteristics - slow improvement in performance with growing average number of retrieved pages. On the contrary, the output stability condition outperforms the other two approaches for the lowest number of retrieved pages, but has no ability to improve the accuracy with an increase of the threshold. Therefore, this condition should be used when the number of additional retrievals should be minimized to absolute minimum.

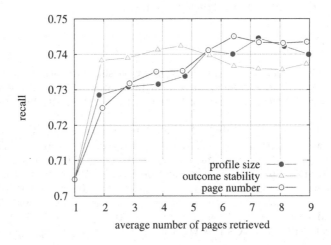

Fig. 3. Comparison of stop conditions. Output stability can improve the performance with the minimal average number of retrieved pages.

5   Conclusions

The proposed system combines ideas from web classification and selective crawling methods to improve the categorization of entity web resources, while at the same time limiting the need for the retrieval of additional information. The categorization process starts with the home page of the entity. The system examines links found on the page and orders them according to their predicted usefulness for entity categorization. Starting with the top link, the system retrieves additional pages expanding the entity profile. The evaluation results on a large set of company web resources show that even a single expansion allows the system to outperform the categorization results computed for all linked pages. Therefore, the system is practically usable for real-time interaction with the user. The proposed ranking approach is based on the fact that web pages share similarities in their structure to match the mental model of a user. For example a company website is likely to have links to alternative language versions of the page at the top and the link to company's products or services should have the corresponding keyword in the link's anchor text. In a broader perspective we demonstrated that a computer system is able to use features designed for human users to retrieve on-line information in an intelligent and orderly way.

In future work on the project we plan to extend the system to extract factual information from entity web resources. Given a description of an attribute format (e.g., regular expression representing company address), the system should crawl the page link graph and retrieve the attribute within an allowed number of steps.

Acknowledgments. This research is partially supported by the Mitacs Elevate and NSERC Engage programs. The authors would like to thank the anonymous reviewers for detailed and constructive comments.

References

1. Angelova, R., Weikum, G.: Graph-based text classification: learn from your neighbors. In: Proceedings of the 29th Annual International ACM SIGIR Conference on Research and Development in Information Retrieval, SIGIR 2006, pp. 485–492. ACM (2006)
2. Baldi, P., Frasconi, P., Smyth, P.: Modeling the Internet and the Web: probabilistic methods and algorithms. Wiley Series in Probability and Statistics. Wiley (2003)
3. Burges, C., Shaked, T., Renshaw, E., Lazier, A., Deeds, M., Hamilton, N., Hullender, G.: Learning to rank using gradient descent. In: Proceedings of the 22nd International Conference on Machine Learning, ICML 2005, pp. 89–96. ACM (2005)
4. Carlson, A., Schafer, C.: Bootstrapping information extraction from semi-structured web pages. In: Daelemans, W., Goethals, B., Morik, K. (eds.) ECML PKDD 2008, Part I. LNCS (LNAI), vol. 5211, pp. 195–210. Springer, Heidelberg (2008)
5. Chakrabarti, S., Dom, B., Indyk, P.: Enhanced hypertext categorization using hyperlinks. SIGMOD Rec. 27(2), 307–318 (1998)
6. Chakrabarti, S., van den Berg, M., Dom, B.: Focused crawling: a new approach to topic-specific web resource discovery. Comput. Netw. 31(11-16), 1623–1640 (1999)
7. Freund, Y., Iyer, R., Schapire, R.E., Singer, Y.: An efficient boosting algorithm for combining preferences. J. Mach. Learn. Res. 4, 933–969 (2003)
8. Guyon, I., Weston, J., Barnhill, S., Vapnik, V.: Gene selection for cancer classification using support vector machines. Mach. Learn. 46(1-3), 389–422 (2002)
9. Hao, Q., Cai, R., Pang, Y., Zhang, L.: From one tree to a forest: a unified solution for structured web data extraction. In: Proceedings of the 34th International ACM SIGIR Conference on Research and Development in Information Retrieval, SIGIR 2011, pp. 775–784. ACM (2011)
10. Herbrich, R., Graepel, T., Obermayer, K.: Support vector learning for ordinal regression. In: International Conference on Artificial Neural Networks, pp. 97–102 (1999)

11. Joachims, T.: Training linear SVMs in linear time. In: Proceedings of the 12th ACM SIGKDD International Conference on Knowledge Discovery and Data Mining, KDD 2006, pp. 217–226. ACM (2006)

12. McCallum, A., Nigam, K.: A comparison of event models for naive bayes text classification. In: AAAI 1998 Workshop on Learning For Text Categorization, pp. 41–48. AAAI Press (1998)

13. Nguyen, H., Fuxman, A., Paparizos, S., Freire, J., Agrawal, R.: Synthesizing products for online catalogs. Proc. VLDB Endow. 4(7), 409–418 (2011)

14. Oh, H.-J., Myaeng, S.H., Lee, M.-H.: A practical hypertext categorization method using links and incrementally available class information. In: Proceedings of the 23rd Annual International ACM SIGIR Conference on Research and Development in Information Retrieval, SIGIR 2000, pp. 264–271. ACM (2000)

15. Pal, A., Tomar, D.S., Shrivastava, S.C.: Effective focused crawling based on content and link structure analysis. International Journal of Computer Science and Information Security 2(1), 140–152 (2009)

16. Pant, G., Srinivasan, P.: Learning to crawl: Comparing classification schemes. ACM Trans. Inf. Syst. 23(4), 430–462 (2005)

17. Qi, X., Davison, B.D.: Web page classification: Features and algorithms. ACM Comput. Surv. 41(2), 12:1–12:31 (2009)

18. Read, J., Pfahringer, B., Holmes, G., Frank, E.: Classifier chains for multi-label classification. Mach. Learn. 85(3), 333–359 (2011)

19. Roth, S.P., Schmutz, P., Pauwels, S.L., Bargas-Avila, J.A., Opwis, K.: Mental models for web objects: Where do users expect to find the most frequent objects in online shops, news portals, and company web pages? Interact. Comput. 22(2), 140–152 (2010)

20. Suchanek, F.M., Kasneci, G., Weikum, G.: Yago: a core of semantic knowledge. In: Proceedings of the 16th International Conference on World Wide Web, WWW 2007, pp. 697–706. ACM (2007)

21. Tang, T.T., Hawking, D., Craswell, N., Griffiths, K.: Focused crawling for both topical relevance and quality of medical information. In: Proceedings of the 14th ACM International Conference on Information and Knowledge Management, CIKM 2005, pp. 147–154. ACM (2005)

22. Tsoumakas, G., Vlahavas, I.: Random k-labelsets: An ensemble method for multi-label classification. In: Kok, J.N., Koronacki, J., Lopez de Mantaras, R., Matwin, S., Mladenič, D., Skowron, A. (eds.) ECML 2007. LNCS (LNAI), vol. 4701, pp. 406–417. Springer, Heidelberg (2007)

23. Xu, J., Li, H.: Adarank: a boosting algorithm for information retrieval. In: Proceedings of the 30th Annual International ACM SIGIR Conference on Research and Development in Information Retrieval, SIGIR 2007, pp. 391–398. ACM (2007)

Navigation by Path Integration
and the Fourier Transform:
A Spiking-Neuron Model

Jeff Orchard[1,2], Hao Yang[2], and Xiang Ji[1,2]

[1] Cheriton School of Computer Science
[2] University of Waterloo
jorchard@uwaterloo.ca

Abstract. In 2005, Hafting et al [1] reported that some neurons in the entorhinal cortex (EC) fire bursts when the animal occupies locations organized in a hexagonal grid pattern in their spatial environment. Previous to that, place cells had been observed, firing bursts only when the animal occupied a particular region of the environment. Both of these types of cells exhibit theta-cycle modulation, firing bursts in the 4-12Hz range. In addition, grid cells fire bursts of action potentials that precess with respect to the theta cycle, a phenomenon dubbed "theta precession". Since then, various models have been proposed to explain the relationship between grid cells, place cells, and theta precession. However, most models have lacked a fundamental, overarching framework. As a reformulation of the pioneering work of Welday et al [2], we propose that the EC is implementing its spatial coding using the Fourier Transform. We show how the Fourier Shift Theorem relates to the phases of velocity-controlled oscillators (VCOs), and propose a model for how various other spatial maps might be implemented. Our model exhibits the standard EC behaviours: grid cells, place cells, and phase precession, as borne out by theoretical computations and spiking-neuron simulations. We hope that framing this constellation of phenomena in Fourier Theory will accelerate our understanding of how the EC – and perhaps the hippocampus – encodes spatial information.

1 Introduction

Some neurons in the entorhinal cortex (EC), called "grid cells", spike preferentially when the animal is at points arranged in a hexagonal grid pattern [1]. Figure 1 shows the output of a sample place cell, taken from [3]. Neurons in the hippocampus, called "place cells", were found to activate when the animal was in a particular location in the environment (see Fig. 1). Both types of cells, place cells and grid cells, are modulated by the theta rhythm, a pattern of activity that oscillates at between 4 and 12 Hz.

Researchers proposed that the grid patterns might arise from an interference pattern between neural oscillators. This is made possible by *velocity-controlled oscillators*, or *VCOs*. A VCO is a neuron or population of neurons whose activity

O. Zaïane and S. Zilles (Eds.): Canadian AI 2013, LNAI 7884, pp. 138–149, 2013.

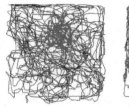 

Place cell spatial map   Grid cell spatial map

Fig. 1. Action potentials (red dots) superimposed on a rat's path, showing the output of a place cell (left) and a grid cell (right). From [3].

oscillates, but at a frequency that is modulated by velocity. As the animal moves, these VCOs take on slightly different frequencies. If the frequency of the VCOs are *linear* functions of velocity, then as the animal moves the phase difference between two VCOs can be written

$$\phi(t) = \int_0^t c_1 v(\tau) - c_2 v(\tau) d\tau = (c_1 - c_2) \int_0^t v(\tau) d\tau = (c_1 - c_2) x(t) \ ,$$

where c_1 and c_2 are scalar constants. Hence the phase difference $\phi(t)$ is proportional to total displacement $x(t)$. In this way, the phase differences between VCOs encodes the rat's position. This is an important point that we will come back to later.

We can think of the state of a VCO as a rotating unit vector, called its *phase vector*. By combining (adding) the phase vectors from two VCOs with different frequencies, the result is a beat interference pattern that generates periods of constructive and destructive interference as their phase difference evolves [4]. Since phase and position are tied together, this interference pattern overlays the animal's spatial environment. Combining three VCOs (that differ in preferred direction by multiples of 120°) tends to create a hexagonal grid interference pattern [5,6].

How are place cells and grid cells related? As recently as 2008, researchers had only a handful of ideas of how grid cells might combine to produce place cells [3]. But a consensus seems to be that place-cell like behaviour results from adding together a number of grid cells [7,8,9,10,11]. A comprehensive review of the various proposed models can be found in [12].

A rather different model, not based on oscillators, used Gaussian surfaces to represent place cells, but encoded these Gaussians by their Fourier coefficients [13]. Their spiking-neuron implementation uses an approximation of the Fourier Shift Theorem (discussed later), moving the Gaussian pattern of excitation around by applying phase shifts to the Fourier coefficients. However, their model does not address grid cells.

In 2011, Welday et al [2] proposed a more complete theory of the mechanisms combining grid cells, place cells, and phase precession. Their model involves a bank of VCOs arranged in a 2-dimensional (2-D) array as shown in the left pane of Fig. 2. In their firing-rate model, each VCO is as a ring oscillator with a wave of

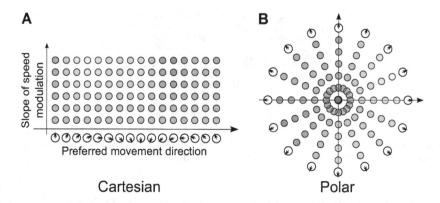

Fig. 2. Cartesian versus polar representation of VCOs. The Cartesian arrangement is derived from part of Fig. 7 in [2]. The polar arrangement consists of a number of "propellers", lines of VCOs that pass through the origin.

activity that cycles at (or near) theta frequency. Hence, each neuron on the ring activates at a particular phase. According to their paper, connecting a read-out node to all the oscillators of a given row produces a place cell. Similarly, choosing only three oscillators from a row, but with preferred directions separated by 120°, yields a grid cell. Finally, choosing all the oscillators with the same preferred direction vector can generate a border cell.

While the ideas presented in that paper have merit, the authors' explanations for their claims are somewhat disconnected and hard to follow. In this paper, we bring together those multiple fragments of theory and re-formulate them into a coherent and elegant framework using Fourier Theory. We also extend the model and hypothesize a separation of labour, where location is encoded using VCO phase, and spatial map patterns are encoded using connection weights.

2   Fourier Model

The bank-of-oscillators model states that a VCO's frequency depends on two parameters: the speed of the animal, and the cosine of the animal's velocity vector with the VCO's preferred direction. Plotting those two factors on axes arranges the VCOs into a 2-D Cartesian space, as shown in Fig. 2A.

Another, perhaps more intuitive way of presenting the same 2-D parameter space is to use polar coordinates, as shown in Fig. 2B. In this view, the direction of displacement from the origin indicates the preferred direction, and the distance from the origin gives the gain of the frequency modulation. Consider a VCO located a position $\mathbf{c}$ in the plane. In this arrangement, the animal moving at velocity $\mathbf{v}$ makes the VCO located at position $\mathbf{c}$ oscillate with frequency $\mathbf{c} \cdot \mathbf{v} + \theta$, where θ is the baseline frequency. This is consistent with the cosine frequency tuning with respect to preferred direction [2].

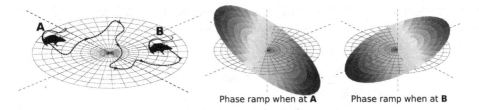

Phase ramp when at **A**   Phase ramp when at **B**

Fig. 3. As the rat moves from position A to position B, the phase ramp changes its slope

Consider the phase difference between the VCO at the origin, and one located at **c**,

$$\phi(t) = \int_0^t (\mathbf{c} \cdot \mathbf{v}(\tau) + \theta) - \theta d\tau \;=\; \mathbf{c} \cdot \int_0^t \mathbf{v}(\tau) d\tau \;=\; \mathbf{c} \cdot \mathbf{x}(t) \;,$$

where $\mathbf{v}(t)$ is the animal's velocity at time t. This is a dot-product, and is linear in **c**. To see this, fix a location **x** and plot the phase difference for all locations **c**; this forms a plane as shown in Fig. 3. Moreover, the slope of the ramp encodes **x**, the location of the animal. As the animal moves from one place to another, the phase ramp tilts to track its location.

These VCOs connect to read-out nodes, including place cells, grid cells, etc. We will refer to these read-out nodes as *spatial-map* nodes, since they draw a map of activity as the animal wanders through its space.

To understand how this ramp can be used to generate spatial maps, we need to know a bit about the Fourier transform. In the following sections, we review the Fourier transform and outline the benefits of thinking about the EC in terms of the this powerful mathematical tool.

2.1   Fourier Theory Basics

We will develop our argument using the Discrete Fourier Transform (DFT). Consider a sampled function f_n with N samples indexed $n = 0, \dots, N - 1$. The DFT of f is

$$F_k = \sum_{n=0}^{N-1} f_n \exp\left(-2\pi i \frac{nk}{N}\right) \;, \quad k = 0, \dots, N - 1 \;. \tag{1}$$

Each complex number F_k is called a Fourier coefficient. We can also denote the transform using $F = \mathrm{DFT}(f)$. In essence, the DFT is a frequency decomposition; it takes a spatial signal and represents it as a sum of wave fronts of various frequencies (and orientations, in 2-D and higher). Each Fourier coefficient occupies a different location in the frequency domain, and each location represents a different wave front. The value of a Fourier coefficient, F_k, represents the contribution of its wave front. The coefficient F_0 has a special name; it is called the DC, and it is located at the origin of the frequency domain.

The Fourier basis functions in (1) are N-periodic. If we also assume that f is periodic (i.e. $f_{-1} = f_{N-1}$, as is convention), then the DFT can equivalently be written,

$$F_k = \sum_{n=-\tilde{N}}^{\tilde{N}} f_n \exp\left(-2\pi i \frac{nk}{N}\right) \, , \, k = -\tilde{N}, \ldots, \tilde{N} \, ,$$

where we assume for simplicity – but without loss of generality – that N is odd, and use the symbol $\tilde{N}$ to represent $\lfloor \frac{N}{2} \rfloor$, where the delimiters $\lfloor \cdot \rfloor$ denote rounding toward zero. We will use this equivalent, centred version of the DFT throughout this paper.

The Fourier Shift Theorem tells us how shifting (translating) a signal influences its Fourier coefficients. Suppose that F_k are the Fourier coefficients of a signal f_n. Consider a shifted version, f_{n-d}, and its Fourier coefficients, G_k. The relationship between G_k and F_k is

$$G_k = \exp\left(-2\pi i \frac{dk}{N}\right) F_k \qquad , \, k = -\tilde{N}, \ldots, \tilde{N} \, .$$

Thus, the Fourier coefficients of the shifted signal can be derived from the coefficients of the original signal multiplied by a phase-shift, where the amount of the phase-shift is a linear function of the frequency index k. The Fourier Shift Theorem even works for non-integer values of d, and in higher dimensions where dk turns into a dot-product between a shift vector, d, and a coordinate in the frequency domain, k.

2.2   Entorhinal Cortex

We propose that the polar arrangement in Fig. 2B is a 2-D frequency domain, and each VCO corresponds to a Fourier coefficient. This formulation splits the production of spatial maps into two components: the spatial pattern of the map, versus movement throughout that map.

Movement throughout the map is taken care of by the phases of the VCOs. As the rat moves around, the VCOs form a phase ramp. This phase ramp is used to shift the spatial map, just like the Fourier Shift Theorem does.

The spatial map itself comes from connection weights. Consider the set of neural connections between a VCO and a spatial-map (read-out) node. Those connections transform the VCO's phase vector and contribute the result to the map node. Since the VCOs generate Fourier basis functions (sines and cosines), this projection to the spatial-map node is tantamount to performing an inverse Fourier transform. Figure 4 illustrates different spatial maps resulting from different selections of VCOs. In that figure, one can think of the inclusion/exclusion of the VCOs as connection weights of all 1s, or all zeros, respectively. Other spatial maps can be realized by choosing different connection weights, as illustrated in Fig. 4D. Though not shown here, border cells (as discussed in [2]) can also be implemented using the same techniques.

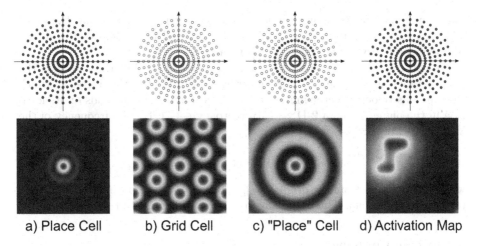

a) Place Cell   b) Grid Cell   c) "Place" Cell   d) Activation Map

Fig. 4. Sample spatial maps (bottom row) and the VCOs used to generate them (top row). For comparison, we included C to show the "place" cell that was proposed in [2]. D shows a general spatial activation map, created by using the Fourier coefficients of an ideal spatial map to set connection weights from a bank of VCOs (18 propellors, 9 rings) to a readout node.

This Fourier interpretation splits the generation of spatial maps into two parts: the VCO phases form a phase ramp, and the neural connections can be interpreted as Fourier coefficients. Together, they constitute an inverse Fourier transform that generates a shifted spatial map, shifted according to the slope of the phase ramp. The beauty is that all spatial maps are shifted at the same time, all driven by the same bank of VCOs.

3   Material and Methods

We implemented a version of the EC Fourier model using spiking leaky integrate-and-fire (LIF) neurons [14]. Here we describe our implementation of the model, outline the challenges, and display results from simulation experiments.

3.1   Oscillators

To build our neural network, we used the Neural Engineering Framework (NEF) [15], a powerful and versatile platform that has proven useful for large-scale cognitive modelling [16]. Explanation of the NEF is beyond the scope of this paper, but we include a brief description.

In the NEF, information is represented in the firing rates of populations of spiking LIF neurons. That data can be extracted and transformed using optimal linear decoders. Moreover, recurrent networks can be designed to implement particular dynamical systems. For example, we used populations of 300 LIF neurons to implement the VCOs as simple harmonic oscillators, storing (x, y, θ)

in each population. Recurrent connection weights were chosen so that the (x, y) state oscillates with frequency θ. We used a normalized version of the simple harmonic oscillator by forcing (x, y) to be a unit-vector.

We constructed arrays of 17 VCOs to form propellers, like those seen in the polar arrangement in Fig. 2B. As the arrangement dictates, the degree to which the animal's velocity vector influences the VCOs frequency depends on its location in the plane. Given a 2-D velocity vector, $\mathbf{v} \in [-1, 1]^2$, the frequency of the VCO at location $\mathbf{c}_n$ is

$$\theta_n = 8 + 1.6 \, \|\mathbf{v}\|_2 - 1.272 \, \mathbf{c}_n \cdot \mathbf{v} \tag{2}$$

where the distance from the origin, $\|\mathbf{c}_n\|$, ranges from -1 at one end of a propeller to 1 at the other end. This is similar in nature to that used in [2].

3.2   Phase Coupling

The stochastic nature of spiking neurons causes imperfect behaviour of the oscillators. If set to the same frequency and started in phase, perfect oscillators will remain in phase. However, slight errors in frequencies will cause them to drift out of phase as time progresses. This random dephasing can disrupt the phases of the oscillators to the point where the phase ramp is overwhelmed by noise.

We designed a phase-coupling method that maintains a linear progression in phase across each propellor, so that the increment in phase from one VCO to the next is the same everywhere. The coupling method is described more completely in [17]. Briefly, each pair of adjacent VCOs is joined by a coupling node, called a *phase-step* node, which computes the phase difference between its two VCOs. All the phase-step nodes are interconnected and arrive at a consensus phase difference (the weighted average). After computing the phase error for each VCO, each phase-step node sends back a correction to keep the VCOs in the correct phase relationship.

While the phase-step nodes keep the phase linear within a 1-D propellor, we still need a way to ensure that all the phase ramps are coplanar. For example, drift could cause one propellor to attain a disproportionately steep slope that makes it tilt out of the plane delineated by the other propellors.

A more complex form of coupling is required to keep the phases coplanar with each other. We need to couple together three phase-step nodes (from three different propellors). The resulting phase adjustments are fed back to the phase-step nodes. Finally, we also used a phase-coupling node to keep the DC nodes of the three propellors in sync.

3.3   Simulation of Rat Motion

We created our network model to test some specific aspects of the Fourier model. In particular, we wanted to see if we would find grid cells that fired spikes on a hexagonal grid of locations. We also wanted to see if these grid cells would exhibit phase precession compared to a global theta cycle. We added a 2-D

VCO node that oscillates at approximately 8 Hz, and used this node's state as the authoritative theta cycle.

To simulate the movement of a rat in a circular environment, we added to our model a random-walk function that adjusts the velocity vector smoothly. The resulting simulated rat trajectories are shown later. One could predict the rat's location by numerically integrating the rat's velocity. However, the rat's own perceived location (as encoded in the phase ramp of the EC VCOs) soon drifted away from the computed position. A real rat seems to avoid this problem by updating its perceived location with sensory information [5]. Though we did not incorporate sensory input in our model here, our companion paper [17] presents an extension that does.

3.4   Network Architecture

As shown in Fig. 5, the network consists of three "wheels" of nodes, along with a velocity node, DC phase-coupling node, a theta-cycle node, and an array of grid-cell nodes. Each wheel has three propellors at angles 0°, 120°, and 240° (though a full model would include more propellors per wheel). The first wheel contains 17 VCO populations per propellor. Each population has 300 LIF neurons and encodes a 3-D vector.

The phase-step wheel also has 3 propellors, but with 16 nodes per propellor (since they model the phase differences between the VCO nodes). Each phase-step population has 500 LIF neurons and encodes a 6-D vector as described in [17]. The coplanar coupling wheel mirrors the phase-step wheel, with each coplanar coupling node having 500 LIF neurons and encoding a 6-D vector.

The grid-cell array has 17 nodes, mirroring the 17 nodes in each of the VCO propellors. Each grid node contains 200 LIF neurons and encodes a 2-D vector of the sum of the phase vectors from the three corresponding VCOs. That is, each grid node receives the phase vectors from a triad of VCOs and simply adds them together.

The DC phase-coupler node has 500 LIF neurons and encodes a 6-D vector that duplicates the phases of the three DC nodes. The velocity node has 100 LIF neurons and encodes a 2-D vector. Finally, the theta-cycle node contains 500 LIF neurons and encodes a 2-D vector that oscillates at approximately 8 Hz. The recurrent connections of the theta-cycle population use a synaptic time constant of $\tau_s = 5$ ms.

Unless otherwise specified, we used the following parameter values for all neurons: synaptic time constant $\tau_s = 5$ ms, refractory period $\tau_{ref} = 2$ ms, membrane time constant $\tau_m = 20$ ms, spiking threshold $J_{th} = 1$, encoding vectors selected randomly (uniformly) from the unit hyper-sphere, neural gain and bias chosen to randomly (uniformly) sample the unit hyper-sphere of the representational space, with a maximum firing rate in the range 200-400 Hz.

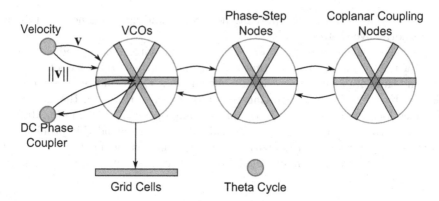

Fig. 5. Network overview. The velocity node modulates the frequency of the VCOs (see equation (2)). The phase-step nodes couple the VCOs to maintain a 1-D phase ramp within each propellor. The coplanar coupling nodes further keep the phase slopes of the different propellors linearly consistent (so that they all rest in a common plane). The DC phase coupler node keeps the absolute phase of the propellors in sync. The grid cells sum triads of VCOs. The theta cycle node is a stand-alone oscillator that maintains a frequency of approximately 8Hz.

4   Results

The simulations were run using the Nengo software package (nengo.ca). The whole model includes 119 nodes, for a total of 68,700 LIF neurons. We ran the model for 300 seconds simulation time. The execution of the model took about 110 minutes to run on a laptop with a 2.9GHz Intel Core i7 processor and 8GB of RAM.

4.1   Grid Cells

Figure 6 shows a sampling of grid cells, with their spikes superimposed overtop of the rat's trajectory. In the figure, the frequency of the grid-cell triad increases from left to right. The red dots of spikes clearly occur on a hexagonal grid with different scales. Not all neurons in the grid nodes exhibited grid firing patterns. However, about 10% did.

4.2   Theta-Phase Precession

If we focus on the timing of the grid-cell spike bursts, we can see that the start of the bursts precess through the theta cycle. Figure 7 plots the spikes as red lines over the theta cycle produced by the "theta" node. The frequency of oscillation for the VCOs – and hence the grid cells – is higher than the nominal 8Hz theta cycle. Thus, we see the bursts of grid-cell activity precess through the lower-frequency theta cycle.

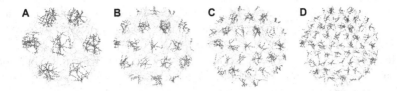

Fig. 6. Spikes from grid cells superimposed on the rat's trajectory. All the grid cells were taken from triads with an orientation of 0°. The neuron in A is from a grid-cell node at position 2 (where the central, or DC, grid-cell node is index 0). The neurons in B, C and D are from grid-cell nodes 3, 4 and 6, respectively.

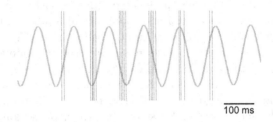

100 ms

Fig. 7. Theta-phase precession of grid-cell spikes

5   Discussion

The model proposed in this paper was inspired by [2], but we re-formulated that approach into a coherent framework that allows for further analysis and deeper understanding. Part of our contribution is to envision the bank of VCOs organized in a polar fashion throughout the frequency domain. Movement of the animal induces frequency changes of those VCOs according to where they rest in the frequency domain. In this polar format, the VCO phases form a ramp. The slope of this ramp encodes the rat's location, and can be used as such when applied to a set of Fourier coefficients.

The connections from the VCOs to the spatial-map nodes represent Fourier coefficients. Thus, projecting the VCO phases through these connections results in a spatial map with the correct positional context. All the spatial maps are shifted in concert with the animal's motion.

Grid cells might emerge as a by-product of a phase coupling mechanism. Some research has shown that the distributed nature of grid-cell encoding offers better accuracy than the same number of sparse place cells [18]. But this theory still does not address why grid cells appear, since the bank of VCOs also offers a distributed representation of location. Another theory, and one that we plan to investigate, is that grid cells are a by-product of the coupling mechanisms that maintain the phase relationships within the bank of VCOs. It seems intuitive that place cells could offer a stable and accurate representation of location as long as the underlying network that feeds into the place cells encodes location in a stable and accurate manner. Coupling between nodes harnesses the redundancy in the network and enables resources to be focussed on lower-dimensional data, such as location. The coplanar-coupling nodes assess the linear consistency among three or more other nodes. In general, a linearity constraint in 2-D will always require input from at least three VCOs (in addition to the implicitly included DC node). We plan to investigate more general implementations of the coplanar constraint and observe whether these mechanisms inherently generate grid-cell behaviours.

The network we have built involves 119 populations, and contains a total of 67,800 LIF neurons. Our implementation is an important step in demonstrating the capabilities and behaviours of our model. However, an obvious question remains, how might such a system get established? What self-organizing principles might apply, and where? Spatial maps of place cells have been learned using Hebbian learning [19]. Grid cells can emerge spontaneously in a topographically connected network with local excitation and lateral inhibition [7,10]. However, these "Turing grids" are not found in adults, leading researchers to suggest that they form during a developmental stage and are used to guide the formation of grid cells in the non-topographical, adult EC network. Even a random selection of grid cells can produce place cells [8,20]. We plan to investigate unsupervised and supervised learning algorithms to derive neural oscillators. One could also look at how such oscillators could take on the proper phase coupling.

6   Conclusion

Although a number of theories have been forwarded regarding the relationships between place cells, grid cells, phase precession, and other spatial-map cells, none have explained all the components with a single overarching framework. Our Fourier model of the entorhinal cortex path integration system organizes the pieces into an architecture with a rich and well-understood foundation. Knowledge about other properties of the Fourier Transform can help to guide further development of the model, and assess how it may (or may not) be extended to explain or predict other observations.

Acknowledgement. We are grateful for the support of the Natural Sciences and Engineering Research Council of Canada (NSERC), the Canada Foundation for Innovation (CFI), and the Ontario Innovation Trust.

References

1. Hafting, T., Fyhn, M., Molden, S., Moser, M., Moser, E.: Microstructure of a spatial map in the entorhinal cortex. Nature 436(7052), 801–806 (2005)
2. Welday, A.C., Shlifer, I.G., Bloom, M.L., Zhang, K., Blair, H.T.: Cosine Directional Tuning of Theta Cell Burst Frequencies: Evidence for Spatial Coding by Oscillatory Interference. Journal of Neuroscience 31(45), 16157–16176 (2011)
3. Moser, E., Kropff, E., Moser, M.: Place cells, grid cells, and the brain's spatial representation system. Annu. Rev. Neurosci. 31, 69–89 (2008)
4. Blair, H., Gupta, K., Zhang, K.: Conversion of a phase- to a rate-coded position signal by a three-stage model of theta cells, grid cells, and place cells. Hippocampus 18(12), 1239–1255 (2008)

5. Burgess, N., Barry, C., O'Keefe, J.: An oscillatory interference model of grid cell firing. Hippocampus 17(9), 801–812 (2007)
6. Krupic, J., Burgess, N., O'Keefe, J.: Neural Representations of Location Composed of Spatially Periodic Bands. Science 337(6096), 853–857 (2012)
7. Fuhs, M.C., Touretzky, D.S.: A spin glass model of path integration in rat medial entorhinal cortex. The Journal of Neuroscience 26(16), 4266–4276 (2006)
8. Solstad, T., Moser, E., Einevoll, G.T.: From grid cells to place cells: a mathematical model. Hippocampus 16(12), 1026–1031 (2006)
9. O'Keefe, J., Burgess, N.: Dual phase and rate coding in hippocampal place cells: theoretical significance and relationship to entorhinal grid cells. Hippocampus 15(7), 853–866 (2005)
10. McNaughton, B., Battaglia, F.P., Jensen, O., Moser, E., Moser, M.: Path integration and the neural basis of the 'cognitive map'. Nature Reviews Neuroscience 7(8), 663–678 (2006)
11. Blair, H., Welday, A.C., Zhang, K.: Scale-Invariant Memory Representations Emerge from Moire Interference between Grid Fields That Produce Theta Oscillations: A Computational Model. Journal of Neuroscience 27(12), 3211–3229 (2007)
12. Zilli, E.A.: Models of grid cell spatial firing published 2005–2011. Frontiers in Neural Circuits 6(16), 1–17 (2012)
13. Conklin, J., Eliasmith, C.: A Controlled Attractor Network Model of Path Integration in the Rat. Journal of Computational Neuroscience 18, 183–203 (2005)
14. Koch, C.: Simplified models of individual neurons. Biophysics of computation (1999)
15. Eliasmith, C., Anderson, C.H.: Neural engineering: Computation, representation, and dynamics in neurobiological systems. MIT Press, Cambridge (2003)
16. Eliasmith, C., Stewart, T.C., Choo, X., Bekolay, T., DeWolf, T., Tang, C., Rasmussen, D.: A Large-Scale Model of the Functioning Brain. Science 338(6111), 1202–1205 (2012)
17. Ji, X., Kushagra, S., Orchard, J.: Sensory updates to combat path-integration drift. In: Zaïane, O., Zilles, S. (eds.) Canadian AI 2013. LNCS (LNAI), vol. 7884, pp. 263–270. Springer, Heidelberg (2013)
18. Mathis, A., Herz, A.: Optimal Population Codes for Space: Grid Cells Outperform Place Cells. Neural Computation 24, 2280–2317 (2012)
19. Rolls, E.T., Stringer, S.M., Elliot, T.: Entorhinal cortex grid cells can map to hippocampal place cells by competitive learning. Network: Computation in Neural Systems 17(4), 447–465 (2006)
20. de Almeida, L., Idiart, M., Lisman, J.E.: The input–output transformation of the hippocampal granule cells: from grid cells to place fields. The Journal of Neuroscience 29(23), 7504–7512 (2009)

Feature Combination for Sentence Similarity

Ehsan Shareghi and Sabine Bergler

CLaC Laboratory, Concordia University
{eh_share,bergler}@cse.concordia.ca

Abstract. The possible combinations of features traditionally used for sentence similarity amount to a very large feature space. Considering all possible combinations and training a support vector machine on the resulting meta-features in a two step process significantly improves performance. The proposed method is trained and tested on the SemEval-2012 Semantic Textual Similarity (STS) Shared Task data, outperforming the task's highest ranking system.

Keywords: Sentence Similarity, Feature Combination, Feature Selection.

1 Introduction

Sentence similarity is a core element of tasks trying to establish how two pieces of text are related, such as Textual Entailment (RTE)[Dagan *et al.*, 2006], and Paraphrase Recognition [Dolan *et al.*, 2004]. SemEval-2012 introduced the Semantic Textual Similarity (STS) Shared Task [Agirre *et al.*, 2012], which differs from RTE in two important points: STS defines a six-point graded similarity scale to measure similarity of two texts, instead of RTE's binary yes/no decision and the similarity relation is considered to be symmetrical, whereas the entailment relation of RTE is inherently unidirectional.

Early sentence similarity systems tested individual methods or resources (like pointwise mutual information [Turney and others, 2001]) but have recently been replaced with hybrid machine learning approaches which combine different feature classes. The complexity of the task (basically being able to competently evaluate and compare any pair of texts) currently gives an edge to basic features, such as n-gram models and WordNet-based metrics [Fellbaum, 2010].

For hybrid systems, the combined feature space becomes too large and feature selection becomes essential, since poor features result in greater computational cost and may lead to overfitting [Ng, 1998]. By the same token, weighting and correlating even the best-quality features for implicit, latent aspects manually is not feasible, since it depends very strongly on the task and the data. We use a combination of simple lexical features (which we refer to as *Basic Features*) from the literature for first pass models and use the meta-feature space of models of all different combinations of *Basic Features* with a second pass support vector machine. This meta-learning approach contrasts with traditional feature selection techniques, like exhaustive search or forward feature selection

O. Zaïane and S. Zilles (Eds.): Canadian AI 2013, LNAI 7884, pp. 150–161, 2013.

[Guyon and Elisseeff, 2003], for instance: while they optimize a subset of features in a single pass, we propose a meta-model, that is trained on models of all possible combinations of the basic features in a second pass SVM, reminiscent of [Chan and Stolfo, 1995].

The proposed approach performs reliably across a range of different test sets with no adjustments and improves results obtained by a single-layer supervised learning approach on the STS SharedTask-2012 [Agirre *et al.*, 2012], outperforming all 88 systems to the 2012 competition.

2 Background

Features used for sentence similarity are either explicit or implicit in the texts. Explicit features concern the surface form of the texts and can themselves be subcategorized into simple string based (e.g. edit distance between two strings [Levenshtein, 1966]) and n-gram or skip-gram based models [Guthrie *et al.*, 2006]. Implicit features, on the other hand, utilize external resources to extract the features. Most well-known approaches use WordNet's hierarchy to characterize either common ancestry between two words of the same part-of-speech such as [Resnik, 1995; Lin, 1998; Jiang and Conrath, 1997] or simply path length [Wu and Palmer, 1994; Leacock and Chodorow, 1998]. More general measures that can relate words of different part-of-speech include [Budanitsky and Hirst, 2006; Banerjee and Pedersen, 2003; Patwardhan, 2003], see Pedersen [Pedersen *et al.*, 2004].

WordNet based approaches have been successfully applied to entailment detection [Inkpen *et al.*, 2006] and to measuring the semantic similarity between two sentences [Mihalcea *et al.*, 2006].

Used by [Morris and Hirst, 1991] for recognizing semantic relationships between words, Roget's Thesaurus is less frequently used because it was not easily accessible electronically. An electronic version was created recently and used for measuring semantic similarity by Jarmasz and Szpakowicz [2003].

Another resource regularly employed to detect implicit similarity measures are co-occurrence measures over reference corpora, of greater importance for phenomena not contained in lexica, such as named entities, technical terms, slang, etc. Latent semantic analysis (LSA) [Landauer *et al.*, 1998], for instance, generates a latent topic-model where correlation of two terms will be higher if they tend to frequently co-occur in the same context (latent topics) but not necessarily in the same documents. LSA has been useful in information retrieval [Deerwester *et al.*, 1990], however, it hasn't been effectively applied to semantic similarity.

A recent extension of LSA, explicit semantic analysis (ESA), [Gabrilovich and Markovitch, 2007], maps the input text into a weighted sequence of Wikipedia concepts (articles' titles) ordered by their relevance to the input text. Since ESA uses predefined concepts represented by the articles' titles, it doesn't have the interpretaion issue associated with LSA output. [Banea *et al.*, 2012] and [Bär *et al.*, 2012] applied ESA successfully to the STS Shared Task-2012.

Other approaches to semantic similarity detection use syntactic dependency relations [Manning and Schütze, 1999] to construct a more comprehensive picture of the meaning of the compared texts, identifying whether a noun is considered the subject or the object of a verb.

3 Related Work

The previously described methods all address partial aspects of semantic relatedness in text and combining different methods promises greater coverage of text meaning. This section reviews three hybrid systems to the STS Shared Task-2012.

The leading UKP [Bär et al., 2012] system uses n-grams, string similarity, WordNet, and ESA, and a support vector regressor [Smola and Schölkopf, 2004]. In addition, the authors use MOSES, a statistical machine translation system [Koehn et al., 2007], to translate each English sentence into Dutch, German, and Spanish and back into English. The system was stable across the different test sets and achieved an average correlation with gold-standard results of 0.6773 [Agirre et al., 2012].

TakeLab [Šaric et al., 2012], in place two, uses n-gram models, two WordNet-based measures, LSA, and dependencies to align subject-verb-object predicate structures. Including named-entities and number matching in the feature space improved performance of their support vector regressor to an average correlation of 0.6753.

The rule-based Sbdlrhmn [AbdelRahman and Blake, 2012] annotates head nouns, main verbs, and named-entities as pertinent information of each sentence. These annotated words are assigned a score if they co-occur in a sentence. The system extracts subject-verb-object triples for every verb in both sentences and aligns them. If predicates align matched annotated words, they are given a score and discarded from the queue. Otherwise, a WordNet similarity score is assigned to the particular unaligned predicate pair. The level of semantic relatedness between two input sentences is determined by a series of if-then rules based on the combinations of scores.

4 Experimental Setup

4.1 Datasets

The STS training dataset consists of 3 different sets of sentence pairs, MSR-PAR (750 pairs), MSR-VID (750 pairs), and SMTeurop (734 pairs), which were gathered from Microsoft Paraphrase corpus, Microsoft Video Description Corpus, and WMT2008 development dataset of the European Parliament corpus, respectively. The test dataset contains 5 sets of sentence pairs, three of which were collected from the same resources as the training sets. The other two test sets, OnWN and SMTnews, are given as surprise sets to examine the reliability of the proposed systems in dealing with unforeseen data. OnWn set comprises

750 pairs of glosses from OntoNotes and WordNet senses. Half of these pairs were collected from senses that were recognized to be equivalent and the remaining from disparate senses. SMTnews set contains 399 sentence pairs from translation systems submitted to WMT2007 and ranked manually.

The task is to measure the similarity between sentence pairs using a score in the range from 0 to 5. The training set gold standard was created using mechanical turk, gathering 5 scores per sentence pair and averaging them.

4.2 Features

After tokenizing, lemmatizing, sentence splitting, and part of speech (POS) tagging, we extract five categories of lexical features: string similarity [Cohen *et al.*, 2003], n-grams [Lin and Och, 2004], WordNet distance [Budanitsky and Hirst, 2006], Roget's Thesaurus [Jarmasz and Szpakowicz, 2003], and ESA[Gabrilovich and Markovitch, 2007].

The following sections discuss the individual features used in each category. To assess their usefulness, single-feature models are trained using cross-validation on the training data, it is these results that are reported in the tables presented throughout Section 4.2. All performance tables also include a model where all features are trained together (marked with a star) and the results for testing on all the different datasets together (last row, captioned ALL).

String Similarity Metrics. In order to calculate similarity for a given pair, 5 common similarity/distance measures were used. In addition, we introduced the ROUGE-W score (RO-W in Table 1) as an effective string similarity measure to the semantic similarity task.

We also added a feature that counts normalized lemma overlap here, even though it is not strictly a string-based technique.

Longest Common Substring[Gusfield, 1997] calculates the distance between two inputs by comparing the length of the longest consecutive sequence of matching characters

Longest Common Subsequence[Allison and Dix, 1986] compare the length of the longest sequence of characters, not necessarily consecutive ones, in order to detect similarities

Jaro[Jaro, 1989] identifies spelling variation between two inputs based on the occurrence of common characters between two text segments at a certain distance

Jaro-Winkler[Winkler, 1990], a variant of *Jaro*, was proposed for name comparison and considers the exact match between the initial characters of the two inputs

Monge-Elkan[Monge and Elkan, 1997] a hybrid method which tokenizes two inputs and finds word pairs with the highest string similarity score and then sums up and normalizes the score and assigns it as the distance between the inputs

ROUGE-W[Lin, 2004] a weighted version of *longest common subsequence* takes into account the number of the consecutive characters in each match, giving higher score for those matches that have larger number of consecutive characters in common. This metric was developed to measure the similarity between machine generated text summaries and a manually generated gold standard.

Table 1. Performance of string-based features, "str*" represents all of the string-based measures being used together

	Lemma	jar	jarWk	lcsq	lcst	monElk	RO-W	str*
VID	0.36	0.19	-0.07	0.48	0.33	0.14	0.75	0.79
PAR	0.39	0.33	0.17	0.18	0.20	0.54	0.52	0.64
SMT	0.55	0.54	0.41	0.30	0.25	0.38	0.53	0.74
ALL	0.41	0.28	0.18	0.22	0.13	0.38	0.62	0.76

Table 1 shows that while the models differ in individual results, none is without merit on this task. Interesingly, the model based on ROUGE-W showed a strong correlation factor of 0.6, which is the highest performing feature in this group.

N-gram Models. We use the ROUGE package, originally developed for automated evaluation of summaries [Lin and Och, 2004], to extract n-gram models. As reported in [Lin, 2004], the most effective for measuring the similarity between small text fragments are:

- **ROUGE-1**, based on Unigrams
- **ROUGE-2**, based on Bigrams
- **ROUGE-SU4**, based on 4-Skip bigrams (including Unigrams)

Note that the small size of the texts compared for similarity paired with the fact that the training data stems from different sources is an important limitation here.

Table 2. Performance of ROUGE-based features. "ROUGE*" represents ROUGE-1,2,SU being used together.

	ROUGE-1	ROUGE-2	ROUGE-SU4	ROUGE*
VID	0.78	0.62	0.67	0.81
PAR	0.63	0.41	0.56	0.63
SMT	0.58	0.39	0.46	0.58
ALL	0.71	0.60	0.61	0.75

WordNet-Based Measures. According to [Budanitsky and Hirst, 2006], the most reliable measures for similarity measurement across different tasks using WordNet are Lin, and Jiang-Conrath.

- **Lin** [Lin, 1998] uses the Brown Corpus of American English to calculate information content of two concepts' least common subsumer (LCS).
- **Jiang-Conrath** [Jiang and Conrath, 1997] uses the conditional probability of encountering a concept given an instance of its parent to calculate the information content.

In order to scale these word-level similarity measures to sentence-level, [Mihalcea *et al.*, 2006] finds for each word w in sentence S_1 the word from sentence S_2 which gives the maximum similarity score.

Roget's Thesaurus-Based Measure. Roget's Thesaurus provides more links between different word types than WordNet. A noun *bank* may be connected to a verb *invest* by determining the common head *Lending*. Roget's has a nine-level ontology: 9 classes, 39 sections, 79 sub-sections, 596 head groups, and 990 heads, paragraphs, parts of speech, semicolon groups, and words. The distance of two terms decreases within the interval of [0,16], as the the common head that subsumes them moves from top to the bottom and becomes more specific. The electronic version of Roget's Thesaurus [Jarmasz and Szpakowicz, 2003] is used for extracting this score.

Table 3. Performance of knowledge-based features. "KB*" represents jcn,lin,Roget's being used together. ESA see below.

	jcn	lin	Roget's	KB*	ESA
VID	0.75	0.71	0.72	0.78	0.83
PAR	0.26	0.24	0.20	0.27	0.33
SMT	0.21	0.21	0.21	0.50	0.45
ALL	0.58	0.58	0.68	0.71	0.73

Explicit Semantic Model. In order to have broader coverage on word types not represented in lexical resources, specifically for named entities, we add ESA generated features to our feature space. The results are shown in the last column of Table 3.

Ablation Assessment. The single feature models we compared led to exclude *Jaro-Winkler*, *longest common substring*, and *Monge-Elkan* from the string-based features.

Jiang-Conrath performs reliably across different datasets. Surprisingly, despite the small size of Roget's coverage compared to WordNet, in the ablation test it demonstrated a strong contribution to the systems performance. Roget's has correlation scores identical or very close to Lin, and in the SMT training subset it achieves the same performance as Jiang-Conrath. In another ablation experiment we combined all the training subsets and Roget's outperformed both Lin and Jiang-Conrath by a margin of 10% with a correlation factor of 0.68.

The average correlation factor achieved by ESA across the training sets was 0.53, which is 2% above the performance reported on other knowledge-based metrics. We thus consider the remaining features to be effective.

4.3 Apparatus

To distribute output in the interval [0,5] we apply the SMOreg support vector regressor from Weka [Hall *et al.*, 2009].

5   Feature Combination

The space of individual features considered is large and even the ones that ablation proved to be effective on the test set are rather similar in some cases. Because the different categories cover different dimensions of the data, we felt a traditional feature selection process was not able to assess tradeoffs properly. After assessing the individual feature strengths in our ablation tests, we wanted to similarly assess the relative strength of different combinations. Picking one or two combinations by the experimenter's gut feeling seemed too arbitrary and we thus generated all combinations. Validating the enormous resulting feature space had to be handled automatically, and we employ a second support vector machine to train the best feature combination model on the dataset.

This two phase supervised learning method for the STS sharedTask is demonstrated in Figure 1. To train the two phase model we first generate all combinations of the *Basic Features* in a *Preprocessing Step*, thus for a feature space of size N we generate $2^N - 1$ combinations.

The *Two Phase Model Training* step trains a separate Support Vector Regressor (SVR), called *Phase One Model*, for each combination with a resulting predicted score ps_i for that phase one combination model M_i. These $2^N - 1$ predicted scores form a new feature vector called *Phase Two Features*. So for each instance in the training set a feature vector of size $2^N - 1$ will be created. Finally a second level SVR, called *Phase Two Model* is trained using the training set and the *Phase Two Feature* space.

After training, we have two types of SVR models: $2^N - 1$ *Phase One Models*, based directly on the Basic Features, and one *Phase Two Model*, based on the predicted scores of *Phase One Models*.

For testing, we proceed as follows: first we send test data as an input to the *Phase One Models* to obtain the predicted scores ps_i for each model for the new feature vector *Phase Two Features*. This new feature vector is fed into the *Phase Two Model* and determines the final output of the system for that particular test instance.

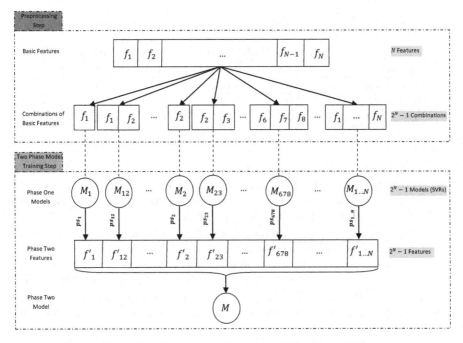

Fig. 1. Two Phase Supervised Learning Method's Architecture

6   Evaluation

The pruned feature space contains 11 Basic Features as described in section 4.2. Two types of experiments were conducted: *Standard Learning*, and *Two Phase Learning*:

Standard Learning the regressor was trained on the training set with all 11 features and tested on the test sets.

Two Phase Learning all combinations of the *Basic Features* create 2047 *Phase Two Features* generated by *Phase One Models* with a predicted score ps_i as input to the *Phase Two Model*.

6.1   Results

The STS Shared Task-2012 used the Pearson Correlation Coefficient (also called correlation factor) as their evaluation metric. The correlation factor results of our experiments are provided in Table 4. Note that the correlation factor ranges from $[-1, 1]$.

Table 4 compares the results of the Standard and the Two Phase learning experiments with the best and baseline performance reported in 2012 for each of the different data sets separately. In all cases the *Two Phase Learning* achieves better results with a margin of 2-7%. The third row shows the result of UKP-run2

Table 4. Pearson correlation for Standard and Two Phase learning

	MSR-PAR	MSR-VID	SMTeurop	SMTnews	OnWN
Standard	0.65	0.87	0.51	0.36	0.66
Two Phase	0.67	0.89	0.56	0.43	0.72
UKP-run2 (best)	0.68	0.87	0.52	0.49	0.66
STS-2012 baseline	0.43	0.29	0.45	0.39	0.58

[Bär *et al.*, 2012] (see Section 3), the best performing system in STS sharedTask-2012. Our system outperforms this system in 3 out of 5 test sets with the average margin of 4%. The last row shows the result achieved by the baseline cosine similarity between the token vectors of each pair.

Most interesting are the extreme points of the system: on OnWN the *Two Phase* model gains 6% over the best system, in STMnews it trailed by 5% behind the best system. Overall, the *Two Phase* model on average outperformed all 88 submitted systems and achieved a mean correlation coefficient of 0.69, which is 2% higher than the best performing system. We speculate that the increase in complexity will be outweighed by better performance. Of note is the fact that the best Phase One Model already outperforms the Standard model, but less than the Two Phase model.

We compare the confidence interval (CI) of the correlation coefficients achieved by each of the methods presented in Table 4 using Fisher's z transformation [Dunn and Clark, 1969], which demonstrates that although there is overlap between the intervals, the Two Phase learning method has an interval of greater possible correlation coefficients especially for OnWN and SMTeurop, see Table 5.

Table 5. Confidence Intervals for correlation coefficients presented in Table 4

	MSR-PAR	MSR-VID	SMTeurop	SMTnews	OnWN
Test Set's Size - N	750	750	459	399	750
Standard Correl.	0.65	0.87	0.51	0.36	0.66
Standard CI	(0.60,0.69)	(0.85,0.88)	(0.44,0.57)	(0.26,0.44)	(0.62,0.70)
Two Phase Correl.	0.67	0.89	0.56	0.43	0.72
Two Phase CI	(0.63,0.71)	(0.87,0.90)	(0.49,0.62)	(0.34,0.50)	(0.68,0.75)
UKP-run2 (best) Correl.	0.68	0.87	0.52	0.49	0.66
UKP-run2 (best) CI	(0.63,0.71)	(0.85,0.88)	(0.45,0.58)	(0.40,0.59)	(0.62,0.70)
STS-2012 baseline Correl.	0.43	0.29	0.45	0.39	0.58
STS-2012 baseline CI	(0.36,0.48)	(0.23,0.35)	(0.37,0.51)	(0.30,0.47)	(0.53,0.62)

We interpret the broadness and the stable relation of our results to the reported systems across the different data sets as a strong indicator that the method has merit. As is the case with all semantic techniques and tasks, evaluation becomes a major task itself with aggregate numbers hiding performance dimensions that might be different for different datasets. This is what motivated our two-level approach.

The experiment has shown that this form of assessing feature combination is successful. It has not overwhelmed the regressor and the results, while marginal on some test sets, show broad improvements on others, which makes it an interesting tool to evaluate further. In our opinion, the recent increase of semantic techniques and resources makes choosing from the great variety of proposed features difficult. We present our technique as a tool to explore the feature space, not as a mature automatic optimization technique.

7 Conclusion

We present here a technique for evaluating possible feature combinations in the context of sentence similarity judgements. We believe that especially in semantic tasks, the different features have to be carefully evaluated in isolation for the given data set. The possible combinations quickly become too complex to prejudge and too many to truly evaluate manually. Our two phase technique, demonstrated in this paper, trains a support vector regressor of the meta-feature space of possible combinations. This technique has been shown to lead to improvements. It is presented here as an interesting methodology for exploration and the added computational complexity is traded off for lower demands on training data coherence.

References

AbdelRahman, S., Blake, C.: Sbdlrhmn: A Rule-based Human Interpretation System for Semantic Textual Similarity Task. In: Proceedings of the 6th International Workshop on Semantic Evaluation (SemEval 2012), in Conjunction with the First Joint Conference on Lexical and Computational Semantics (2012)

Agirre, E., Cer, D., Diab, M., Gonzalez-Agirre, A.: Semeval-2012 Task 6: A Pilot on Semantic Textual Similarity. In: Proceedings of the 6th International Workshop on Semantic Evaluation (SemEval 2012), in Conjunction with the First Joint Conference on Lexical and Computational Semantics (2012)

Allison, L., Dix, T.I.: A Bit-String Longest-Common-Subsequence Algorithm. Information Processing Letters 23(5) (1986)

Banea, C., Hassan, S., Mohler, M., Mihalcea, R.: UNT: A Supervised Synergistic Approach to Semantic Text Similarity. In: Proceedings of the 6th International Workshop on Semantic Evaluation (SemEval 2012), in Conjunction with the First Joint Conference on Lexical and Computational Semantics (2012)

Banerjee, S., Pedersen, T.: Extended Gloss Overlaps as a Measure of Semantic Relatedness. In: International Joint Conference on Artificial Intelligence, vol. 18. Lawrence Erlbaum Associates Ltd. (2003)

Bär, D., Biemann, C., Gurevych, I., Zesch, T.: UKP: Computing Semantic Textual Similarity by Combining Multiple Content Similarity Measures. In: Proceedings of the 6th International Workshop on Semantic Evaluation (SemEval 2012), in Conjunction with the First Joint Conference on Lexical and Computational Semantics (2012)

Budanitsky, A., Hirst, G.: Evaluating WordNet-based Measures of Lexical Semantic Relatedness. Computational Linguistics 32(1) (2006)

Chan, P.K., Stolfo, S.J.: A comparative evaluation of voting and meta-learning on partitioned data. In: Machine Learning International Conference. Citeseer (1995)

Cohen, W.W., Ravikumar, P., Fienberg, S.E., et al.: A Comparison of String Distance Metrics for Name-Matching Tasks. In: Proceedings of the International Joint Conference on Artificial Intelligence Workshop on Information Integration on the Web, IIWeb 2003 (2003)

Dagan, I., Glickman, O., Magnini, B.: The Pascal Recognising Textual Entailment Challenge. In: Machine Learning Challenges. Evaluating Predictive Uncertainty, Visual Object Classification, and Recognising Tectual Entailment (2006)

Deerwester, S., Dumais, S.T., Furnas, G.W., Landauer, T.K., Harshman, R.: Indexing by Latent Semantic Analysis. Journal of the American society for information science 41(6) (1990)

Dolan, B., Quirk, C., Brockett, C.: Unsupervised Construction of Large Paraphrase Corpora: Exploiting Massively Parallel News Sources. In: Proceedings of the 20th International Conference on Computational Linguistics. Association for Computational Linguistics (2004)

Dunn, O.J., Clark, V.: Correlation coefficients measured on the same individuals. Journal of the American Statistical Association 64(325) (1969)

Fellbaum, C.: WordNet. Theory and Applications of Ontology: Computer Applications (2010)

Gabrilovich, E., Markovitch, S.: Computing Semantic Relatedness Using Wikipedia-based Explicit Semantic Analysis. In: Proceedings of the 20th International Joint Conference on Artificial Intelligence (2007)

Gusfield, D.: Algorithms on Strings, Trees, and Sequences: Computer Science and Computational Biology. Cambridge Univ. Press (1997)

Guthrie, D., Allison, B., Liu, W., Guthrie, L., Wilks, Y.: A Closer Look at Skip-Gram Modelling. In: Proceedings of the 5th International Conference on Language Resources and Evaluation (2006)

Guyon, I., Elisseeff, A.: An Introduction to Variable and Feature Selection. The Journal of Machine Learning Research 3 (2003)

Hall, M., Frank, E., Holmes, G., Pfahringer, B., Reutemann, P., Witten, I.H.: The WEKA Data Mining Software: an Update. ACM SIGKDD Explorations Newsletter 11(1) (2009)

Inkpen, D., Kipp, D., Nastase, V.: Machine Learning Experiments for Textual Entailment. In: Proceedings of the Second Recognizing Textual Entailment Challenge (2006)

Jarmasz, M., Szpakowicz, S.: Roget's Thesaurus and Semantic Similarity. In: Proceedings of the Conference on Recent Advances in Natural Language Processing (2003)

Jaro, M.A.: Advances in Record-Linkage Methodology as Applied to Matching the 1985 Census of Tampa, Florida. Journal of the American Statistical Association (1989)

Jiang, J.J., Conrath, D.W.: Semantic Similarity Based on Corpus Statistics and Lexical Taxonomy. In: Proceedings of the 10th International Conference on Research on Computational Linguistics (1997)

Koehn, P., Hoang, H., Birch, A., Callison-Burch, C., Federico, M., Bertoldi, N., Cowan, B., Shen, W., Moran, C., Zens, R., et al.: Moses: Open Source Toolkit for Statistical Machine Translation. In: Proceedings of the 45th Annual Meeting of the ACL on Interactive Poster and Demonstration Sessions. Association for Computational Linguistics (2007)

Landauer, T.K., Foltz, P.W., Laham, D.: An Introduction to Latent Semantic Analysis. Discourse Processes 25(2-3) (1998)

Leacock, C., Chodorow, M.: Combining Local Context and WordNet Similarity for Word Sense Identification. WordNet: An Electronic Lexical Database 49(2) (1998)

Levenshtein, V.I.: Binary Codes Capable of Correcting Deletions, Insertions, and Reversals. Soviet Physics Doklady 10(8) (1966)

Lin, C.Y., Och, F.J.: Automatic Evaluation of Machine Translation Quality Using Longest Common Subsequence and Skip-Bigram Statistics. In: Proceedings of the 42nd Annual Meeting on Association for Computational Linguistics. Association for Computational Linguistics (2004)

Lin, D.: An Information-Theoretic Definition of Similarity. In: Proceedings of the 15th International Conference on Machine Learning, vol. 1 (1998)

Lin, C.Y.: ROUGE: A Package for Automatic Evaluation of Summaries. In: Text Summarization Branches Out: Proceedings of the ACL 2004 Workshop (2004)

Manning, C.D., Schütze, H.: Foundations of Statistical Natural Language Processing. MIT Press (1999)

Mihalcea, R., Corley, C., Strapparava, C.: Corpus-based and Knowledge-based Measures of Text Semantic Similarity. In: Proceedings of the National Conference on Artificial Intelligence (2006)

Monge, A., Elkan, C.: An Efficient Domain-Independent Algorithm for Detecting Approximately Duplicate Database Records. In: Proceedings of the SIGMOD Workshop on Data Mining and Knowledge Discovery. Citeseer (1997)

Morris, J., Hirst, G.: Lexical Cohesion Computed by Thesaural Telations as an Indicator of the Structure of Text. Computational Linguistics 17(1) (1991)

Ng, A.Y.: On Feature Selection: Learning with Exponentially many Irrelevant Features as Training Examples. In: Proceedings of the 15th International Conference on Machine Learning (1998)

Patwardhan, S.: Incorporating Dictionary and Corpus Information into a Context Vector Measure of Semantic Relatedness. Master's thesis, University of Minnesota (2003)

Pedersen, T., Patwardhan, S., Michelizzi, J.: WordNet: Similarity: Measuring the Relatedness of Concepts. In: Demonstration Papers at North American Chapter of the Association for Computational Linguistics: Human Language Technologies. Association for Computational Linguistics (2004)

Resnik, P.: Using Information Content to Evaluate Semantic Similarity in a Taxonomy. In: Proceedings of the 14th International Joint Conference on Artificial Intelligence (1995)

Šaric, F., Glavaš, G., Karan, M., Šnajder, J., Bašic, B.D.: TakeLab: Systems for Measuring Semantic Text Similarity. In: Proceedings of the 6th International Workshop on Semantic Evaluation (SemEval 2012), in Conjunction with the First Joint Conference on Lexical and Computational Semantics (2012)

Smola, A.J., Schölkopf, B.: A Tutorial on Support Vector Regression. Statistics and Computing 14(3) (2004)

Turney, P.D.: Mining the web for synonyms: PMI-IR versus LSA on TOEFL. In: Flach, P.A., De Raedt, L. (eds.) ECML 2001. LNCS (LNAI), vol. 2167, pp. 491–502. Springer, Heidelberg (2001)

Winkler, W.E.: String Comparator Metrics and Enhanced Decision Rules in the Fellegi-Sunter Model of Record Linkage. In: Proceedings of the Section on Survey Research Methods. American Statistical Association (1990)

Wu, Z., Palmer, M.: Verbs Semantics and Lexical Selection. In: Proceedings of the 32nd Annual Meeting on Association for Computational Linguistics. Association for Computational Linguistics (1994)

Exhaustive Search
with Belief Discernibility Matrix and Function

Salsabil Trabelsi[1], Zied Elouedi[1], and Pawan Lingras[2]

[1] Larodec, Institut Superieur de Gestion de Tunis, Tunisia
[2] Saint Mary's University Halifax, Canada

Abstract. This paper proposes a new feature selection method based on rough sets to take away the unnecessary attributes for the classification process from partially uncertain decision system. The uncertainty exists only in the decision attributes (classes) and is represented by the belief function theory. The simplification of the uncertain decision table to generate more significant attributes is based on computing all possible reducts. To obtain these reducts, we propose a new definition of the concepts of discernibility matrix and function under the belief function framework. Experimentations have been done to evaluate this exhaustive solution.

keywords: Uncertainty, belief function theory, rough sets, attribute selection, discernibility matrix and function.

1 Introduction

The method of discernibility matrix and function was proposed by Skowron and Rauszer [14] to solve the problem of relevant feature selection using rough set theory followed by several other attempts [7, 10–12]. It is widely applied in the pre-processing stage of the modeling process in machine learning [3, 6, 15] where it is easy to represent and interpret knowledge. It is convenient to calculate all the possible reducts and the core of data [21]. However, the original discernibility matrix and function do not deal with uncertainty, imprecision or incompleteness in data. This kind of data exists in many real-world databases like in medicine where diseases and symptom of some patients may be partially or totally uncertain.

In this paper, we develop a new attribute selection method based on new definitions of discernibility matrix and function to compute the possible reducts and core from partially uncertain data based on rough sets. The uncertainty exists only in the decision attribute and is represented by the belief function theory [13]. It is considered as a useful theory for representing and managing total or partial uncertain knowledge because of its relative flexibility. The belief function theory is widely applied in artificial intelligence and to real life problems for decision making and classification. In this paper, we use the Transferable Belief Model (TBM) - one interpretation of belief function theory [17]. To remove the superfluous and the redundant attributes from the uncertain decision table, we

O. Zaïane and S. Zilles (Eds.): Canadian AI 2013, LNAI 7884, pp. 162–173, 2013.

adapt the concept of discernibility matrix and function in the new context to be called belief discernibility matrix and function. The latter is an exhaustive method able to compute the possible reducts from the uncertain decision table. Previously, we have also proposed feature selection methods from this kind of uncertainty [18, 19]. However, these solutions are heuristic methods. The approach described in this paper is experimentally compared with the heuristics proposed in [18]. While there is a significant amount of research on feature selection, a direct comparison with them is not possible because those methods do not handle uncertainty in the decision attribute.

The paper is organized as follows: Section 2 provides an overview of the rough set theory. Section 3 introduces the belief function theory as understood in the Transferable Belief Model (TBM). Section 4 details the proposed attribute selection method based on rough sets under uncertainty. Section 5 describes a classification system used in the experimentation and able to generate uncertain decision rules from partially uncertain data called Belief Rough Set Classifier proposed originally in [20] where the feature selection is one of the important steps in the construction procedure. We further report the experimental results obtained from modified uncertain databases to evaluate the performance of our solution based on two evaluation criteria: the time requirement and the classification accuracy. Finally, we draw some conclusions in Section 6.

2   Rough Set Theory

Let us define some basic notions related to decision tables and rough sets [9, 10]. A decision table (DT) is defined as A= (U, C, {d}), where $U = \{x_1, x_2, \ldots x_n\}$ is a nonempty finite set of n objects called *the universe*, $C = \{c_1, c_2, \ldots c_k\}$ is a finite set of k *condition* attributes and $d \notin C$ is a distinguished attribute called *decision*. The value set of d is called $\Theta = \{d_1, d_2, \ldots d_s\}$. The notation $c(x_j)$ is used to represent the value of a condition attribute $c \in C$ for $x_j \in U$. The rough sets adopt the concept of indiscernibility relation [10] to partition the object set U into disjoint subsets, denoted by U/B or IND_B. The partition that includes x_j is denoted $[x_j]_B$. For every set of attributes $B \subseteq C$, the equivalence relation denoted by IND_B and called the B-indiscernibility relation, is defined by

$$IND_B = U/B = \{[x_j]_B | x_j \in U\} \tag{1}$$

where

$$[x_j]_B = \{x_i | \forall c \in B \, c(x_i) = c(x_j)\} \tag{2}$$

Let $B \subseteq C$ and $X \subseteq U$. We can approximate X by constructing the $B - lower$ and $B - upper$ *approximations* of X, denoted $\underline{B}(X)$ and $\bar{B}(X)$, respectively, where

$$\underline{B}(X) = \{x_j | [x_j]_B \subseteq X\} \; and \; \bar{B}(X) = \{x_j | [x_j]_B \cap X \neq \emptyset\} \tag{3}$$

2.1 Reduct and Core

For feature selection, a reduct [7, 11, 12] is a minimal subset of attributes from C that preserves the positive region and the ability to perform classifications as the entire attributes set C. A subset $B \subseteq C$ is a reduct of C with respect to d, iff B is minimal and:

$$Pos_B(\{d\}) = Pos_C(\{d\}) \tag{4}$$

where $Pos_C(\{d\})$, called a positive region of the partition $U/\{d\}$ with respect to C.

$$Pos_C(\{d\}) = \bigcup_{X \in U/\{d\}} \underline{C}(X) \tag{5}$$

The core is the most important subset of attributes, it is included in every reduct.

$$Core(\{d\}) = \bigcap RED(\{d\}) \tag{6}$$

where $RED(\{d\})$ is the set of all reducts of DT relative to d.

2.2 Discernibility Matrix and Function

In order to easily compute reducts and cores, we can use the discernibility matrix [14, 15] which is defined below. Let DT be a decision table with n objects. The discernibility matrix $M(DT)$ is a symmetric $n \times n$ matrix with entries $M_{i,j}$. Each entry thus consists of the set of attributes upon which objects x_i and x_j differ. For $i, j = 1, ..., n$

$$M_{i,j} = \{c \in C | c(x_i) \neq c(x_j) \ and \ d(x_i) \neq d(x_j)\} \tag{7}$$

Thus, entry $M_{i,j}$ is the set of all attributes which discern objects x_i and x_j that do not belong to the same equivalence class $IND_{\{d\}}$. A discernibility function $f(DT)$ for a decision table DT is a boolean function of k boolean variables $c_1^*...c_k^*$ (corresponding to the attributes $c_1...c_k$) defined as follows, where $M_{i,j}^* = \{c^* | c \in M_{i,j}\}$

$$f(DT) = \wedge\{\vee M_{i,j}^* | 1 \leq j \leq i \leq n, M_{i,j} \neq \emptyset\} \tag{8}$$

where $\wedge$ and $\vee$ are two logical operators for conjunction and disjunction. The set of all prime implicants of $f(DT)$ determines the sets of all reducts of DT.

3 Belief Function Theory

In this section, we briefly review the main concepts underlying the belief function theory as interpreted in the TBM [16]. Let Θ be a finite set of elementary events of a given problem, called the frame of discernment. All the subsets of Θ belong to the power set of Θ, denoted by 2^Θ.

The impact of a piece of evidence on the subsets of the frame of discernment Θ is represented by a basic belief assignment (bba). The bba is a function m : $2^\Theta \rightarrow [0, 1]$ such that:

$$\sum_{E \subseteq \Theta} m(E) = 1 \qquad (9)$$

where $m(E)$, named a basic belief mass (bbm), shows the part of belief exactly committed to the element E. The bba's induced from distinct pieces of evidence are combined by the conjunctive rule of combination [16].

$$(m_1 \cap\!\!\!\bigcirc m_2)(E) = \sum_{F,G \subseteq \Theta : F \cap G = E} m_1(F) \times m_2(G) \qquad (10)$$

To make decisions in the TBM, belief functions can be represented by probability functions called the pignistic probabilities denoted $BetP$, which are defined as [16]:

$$BetP(\{a\}) = \sum_{F \subseteq \Theta} \frac{|\{a\} \cap F|}{|F|} \frac{m(F)}{(1 - m(\emptyset))} \text{ for all } a \in \Theta \qquad (11)$$

4 Attribute Selection Method Using Belief Discernibility Matrix and Function

Discernibility matrix and function can provide reducts and cores of decision tables, but the original discernibility matrix and function cannot be applied to uncertain data. In this section, we present a new method for attribute selection from partially uncertain decision system based on rough sets under the belief function framework. We will remove the superfluous and redundant attributes for rules discovery and keep only the features from the reduct by defining a new discernibility matrix and function under the uncertain context. First, we will give an overview of uncertain decision table followed by a description of the proposed approach.

4.1 Uncertain Decision Table

Our uncertain decision table denoted UDT contains n objects x_j, characterized by a set of certain condition attributes $C = \{c_1, c_2, ..., c_k\}$ and uncertain decision attribute ud. We propose to represent the uncertainty of each object x_j by a bba m_j expressing belief on decision defined on the frame of discernment $\Theta = \{ud_1, ud_2, ..., ud_s\}$ representing the possible values of ud.

Example. Let us use Table 1 to describe our uncertain decision system. It contains five objects, three certain condition attributes $C = \{a, b, c\}$ and an uncertain decision attribute ud with possible value $\{yes, no\}$ representing Θ. For example, for the object x_2, belief of 0.6 is exactly committed to the decision $ud_2 = no$, whereas the remaining mass 0.4 is assigned to the entire frame of discernment Θ (ignorance).

Table 1. Uncertain Decision Table (UDT)

U	a	b	c	ud
x_1	0	1	1	$m_1(\{yes\}) = 0.95 \quad m_1(\Theta) = 0.05$
x_2	1	0	2	$m_2(\{no\}) = 0.6 \quad m_2(\Theta) = 0.4$
x_3	1	0	2	$m_3(\{no\}) = 1$
x_4	1	1	1	$m_4(\{no\}) = 0.95 \quad m_4(\Theta) = 0.05$
x_5	0	0	1	$m_5(\{yes\}) = 1$

4.2 Belief Discernibility Matrix and Function

In order to compute the possible reducts from our uncertain decision table, we propose to adapt the concepts of discernibility matrix and function under the belief function framework which are originally based on certain decision attribute (see subsection 2.2) to be called belief discernibility matrix M'(UDT) and function $f'(UDT)$. In this case, the belief discernibility matrix will be based on a distance measure to identify the similarity or dissimilarity between decision values of the objects x_i and x_j. The idea is to use a distance measure between two bba's m_i and m_j. The threshold value provides a bit of flexibility in our approach. Hence, belief discernibility matrix M'(UDT) is a $n \times n$ matrix with entries $M'_{i,j}$ defined as follows:

For $i, j = 1, ..., n$

$$M'_{i,j} = \{c \in C \mid c(x_i) \neq c(x_j) \text{ and } dist(m_i, m_j) \geq threshold\} \qquad (12)$$

where $dist$ is a distance measure between two bba's proposed in [2] which satisfies more properties than many other distance measures proposed in [1, 4, 5] as defined below.

$$dist(m_1, m_2) = \sqrt{\frac{1}{2}(\| \vec{m_1} \|^2 + \| \vec{m_2} \|^2 - 2 < \vec{m_1}, \vec{m_2} >)} \qquad (13)$$

$$0 \leq dist(m_1, m_2) \leq 1 \qquad (14)$$

where $< \vec{m_1}, \vec{m_2} >$ is the scalar product defined by:

$$< \vec{m_1}, \vec{m_2} > = \sum_{i=1}^{|2^\Theta|} \sum_{j=1}^{|2^\Theta|} m_1(A_i) m_2(A_j) \frac{|A_i \cap A_j|}{|A_i \cup A_j|} \qquad (15)$$

with $A_i, A_j \in 2^\Theta$ for $i, j = 1, 2, \cdots, |2^\Theta|$. $\| \vec{m_1} \|^2$ and $\| \vec{m_2} \|^2$ are respectively the square norm of $\vec{m_1}$ and $\vec{m_2}$.

The belief discernibility function $f'(UDT)$ for the uncertain decision table UDT is equivalent to the certain discernibility function $f(DT)$. The only difference is that it is computed from the belief discernibility matrix. Because the latter has the same structure as the certain discernibility matrix. So, the belief discernibility function is a boolean function of m boolean variables $c_1^*...c_m^*$ (corresponding to the attributes $c_1...c_m$) defined as below, where $M_{ij}'^* = \{c^*|c \in M_{i,j}'\}$

$$f'(UDT) = \wedge\{\vee M_{i,j}'^* | 1 \leq j \leq i \leq n, M_{i,j}' \neq \emptyset\} \tag{16}$$

where $\wedge$ and $\vee$ are two logical operators for conjunction and disjunction. The set of all prime implicants of $f'(UDT)$ determines the sets of all reducts of UDT.

Example. To apply our feature selection method to the uncertain decision table (see Table 1), we start by computing the belief discernibility matrix (see Table 2). To obtain Table 2, we have used Equation (12) with a threshold value equal to 0.1. For example, $M_{1,5}' = \emptyset$ because the two objects x_1 and x_5 have $dist(m_1, m_5) = 0.07 \leq 0.1$. The decision values of the two objects are considered similar.

Table 2. Belief discernibility matrix (M'(UDT))

U	x_1	x_2	x_3	x_4	x_5
x_1					
x_2	a,b,c				
x_3	a,b,c				
x_4	a	b,c			
x_5		a,c	a,c	a,b	

Next, we compute the possible reducts by computing the discernibility function. $f'(UDT) = (a \vee b \vee c) \wedge (a) \wedge (a \vee c) \wedge (b \vee c) \wedge (a \vee b) = (a \wedge b) \vee (a \wedge c)$.

To simplify our uncertain decision table, we find two possible reducts: {a and b} or {a and c} (see Tables 3 and 4). The two reducts have the ability to perform the same classification as the entire attributes set C.

Table 3. Reduct1

U	a	b	ud
x_1	0	1	$m_1(\{yes\}) = 0.95$ $\quad m_1(\Theta) = 0.05$
x_2	1	0	$m_2(\{no\}) = 0.6$ $\quad m_2(\Theta) = 0.4$
x_3	1	0	$m_3(\{no\}) = 1$
x_4	1	1	$m_4(\{no\}) = 0.95$ $\quad m_4(\Theta) = 0.05$
x_5	0	0	$m_5(\{yes\}) = 1$

5 Experimentation

Several tests have been done on real-world databases to evaluate the proposed feature selection method in comparison with another heuristic feature selection method proposed originally in [18]. The comparison is based on two evaluation criteria: the time requirement (the number of seconds needed to find the reduct) and the classification accuracy (Percent of Correct Classification (PCC)) of the

Table 4. Reduct2

U	a	c	ud
x_1	0	1	$m_1(\{yes\}) = 0.95 \quad m_1(\Theta) = 0.05$
x_2	1	2	$m_2(\{no\}) = 0.6 \quad m_2(\Theta) = 0.4$
x_3	1	2	$m_3(\{no\}) = 1$
x_4	1	1	$m_4(\{no\}) = 0.95 \quad m_4(\Theta) = 0.05$
x_5	0	1	$m_5(\{yes\}) = 1$

generated decision rules by incorporating the two methods into a classification system. For this reason, in the first step of the belief rough set classifier described in the following subsection, we apply the two feature selection methods. The latter is able to generate uncertain decision rules used for classification process where the feature selection is one of the important steps.

5.1   Belief Rough Set Classifier

Belief rough set classification, proposed in [20], is able to learn decision rules for the classification process from the uncertain decision system (see Table 1). It was proposed originally in [20]. The induced decision rules are called belief decision rules where the decision is represented by a bba. The simplification of the uncertain decision table to generate the significant belief decision rules is an important phase in generation of the belief rough set classifier. It is described as follows:

1. **Step 1. Eliminate the Superfluous Condition Attributes:** We remove the superfluous condition attributes that are not in reduct.
2. **Step 2. Eliminate the Redundant Objects:** After removing the superfluous condition attributes, we will find redundant objects. They may not have the same bba on decision attribute. So, we use their combined bba's using a rule of combination.
3. **Step 3. Eliminate the Superfluous Condition Attribute Values:** In this step, we compute the reduct value for each belief decision rule R_j of the form: **If** $C(x_j)$ **then** m_j.
4. **Step 4. Generate Belief Decision Rules:** After the simplification of the uncertain decision table, we can generate shorter and significant belief decision rules. With simplification, we can improve the time and the performance of classification of unseen objects.

5.2   Experimental Results

– In the first set of experiments, we tested our methods on standard real-world databases obtained from the U.C.I. repository[1]. A brief description of these databases is presented in Table 5. These databases are of varying sizes (number of instances, number of attributes and number of decision values). The

---

[1] http://www.ics.uci.edu/~mlearn/MLRepository.html

databases were artificially modified in order to include uncertainty in deci-sion attribute. We took different degrees of uncertainty based on increasing values of probabilities P used to transform the actual decision value d_i of each object x_j to a bba $m_j(\{d_i\}) = 1 - P$ and $m_j(\Theta) = P$. A larger P gives a larger degree of uncertainty.

- Low degree of uncertainty: we take $0 < P \leq 0.3$
- Middle degree of uncertainty: we take $0.3 < P \leq 0.6$
- High degree of uncertainty: we take $0.6 < P \leq 1$

Each database is divided into ten parts. Nine parts are used as the training set, the last is used as the testing set. The procedure is repeated ten times, each time another part is chosen as the testing set. This method, called a cross-validation, permits a more reliable estimation of the evaluation crite-rion. In this paper, we report the average of the evaluation criteria.

Table 5. Description of databases

Databases	#instances	#attributes	#decision values
W. Breast Cancer	690	8	2
Balance Scale	625	4	3
C. Voting records	497	16	2
Zoo	101	17	7
Nursery	12960	8	3
Solar Flares	1389	10	2
Lung Cancer	32	56	3
Hayes-Roth	160	5	3
Car Evaluation	1728	6	4
Lymphography	148	18	4
Spect Heart	267	22	2
Tic-Tac-Toe Endgame	958	9	2

Table 6 reports the experimental results relative to the classification accuracy for different degrees of uncertainty. From this table, we see that the proposed exhaustive feature selection method is more accurate than the heuristic search method for attribute selection. It is true for all the databases and for all degrees of uncertainty. For example, the mean PCC for Balance Scale database goes from 83.23% with exhaustive method to 77.85% with heuristic method. We can also conclude that when the degree of uncertainty increases there is a slight decline in accuracy.

The Table 7 gives the experimental results relative to the second evalua-tion criterion, the time requirement needed to simplify the databases. Note that the time requirement is almost the same for different degrees of un-certainty. We conclude from the table that the heuristic feature selection method is little faster than the exhaustive search method for attribute selec-tion. It is true for all the databases. For example, the time requirement for

Table 6. The PCC relative for exhaustive and heuristic methods

Databases	Exhaustive method PCC (%)				Heuristic method PCC (%)			
	Low	Middle	High	Mean	Low	Middle	High	Mean
W. Breast Cancer	86.87	86.58	86.18	86.54	83.41	83.39	83.17	83.32
Balance Scale	83.46	83.21	83.03	83.23	77.96	77.83	77.76	77.85
C. Voting records	98.94	98.76	98.52	98.74	97.91	97.76	97.71	97.79
Zoo	96.52	96.47	95.87	95.95	90.41	90.37	90.22	90.33
Nursery	96.68	96.21	96.07	96.32	94.34	94.13	94.11	94.19
Solar Flares	88.67	88.61	88.56	88.61	85.72	85.61	85.46	85.59
Lung Cancer	75.77	75.50	75.33	75.53	66.43	66.28	66.08	66.26
Hayes-Roth	97.96	97.15	96.75	96.95	83.66	83.31	83.14	83.37
Car Evaluation	84.46	84.17	84.01	84.21	73.39	73.22	73.17	73.26
Lymphography	83.24	83.03	82.67	82.64	78.85	78.67	78.34	78.62
Spect Heart	85.34	85.28	85.07	85.23	83.54	83.21	83.17	83.30
Tic-Tac-Toe Endgame	86.26	86.21	86.18	86.21	83.93	83.72	83.47	83.70

W. Breast Cancer database goes from 154 seconds with exhaustive method to 122 seconds with heuristic method.

- In the second set of experiments, we tested our methods on a naturally uncertain web usage database. The latter was obtained from web access logs of the introductory computing science course at Saint Mary's University. The course is "Introduction to Computing Science and Programming" offered in the first term of the first year. Lingras and West [8] showed that visits from students attending the first course could fall into one of the following three categories (decision values):

  - Studious: These visitors download the current set of notes. Since they download a limited/current set of notes, they probably study class-notes on a regular basis.
  - Crammers: These visitors download a large set of notes. This indicates that they have stayed away from the class-notes for a long period of time. They are planning for pretest cramming.
  - Workers: These visitors are mostly working on class or lab assignments or accessing the discussion board.

The web logs were preprocessed to create an appropriate representation of each user, corresponding to a visit. The abstract representation of a web user is a critical step that requires a good knowledge of the application domain. Based on some observations, it was decided to use the following attributes for representing each visitor [8]:

- On campus/Off campus access.
- Day time/Night time access: 8 a.m. to 8 p.m. were considered to be the daytime.

Table 7. The time requirement relative for exhaustive and heuristic methods

Databases	Exhaustive method Time (s)	Heuristic method Time(s)
W. Breast Cancer	154	122
Balance Scale	129	97
C. Voting records	110	83
Zoo	101	75
Nursery	380	269
Solar Flares	157	113
Lung Cancer	48	37
Hayes-Roth	91	46
Car Evaluation	178	149
Lymphography	102	79
Spect Heart	109	87
Tic-Tac-Toe Endgame	139	109

- Access during lab/class days or non-lab/class days: All the labs and classes were held on Tuesdays and Thursdays. The visitors on these days are more likely to be workers.
- Number of hits.
- Number of class-notes downloads.

Total visits were 7965. Instead of representing an object as belonging to a cluster ud_i (decision value), it is also associated a degree of belief in the object belonging to the cluster ud_i.

In the Table 8, we conclude that the results relative to the artificial databases are also the same for the real and uncertain web usage database. The PCC goes from 84.16% with heuristic method to 85.07% with exhaustive method. The time requirement is slightly increased.

Table 8. Experimental results for the web usage database

Approaches	PCC (%)	Time requirement (seconds)
Exhaustive method	85.07	188
Heuristic method	84.16	157

6   Conclusion and Future Work

In this paper, we have proposed a new attribute selection method to remove the superfluous attributes from uncertain decision table by defining a new discernibility matrix and its corresponding discernibility function. We handle uncertainty in decision attributes using the belief function. An exhaustive method is able to calculate all possible reducts from uncertain data.

The experimental results show the efficiency of the method especially for the classification accuracy. As a future work, we suggest proposing a new feature selection method for big and incremental data. We also suggest adapting the concepts of discernibility matrix and function to select relevant features from data characterized by uncertain condition attribute values.

References

1. Bauer, M.: Approximations algorithm and decision making in the Dempster-Shafer theory of evidence - an empirical study. International Journal of Approximate Reasoning 17(2-3), 217–237 (1997)
2. Bosse, E., Jousseleme, A.L., Grenier, D.: A new distance between two bodies of evidence. Information Fusion 2, 91–101 (2001)
3. Deng, D., Huang, H.: A new discernibility matrix and function. In: Wang, G.-Y., Peters, J.F., Skowron, A., Yao, Y. (eds.) RSKT 2006. LNCS (LNAI), vol. 4062, pp. 114–121. Springer, Heidelberg (2006)
4. Elouedi, Z., Mellouli, K., Smets, P.: Assessing sensor reliability for multisensor data fusion within the transferable belief model. IEEE Trans. Syst. Man Cybern. 34(1), 782–787 (2004)
5. Fixen, D., Mahler, R.P.S.: The modified Dempster-Shafer approach to classification. IEEE Trans. Syst. Man Cybern. 27(1), 96–104 (1997)
6. Hu, X., Cercone, N.: Learning in Relational Databases: a Rough Set Approach. Computational Intelligence 2, 323–337 (1995)
7. Modrzejewski, M.: Feature selection using rough sets theory. In: Proceedings of the 11th International Conference on Machine Learning, pp. 213–226 (1993)
8. Lingras, P., West, C.: Interval Set Clustering of Web Users with Rough K-means. Journal of Intelligent Information Systems 23(1), 5–16 (2004)
9. Pawlak, Z.: Rough Sets. International Journal of Computer and Information Sciences 11, 341–356 (1982)
10. Pawlak, Z., Zdzislaw, A.: Rough Sets: Theoretical Aspects of Reasoning About Data. Kluwer Academic Publishing, Dordrecht (1991) ISBN 0-7923-1472-7
11. Pawlak, Z., Rauszer, C.M.: Dependency of attributes in Information systems. Bull. Polish Acad. Sci., Math. 33, 551–559 (1985)
12. Rauszer, C.M.: Reducts in Information systems. Fundamenta Informaticae (1990)
13. Shafer, G.: A mathematical theory of evidence. Princeton University Press, Princeton (1976)
14. Skowron, A., Rauszer, C.: The Discernibility Matrices and Functions in Infor- mation Systems. In: Slowiski, R. (ed.) Intelligent Decision Support, Handbook of Applications and Advances of the Rough Set Theory, pp. 311–362. Kluwer Academic Publishers, Dordrecht (1992)
15. Skowron, A.: Rough Sets in KDD. Special Invited Speaking, WCC 2000 in Beijing (August 2000)
16. Smets, P., Kennes, R.: The transferable belief model. Artificial Intelligence 66, 191–236 (1994)
17. Smets, P.: The transferable belief model for quantified belief representation. In: Gabbay, D.M., Smets, P. (eds.) Handbook of Defeasible Reasoning and Uncertainty Management Systems, vol. 1, pp. 207–301. Kluwer, Doordrecht (1998)
18. Trabelsi, S., Elouedi, Z.: Heuristic method for attribute selection from partially uncertain data using rough sets. International Journal of General Systems 39(3), 271–290 (2010)

19. Trabelsi, S., Elouedi, Z., Lingras, P.: Heuristic for Attribute Selection Using Belief Discernibility Matrix. In: Li, T., Nguyen, H.S., Wang, G., Grzymala-Busse, J., Janicki, R., Hassanien, A.E., Yu, H. (eds.) RSKT 2012. LNCS (LNAI), vol. 7414, pp. 129–138. Springer, Heidelberg (2012)
20. Trabelsi, S., Elouedi, Z., Lingras, P.: Belief Rough Set Classifier. In: Gao, Y., Japkowicz, N. (eds.) AI 2009. LNCS (LNAI), vol. 5549, pp. 257–261. Springer, Heidelberg (2009)
21. Yao, Y., Zhao, Y.: Discernibility Matrix Simplification for Constructing Attribute Reducts. Information Sciences 179(5), 867–882 (2009)

Cost-Sensitive Boosting Algorithms
for Imbalanced Multi-instance Datasets

Xiaoguang Wang[1], Stan Matwin[2], Nathalie Japkowicz[1,3], and Xuan Liu[1]

[1] School of Electrical Engineering and Computer Science, University of Ottawa, Canada
Bwang009@eecs.uottawa.ca, Xliu107@uottawa.ca
[2] Faculty of Computer Science, Dalhousie University, Canada
Stan@cs.dal.ca
[3] Department of Computer Science, Northern Illinois University, USA
Nat@eecs.uottawa.ca

Abstract. Multi-instance learning is different than standard propositional classification, because it uses a set of bags containing many instances as input. The instances in each bag are not labeled, but the bags themselves are labeled positive or negative. Our research shows that classification of multi-instance data with imbalanced class distributions significantly decreases the performance normally achievable by most multi-instance algorithms, which is the same as the performance of most standard, single-instance classifier learning algorithms. In this paper, we present and analyze this multi-instance class imbalance problem, and propose a novel solution framework. We focus on how to utilize the extended AdaBoost techniques applicable to most multi-instance classifier learning algorithms. Cost-sensitive boosting algorithms are developed by introducing cost items into the learning framework of AdaBoost, to enable classification of imbalanced multi-instance datasets.

Keywords: multi-instance classification, class imbalance problem, AdaBoost, cost-sensitive learning.

1 Introduction

Multi-instance learning (MIL) differs from traditional supervised learning algorithms, in that a multi-instance dataset consists of bags of individual instances with unknown classifications, and only the bags are labeled. Each bag can contain several instances, but the number of instances in each is different, and the same instance can belong to several bags.

While MIL has been used in many applications, including drug activity recognition [3], text-categorization [15] and computer vision recognition [14], [20], there is a vast amount of research about, and many different approaches to, solving the MIL problem. For example, Diverse Density (DD) [4] and the Expectation-Maximization version [10] were proposed as general frameworks for solving multi-instance learning problems. The k Nearest Neighbour approach, known as Citation kNN, was adapted for MIL problems in [8]. Andrews et al. [15] proposed two approaches to modify Support Vector Machines: mi-SVM for instance-level classification, and MI-SVM for

O. Zaïane and S. Zilles (Eds.): Canadian AI 2013, LNAI 7884, pp. 174–186, 2013.

bag-level classification. For tree methods, Blockeel et al. [17] proposed a multi-instance tree method (MITI), and Bjerring et al. [21] extended this in their work by adopting MITI to learn rules (MIRI).

A dataset is imbalanced if the classes are not represented approximately equally. In a two-class imbalanced dataset, there are often far more negative examples than positive examples, and in this situation a default classifier will always predict 'negative'. In practice, one would want to penalize errors on positive examples more strongly than errors on negative examples. There have been attempts to deal with imbalanced datasets in real life domains, such as text classification [16], image classification [7], disease detection [18] and others [9], [13]. However, the data imbalance problem still exists for multi-instance classification without specialized solution provided. Although there are many published works about multi-instance classification, there is very little related discussion about imbalanced multi-instance classification problems. The multi-instance data imbalanced problem is presented in this paper, and cost-sensitive boosting algorithms are developed by introducing cost items into the learning framework of AdaBoost, for classification of imbalanced multi-instance datasets.

The paper examines two class classification problems, and the algorithms discussed can be extended to multi-class classification. The rest of the paper is organized as follows: Section 2 presents the class imbalance problem for multi-instance datasets and related concepts. In Section 3, the AdaBoost algorithm and its cost-sensitive adaptations for the single-instance class imbalance problem are discussed. Cost-sensitive boosting algorithms for the multi-instance class imbalance problem are presented in Section 4. Section 5 illustrates the efficiency of our algorithm as determined by experimentation, and offers some final remarks. Finally, Section 6 presents the conclusion and future work.

2   The Class Imbalance Problem of Multi-instance Datasets

The multi-instance learning problem can be defined as:
Given:

- A set of bags $\chi_i, i = 1, \ldots, N$, where each bag can consist of an arbitrary number of instances and a given label: $\chi_i = \{x_i^1, x_i^2, \ldots, x_i^{n_i}; y_i\}, i = 1, \ldots, N, y_i \in \{-1, +1\}$, where each instance $x_i^{n_i}$ is an M-tuple of attribute values belonging to a certain domain or instance space $\mathbb{R}$.
- The existence of an unknown function f that classifies individual instances as $+1$ or -1, and for which it holds that $c(\chi_i) = +1$ if and only if $\exists x_i^{n_i} \in \chi_i : f(x_i^{n_i}) = +1$. (multi-instance constraint, MIC)

From this definition we derive the standard assumption of MI learning, which is that a bag is negative if and only if all instances in the bag are negative; if the bag contains one or more positive instances, the bag is positive.

Although specific discussions about the multi-instance class imbalance problem are rarely found in previous work, the issue occurs frequently in real life application areas such as computer vision recognition and text mining. In our related research [23], multi-instance classification algorithms are used in underwater mine like object sonar image processing. Each target to be classified has many images from different angles and distances and these images build a multi-instance dataset. In real world environment the number of non-mine like objects is much greater than the number of mine like objects, so the class imbalance problem is an important factor affecting the performance of the classifiers. Our research shows that for most multi-instance dataset with imbalanced class distributions, classification of these datasets significantly decrease the performance normally achievable by most multi-instance algorithms but we can hardly find solutions to deal with this problem from these algorithms. This motivated us to investigate the problem of multi-instance class imbalance more thoroughly.

As we already know, a single-instance dataset is defined as imbalanced if at least one class is under-represented relative to others. For multi-instance datasets, the problem is similar but the circumstances are more complex. The class imbalance situation occurs not only at the instance-level, but also at the bag-level. Figure 1 shows the imbalanced multi-instance classification problem with the separating plane and the margin. Since the final margin of multi-instance classification is at bag-level, the default classifier would tend to penalize errors on positive bags more strongly than errors on negative bags. In Figure 1, there are far more majority bags than minority bags, and the margin learned by the default classifier is 'pushed' closer to the minority bags from the ideal margin.

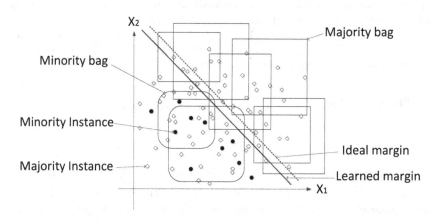

Fig. 1. The imbalanced multi-instance classification problem with the separating plane and the margin. Black dots denote instances of minority class while diamond dots denote instances of majority class. Rectangle frames with round corner denote minority bags and rectangle frames denote majority bags. The solid line denotes the learned margin by classifier and the dotted line denotes the ideal margin of two classes.

For the single-instance data imbalance problem, the machine learning community has addressed the issue of class imbalances in two different ways to solve the skewed vector space problem. The first method, which is classifier-independent, is to balance the distributions by considering the representative proportions of class examples in the distribution of the original data. The simplest way to balance a dataset is to under-sample or over-sample (randomly or selectively) the majority class, while maintaining the original minority class population [16]. One of the most common pre-processing methods to balance a dataset, Synthetic Minority Over-sampling Technique (SMOTE) [13], over-samples the minority class by taking each minority class sample and introducing synthetic examples along the line segments joining any or all of the k minority class nearest neighbors. Evidence shows that synthetic sampling methods are effective when dealing with learning from imbalanced data [9], [13], [16].

Working with classifiers to adapt datasets is another way to deal with the single-instance imbalanced data problem. The theoretical foundations and algorithms of cost-sensitive methods naturally apply to imbalanced learning problems [7], [11]. Thus, for imbalanced learning domains, cost-sensitive techniques provide a viable alternative to sampling methods. Recent research ([9], [11], [16]) suggests that assigning distinct costs to the training examples is a fundamental approach of this type, and various experimental studies of this ([5], [7], [18]) have been performed using different kinds of classifiers.

3   AdaBoost and Cost-Sensitive Adaptations

Boosting has been proven to be an effective method of combining multiple models in order to enhance the predictive accuracy of a single model [1], [6]. AdaBoost is a version of boosting that uses the confidence-rated predictions described in [1], [6]. It applies a base learner to induce multiple individual classifiers in sequential trials, and a weight is assigned to each example. After each trial, the vector of weights is adjusted to reflect the importance of each training example in the next induction trial. This adjustment effectively increases the weights of misclassified examples, and decreases the weights of correctly classified examples. Finally, the individual classifiers are combined to form a composite classifier.

Take as input the training set $(x_1, y_1), ..., (x_m, y_m)$; $x_i \in \chi, y_i \in \{-1, +1\}$, where each x_i is an n-tuple of attribute values belonging to a certain domain or instance space X, and y_i is a label in a label set Y. The key process of the AdaBoost.M1 method [1] is to iteratively update the distribution function over the training data. This means that for every iteration $t = 1, ..., T$, where T is a given number of the total number of iterations, the distribution function D_t is updated sequentially, and used to train a new hypothesis:

$$D_{t+1}(i) = \frac{D_t(i) \exp(-\alpha_t y_i h_t(x_i))}{Z_t} \tag{1}$$

where $\alpha_t = \frac{1}{2} ln\left(\frac{1-\varepsilon_t}{\varepsilon_t}\right)$ is the weight updating parameter, $h_t(x_i)$ is the prediction output of hypothesis h_t on the instance x_i, ε_t is the error of hypothesis h_t over the training data, and Z_t is a normalization factor.

Schapire and Singer [6] used a generalized version of Adaboost. As shown in [6], the training error of the final classifier is bounded as:

$$\frac{1}{m}|\{i: H(x_i) \neq y_i\}| \leq \prod_t Z_t \tag{2}$$

where

$$Z_t = \sum_i D_t(i) \exp\left(-\alpha_t y_i h_t(x_i)\right) \leq \sum_i D_t(i) \left(\frac{1 + y_i h_t(x_i)}{2} e^{-\alpha} + \frac{1 + y_i h_t(x_i)}{2} e^{\alpha}\right) \tag{3}$$

Minimizing Z_t on each round, α_t is induced as:

$$\alpha_t = \frac{1}{2} \ln \left(\frac{\sum_{i, y_i = h_t(x_i)} D_t(i)}{\sum_{i, y_i \neq h_t(x_i)} D_t(i)}\right) \tag{4}$$

The weighting strategy of AdaBoost identifies samples on their classification outputs as correctly classified or misclassified. However, it treats samples of different classes equally. The weights of misclassified samples from different classes are increased by an identical ratio, and the weights of correctly classified samples from different classes are decreased by an identical ratio.

Since boosting is suitable for cost-sensitive adaption, motivated by [6]'s analysis and methods for choosing α_t, several cost-sensitive boosting methods for imbalanced learning have been proposed in recent years. Three cost-sensitive boosting methods, AdaC1, AdaC2 and AdaC3, were proposed in [18], which introduced cost items into the weight updating strategy of AdaBoost. AdaCost [5] is another cost-sensitive boosting algorithm that follows a similar methodology.

4   Proposed Methods

Similar to the methods for managing the single-instance class imbalance problem in Refs. [5], [7], [11], [18], the learning objective in dealing with the multi-instance class imbalance problem is to improve the identification performance on the minority class. In our research, the first strategy is to target the multi-instance imbalanced learning problem by using different cost matrices that describe the costs for misclassifying any particular data example. When used for single-instance learning, this method reportedly improved classification performance on class imbalanced datasets significantly [11], [12].

For multi-instance class imbalance datasets, the optimal prediction for a bag χ is the class i that minimizes

$$L(\chi, i) = \sum_j P(j|\chi) C(i, j) \tag{5}$$

where C denotes the cost matrix [11], and (i, j) is the cost of predicting class i when the true class is j. $P(j|\chi)$ denotes the probability of each class j being the true class of bag χ.

Our second strategy is to apply cost-minimizing techniques to the combination schemes of ensemble methods. This learning objective expects that the weighting strategy of a boosting algorithm will preserve a considerable weighted sample size of the minority class. A preferred boosting strategy is one that can distinguish different types of samples, and boost more weights on those samples associated with higher identification importance.

To denote the different identification importance among bags, each bag is associated with a cost item; the higher the value, the higher the importance of correctly identifying the sample. For an imbalanced multi-instance dataset, there are many more bags with class label $y = -1$ than bags with class label $y = +1$. Using the same learning framework as AdaBoost, the cost items can be fed into the weight update formula of AdaBoost (Eq. (1)) to bias the weighting strategy. The proposed methods are similar to those proposed in Ref. [18]. Figure 2 shows the proposed algorithms.

Given: A multi-instance training dataset with a set of bags $\chi_i, i = 1, ..., N$, where each bag can consist of an arbitrary number of instances and a given label: $\chi_i = \{x_i^1, x_i^2, ..., x_i^{n_i}; y_i\}, i = 1, ..., N, y_i \in \{-1, +1\}$, and each instance $x_i^{n_i}$ is an M-tuple of attribute values belonging to a certain domain or instance space $\mathbb{R}$.

Initialize $D_1(i) = 1/m$.

For $t = 1, ..., T$ && the constraint condition η is satisfied:

- Train a weak learner using distribution D_t.
- Get a weak hypothesis $h_t: \chi \to \mathbb{R}$.
- Choose $\alpha_t \in \mathbb{R}$.
- Update:

$$D_{t+1}(i) = \frac{D_t(i)K_t(\chi_i, y_i)}{Z_t} \tag{6}$$

where Z_t is a normalization factor (chosen so that D_{t+1} will be a distribution).

Output the final hypothesis:

$$H(\chi) = sign\left(\sum_{t=1}^{T} \alpha_t h_t(\chi)\right) \tag{7}$$

Fig. 2. Cost-sensitive Adaboost for Multi-Instance Learning Algorithm

For the original adaboost, $K_t(\chi_i, y_i) = \exp(-\alpha_t y_i h_t(\chi_i))$, our proposed algorithms introduced four cost items into the weight update formula of AdaBoost: inside the exponent, outside the exponent, or in two ways both inside and outside the exponent. Each modification can be a new boosting algorithm, denoted as Ab1, Ab2, Ab3 and Ab4 respectively. Ab1, Ab2 and Ab3 are similar to AdaC1, AdaC2 and AdaC3 respectively for single-instance learning in Ref. [18]. The difference is, in our algorithms the training samples are bags of instances, not instances.

The modifications of $K_t(\chi_i, y_i)$ are then given by:

- Ab1:

$$K_t(\chi_i, y_i) = \exp(-C_i \alpha_t y_i h_t(\chi_i)) \tag{8}$$

- Ab2:

$$K_t(\chi_i, y_i) = C_i \exp(-\alpha_t y_i h_t(\chi_i)) \tag{9}$$

- Ab3:

$$K_t(\chi_i, y_i) = C_i \exp\left(-C_i \alpha_t y_i h_t(\chi_i)\right) \tag{10}$$

- Ab4:

$$K_t(\chi_i, y_i) = C_i^2 \exp\left(-C_i^2 \alpha_t y_i h_t(\chi_i)\right) \tag{11}$$

Now we induce the weight update parameter α_t and constraint condition η in Figure 2 for Ab4. From Eq. (11) we get:

$$D_{t+1}(i) = \frac{C_i^2 D_t(i) \exp\left(-C_i^2 \alpha_t y_i h_t(\chi_i)\right)}{Z_t} = \frac{C_i^{2t} \exp\left(-C_i^2 y_i f(\chi_i)\right)}{m \prod_t Z_t} \tag{12}$$

where

$$f(\chi) = \sum_t \alpha_t h_t(\chi) \tag{13}$$

and

$$Z_t = \sum_i C_i^2 D_t(i) \exp\left(-C_i^2 \alpha_t y_i h_t(\chi_i)\right) \tag{14}$$

The overall training error is bounded as:

$$\frac{1}{m}|\{i: H(\chi_i) \neq y_i\}| \leq \frac{1}{m}\sum_i C_i^2 \exp(-C_i^2 y_i f(\chi_i)) = \prod_t Z_t \sum_i \frac{C_i^2}{C_i^{2t}} D_{t+1}(i) \tag{15}$$

According to Ref. [6], for weak hypotheses $C_i^2 \alpha_t y_i h_t(\chi_i) \in [-1, +1]$ with range $[-1, +1]$, α can be obtained by approximating Z as follows:

$$Z_t = \sum_i C_i^2 D_t(i) \exp\left(-C_i^2 \alpha_t y_i h_t(\chi_i)\right) \leq \sum_i C_i^2 D_t(i)\left(\frac{1 + C_i^2 y_i h_t(\chi_i)}{2}e^{-\alpha_t} + \frac{1 - C_i^2 y_i h_t(\chi_i)}{2}e^{\alpha_t}\right) \tag{16}$$

Let

$$G(\alpha_t) = \sum_i C_i^2 D_t(i)\left(\frac{1 + C_i^2 y_i h_t(\chi_i)}{2}e^{-\alpha_t} + \frac{1 - C_i^2 y_i h_t(\chi_i)}{2}e^{\alpha_t}\right) \tag{17}$$

Our purpose is for α_t to minimize $G(\alpha_t)$, so we can obtain:

$$G'(\alpha_t) = \frac{dG}{d\alpha_t} = 0 \tag{18}$$

Next, we can analytically obtain α_t from Eq. (18), giving:

$$\alpha_t = \frac{1}{2}\ln\left(\frac{\sum_i C_i^2 D_t(i) + \sum_{i, y_i = h_t(\chi_i)} C_i^4 D_t(i) - \sum_{i, y_i \neq h_t(\chi_i)} C_i^4 D_t(i)}{\sum_i C_i^2 D_t(i) - \sum_{i, y_i = h_t(\chi_i)} C_i^4 D_t(i) + \sum_{i, y_i \neq h_t(\chi_i)} C_i^4 D_t(i)}\right) \tag{19}$$

The sample weight updating goal of AdaBoost is to decrease the weight of the training samples that are correctly classified, and increase the weights of the opposite samples [1], [6]. Therefore, α_t should be a positive value, and the training error should be less than random guessing, based on the current data distribution.

To ensure that α_t is positive, we get

$$\sum_{i,y_i=h_t(\chi_i)} C_i^4 D_t(i) > \sum_{i,y_i \neq h_t(\chi_i)} C_i^4 D_t(i) \tag{20}$$

This is the constraint condition η in Figure 2.

Similarly, we can analytically choose α_t and constraint condition η for the other three modifications of Eq. (1) ([6], [18]). Table 1 lists all α_t and η of Ab1 to Ab4.

Table 1. Parameter α_t and η chosen for Figure 2

	α_t	η
Ab1	$\frac{1}{2} ln \left(\frac{1 + \sum_{i,y_i=h_t(\chi_i)} C_i D_t(i) - \sum_{i,y_i \neq h_t(\chi_i)} C_i D_t(i)}{1 - \sum_{i,y_i=h_t(\chi_i)} C_i D_t(i) + \sum_{i,y_i \neq h_t(\chi_i)} C_i D_t(i)} \right)$	$\sum_{i,y_i=h_t(\chi_i)} C_i D_t(i) > \sum_{i,y_i \neq h_t(\chi_i)} C_i D_t(i)$
Ab2	$\frac{1}{2} ln \left(\frac{\sum_{i,y_i=h_t(\chi_i)} C_i D_t(i)}{\sum_{i,y_i \neq h_t(\chi_i)} C_i D_t(i)} \right)$	$\sum_{i,y_i=h_t(\chi_i)} C_i D_t(i) > \sum_{i,y_i \neq h_t(\chi_i)} C_i D_t(i)$
Ab3	$\frac{1}{2} ln \left(\frac{\sum_i C_i D_t(i) + \sum_{i,y_i=h_t(\chi_i)} C_i^2 D_t(i) - \sum_{i,y_i \neq h_t(\chi_i)} C_i^2 D_t(i)}{\sum_i C_i D_t(i) - \sum_{i,y_i=h_t(\chi_i)} C_i^2 D_t(i) + \sum_{i,y_i \neq h_t(\chi_i)} C_i^2 D_t(i)} \right)$	$\sum_{i,y_i=h_t(\chi_i)} C_i^2 D_t(i) > \sum_{i,y_i \neq h_t(\chi_i)} C_i^2 D_t(i)$
Ab4	$\frac{1}{2} ln \left(\frac{\sum_i C_i^2 D_t(i) + \sum_{i,y_i=h_t(\chi_i)} C_i^4 D_t(i) - \sum_{i,y_i \neq h_t(\chi_i)} C_i^4 D_t(i)}{\sum_i C_i^2 D_t(i) - \sum_{i,y_i=h_t(\chi_i)} C_i^4 D_t(i) + \sum_{i,y_i \neq h_t(\chi_i)} C_i^4 D_t(i)} \right)$	$\sum_{i,y_i=h_t(\chi_i)} C_i^4 D_t(i) > \sum_{i,y_i \neq h_t(\chi_i)} C_i^4 D_t(i)$

We also used AdaCost [5] as a cost-sensitive boosting algorithm to deal with the multi-instance class imbalance problem. In Figure 2, AdaCost is developed when dealing with multi-instance classification by introducing:

$$K_t(\chi_i, y_i) = \exp\left(-\alpha_t y_i h_t(\chi_i)\beta(i)\right) \tag{21}$$

where

$$\beta(i) = \beta(\text{sign}(y_i h_t(\chi_i)), C_i) \tag{22}$$

Like Ab1, AdaCost introduces cost sensitivity inside the exponent of the weight updating formula of Adaboost. However, instead of applying the cost items directly, AdaCost employs a cost-adjustment function that aggressively increases the weights of costly misclassifications, while conservatively decreasing the weights of high-cost examples that are correctly classified.

5   Experiments

In this section, we explain our experiments to investigate and compare the following boosting and non-boosting algorithms: Base learner, Cost Sensitive, Adaboost, AdaCost, Ab1, Ab2, Ab3 and Ad4 with two weak learners using tree methods (MITI [17] and MIRI [21] respectively). Tree methods were chosen as the weak learners because they are 1) stable in multi-instance learning [17], [21], and 2) suitable to be weak learners in many related works [12], [18].

For these experiments parameter T, which governs the number of classifiers generated, was set to ten in each boosting algorithm. For the cost sensitive methods, the original costs were chosen according to the bag number of each class. For the cost sensitive boosting methods, the iteration rounds of boosting could be terminated through one of two conditions: a) the prefixed number T, or b) the constraint condition η in Figure 2. The Ten-fold Cross-Validation method was used in all experiments.

5.1   Details of Datasets

The first seven datasets used in our experiments are those employed in [19] and [21]. The original datasets are not imbalanced, so to make them all imbalanced we chose only a portion of the bags in one class.

Table 2 shows the details of the datasets used in our experiment. These datasets can be retrieved from http://www.eecs.uottawa.ca/~bwang009/.

Table 2. Details of Datasets ('#' denotes 'number of' and '%' denotes of 'percentage of').

Dataset	Size	#attribute	#minority bags	%minority bags	#minority instances	%minority instances
Elephant	125	230	25	20	150	19.69
Fox	121	230	21	17.36	134	20.71
Tiger	126	230	26	20.63	164	30.15
Mutagenesis_atom	167	10	42	25.15	365	34.02
Mutagenesis_bond	160	16	35	21.88	603	20.41
Mutagenesis_chain	152	24	27	17.76	514	12.49
Process	142	200	29	20.42	281	9.78

5.2   Experimental Results

When learning extremely imbalanced data, a trivial classifier that predicts every case as the majority class can still achieve very high accuracy. Thus, the overall classification accuracy is often not an effective measure of performance. We chose Gmean [2] as the measure for our algorithms and experiments. The definition of Gmean is found in Eq. (23) and the confusion matrix is defined in Table 3.

Table 3. Confusion Matrix

	Predicted Positive Class	Predicted Negative Class
Actual Positive class	TP (True Positive)	FN (False Negative)
Actual Negative class	FP (False Positive)	TN (True Negative)

$$\text{Gmean} = (\frac{TN}{TN + FP} \times \frac{TP}{TP + FN})^{1/2} \tag{23}$$

Table 4. Experiment results on MITI (Gmean)

Dataset	Base	CS	AdaCost	Adaboost	Ab1	Ab2	Ab3	Ab4
Elephant	0.4313	0.6	0.5485	0.5238	0.5426	0.5514	0.5514	0.6099
Fox	0.2149	0.4337	0.4445	0	0.296	0.4655	0.5555	0.4928
Tiger	0.5463	0.7317	0.7038	0.6928	0.7805	0.7114	0.7671	0.6923
Mutagenesis_atom	0.5831	0.6309	0.6973	0.6234	0.5995	0.7059	0.657	0.6928
Mutagenesis_bond	0.5493	0.7538	0.7928	0.7085	0.6943	0.7335	0.7899	0.7783
Mutagenesis_chain	0.4251	0.6742	0.7272	0.503	0.7542	0.7242	0.7933	0.8327
Process	0.7096	0.7913	0.8156	0.8194	0.8593	0.8194	0.8043	0.842

Table 5. Experiment results on MIRI (Gmean)

Dataset	Base	CS	AdaCost	Adaboost	Ab1	Ab2	Ab3	Ab4
Elephant	0.5158	0.6536	0.5786	0.5571	0.5514	0.56	0.5817	0.6033
Fox	0.2116	0.4337	0.5014	0.4276	0.4756	0.3546	0.4499	0.4024
Tiger	0.5004	0.6844	0.6785	0.6725	0.7114	0.7981	0.7894	0.7442
Mutagenesis_atom	0.7005	0.7709	0.7358	0.7579	0.7554	0.7926	0.7687	0.7916
Mutagenesis_bond	0.7219	0.7746	0.7517	0.767	0.7783	0.7838	0.7697	0.7453
Mutagenesis_chain	0.631	0.7055	0.7694	0.6713	0.811	0.7976	0.8348	0.7797
Process	0.8358	0.8396	0.8993	0.8119	0.8554	0.8706	0.8502	0.8316

Tables 4 and 5 present the experimental results of the base learners and all the presented algorithms using the base learners. MITI [17] is chosen as the base learner in Table 4, and MIRI [21] is the base learner in Table 5. In the Tables, 'Base' denotes the base learner, and 'CS' indicates the cost sensitive method.

As Friedman's test [23] is a non-parametric statistical test for multiple classifiers and multiple domains, we performed it on the results in Tables 4 and 5. The null hypothesis for this test is that all the classifiers perform equally, and rejection of the null hypothesis means that there is at least one pair of classifiers with significantly different performance.

For Table 4 the test results are Friedman chi-squared = 25.0017, df = 7, and p-value = 0.0007583, and for Table 5 the results are Friedman chi-squared = 22.2381, df = 7, and p-value = 0.002311. The critical value for the chi-square distribution is 14.07 for a 0.05 level of significance for a single-tailed test. As 25.0017 and 22.2381 are both larger than 14.07, we can reject the hypothesis for both Tables 4 and 5.

We applied Nemenyi's post-hoc test [23] to determine which classifier has the best performance. First we ranked the Gmean values for each dataset with different classifiers. The sum of the ranks for all datasets is represented as $R_{.i}$, where i represents a classifier. Then we used the following formula to calculate the q value between different classifiers:

$$q_{ij} = (\bar{R}_i - \bar{R}_j) / \sqrt{\frac{k(k+1)}{6n}} \qquad (24)$$

where k is the number of classifiers and n is the number of datasets. We then determined if one algorithm is better than another by comparing their q values with the critical value $q_\alpha = 3.19$ (gotten from [22]), as shown in Table 6. The result of 6-1-0 for Ab3 with MITI as the base learner means that the algorithm wins six times, is equal once, and loses zero times. If we set the scores as win=1, equal=0 and lose=-1, the total score of each algorithm in Tables 4 and 5 can be calculated. The result is shown in Table 6.

Table 6. Experiment result using the statistical test method (sorted by score from high to low)

	Ab3	Ab2	Ab4	AdaCost	Ab1	CS	Adaboost	Base
MITI	6-1-0	4-1-2	6-1-0	4-1-2	2-1-4	2-1-4	1-0-6	0-0-7
MIRI	6-1-0	6-1-0	2-3-2	2-3-2	2-3-2	2-3-2	1-0-6	0-0-7
Score	12	8	6	2	-2	-2	-10	-14

The experimental results show that all the proposed algorithms can improve the performance of the base learner. CS did not overcome the cost sensitive boosting methods, since it does not adopt a weight updating strategy. AdaCost did not outperform Ab3, Ab2 and Ab4, since in Ref. [5] AdaCost requires that β for the minority class be non-increasing with respect to c_n, which means the reward for correct classification is low when the cost is high [7]. The performance of Ab3 is the best in all experiments, and Ab2 and Ab4 are also competitive. The weighted updating strategy of cost-sensitive boosting algorithms increases the weights on the misclassified bags from the minority class more than it does on those from the majority class. Similarly, it decreases the weight on correctly classified bags from the minority class less than on those from the majority class.

6   Conclusions and Future Research

We have presented and analyzed the multi-instance class imbalance problem, and provided a novel framework for the design of cost-sensitive boosting algorithms for this problem. We compared the cost sensitive method, the Adacost algorithm, and the four proposed weight updating cost-sensitive boosting algorithms, along with two multi-instance tree methods.

Experimental evidence derived from standard datasets was presented to support the cost-sensitive optimality of the proposed algorithms. We found that cost-sensitive boosting consistently outperformed all other methods tested. In the future, we plan to investigate whether the proposed methods are sensitive to the cost setup. On the given datasets in our experiments, the updating strategies of Ab3 and Ab2 are more suitable than that of Ab4. Since the imbalance ratio is not very large for the chosen datasets in these experiments, in future work it would be worthwhile to investigate whether Ab4 can provide better result than Ab3 and Ab2 on highly class imbalanced multi-instance datasets. We also intend to research the application of the cost-sensitive boosting algorithms for multi-instance classification, and investigate other related algorithms.

Acknowledgement. The authors acknowledge the support of the Defence R&D Canada Centre for Operational Research and Analysis for this work. This work was funded by Defence R&D Canada Centre for Operational Research and Analysis under PWGSC Contract #W7714-115078/001/SV.

References

1. Freund, Y., Schapire, R.E.: Experiments with a new boosting algorithm. In: Machine Learning: Proceedings of the Thirteenth International Conference, pp. 148–156 (1996)
2. Kubat, M., Matwin, S.: Addressing the curse of imbalanced training sets: One-sided selection. In: Proceddings of the Fourteenth International Conference on Machine Learning, pp. 179–186 (1997)
3. Dietterich, T., Lathrop, R., Lozano-P´erez, T.: Solving the multiple instance problem with the axis-parallel rectangles. Artificial Intelligence 89(1-2), 31–71 (1997)
4. Maron, O., Lozano-Pérez, T.: A framework for multiple instance learning. In: Proc. of the 1997 Conf. on Advances in Neural Information Processing Systems 10, pp. 570–576 (1998)
5. Fan, W., Stolfo, S.J., Zhang, J., Chan, P.K.: AdaCost: Misclassification Cost-Sensitive Boosting. In: Proc. Int'l Conf. Machine Learning, pp. 97–105 (1999)
6. Schapire, R.E., Singer, Y.: Improved boosting algorithms using confidence-rated predictions. Machine Learning 37(3), 297–336 (1999)
7. Ting, K.M.: A Comparative Study of Cost-Sensitive Boosting Algorithms. In: Proc. Int'l Conf. Machine Learning, pp. 983–990 (2000)
8. Wang, J., Zucker, J.D.: Solving the multiple-instance problem: A lazy learning approach. In: ICML (2000)
9. Japkowicz, N.: Learning from Imbalanced Data Sets: A Comparison of Various Strategies. In: Proc. Am. Assoc. for Artificial Intelligence (AAAI) Workshop Learning from Imbalanced Data Sets, pp. 10-15 (Technical Report WS-00-05) (2000)
10. Zhang, Q., Goldman, S.A.: EM-DD: An improved multiple instance learning technique. In: Neural Information Processing Systems 14 (2001)
11. Elkan, C.: The Foundations of Cost-Sensitive Learning. In: Proc. Int'l Joint Conf. Artificial Intelligence, pp. 973–978 (2001)
12. Ting, K.M.: An Instance-Weighting Method to Induce Cost-Sensitive Trees. IEEE Trans. Knowledge and Data Eng. 14(3), 659–665 (2002)
13. Chawla, N.V., Bowyer, K.W., Hall, L.O., Kegelmeyer, W.P.: SMOTE: Synthetic Minority Over-sampling Technique. Journal of Artificial Intelligence Research 16, 321–357 (2002)
14. Zhang, M.L., Goldman, S.: Em-dd: An improved multi-instance learning technique. In: NIPS (2002)
15. Andrews, S., Tsochandaridis, I., Hofman, T.: Support vector machines for multiple instance learning. Adv. Neural. Inf. Process. Syst. 15, 561–568 (2003)
16. Batista, G.E.A.P.A., Prati, R.C., Monard, M.C.: A Study of the Behavior of Several Methods for Balancing Machine Learning Training Data. ACM SIGKDD Explorations Newsletter 6(1), 20–29 (2004)
17. Blockeel, H., Page, D., Srinivasan, A.: Multi-instance tree learning. In: ICML (2005)
18. Sun, Y., Kamel, M.S., Wong, A.K.C., Wang, Y.: Cost-Sensitive Boosting for Classification of Imbalanced Data. Pattern Recognition 40(12), 3358–3378 (2007)

19. Foulds, J., Frank, E.: Revisiting multiple-instance learning via embedded instance selection. In: Wobcke, W., Zhang, M. (eds.) 21st Australasian Joint Conference on Artificial Intelligence Auckland, New Zealand, pp. 300–310 (2008)
20. Leistner, C., Saffari, A., Bischof, H.: MIForests: Multiple-instance learning with randomized trees. In: Daniilidis, K., Maragos, P., Paragios, N. (eds.) ECCV 2010, Part VI. LNCS, vol. 6316, pp. 29–42. Springer, Heidelberg (2010)
21. Bjerring, L., Frank, E.: Beyond trees: Adopting MITI to learn rules and ensemble classifiers for multi-instance data. In: Wang, D., Reynolds, M. (eds.) AI 2011. LNCS (LNAI), vol. 7106, pp. 41–50. Springer, Heidelberg (2011)
22. Japkowicz, N., Shah, M.: Evaluating Learning Algorithms: A Classification Perspective. Cambridge University Press (2011)
23. Wang, X., Shao, H., Japkowicz, N., Matwin, S., Liu, X., Bourque, A., Nguyen, B.: Using SVM with Adaptively Asymmetric Misclassification Costs for Mine-Like Objects Detection. In: ICMLA (2012)

Boundary Set Based Existence Recognition
and Construction of Hypertree Agent Organization

Yang Xiang and Kamala Srinivasan

School of Computer Science, University of Guelph, Canada

Abstract. Some of the essential tasks of a multiagent system (MAS) include distributed probabilistic reasoning, constraint reasoning, and decision making. Junction tree (JT) based agent organizations have been adopted by some MAS frameworks for their advantages of efficient communication and sound inference. In addition, JT organizations have the potential capacity to support a high degree of agent privacy. This potential, however, has not been fully realized. We present two necessary and sufficient conditions on the existence of JT organization given a MAS. Following these conditions, we propose a new algorithm suite, based on elimination in the so called *boundary set* of a MAS, that recognizes JT organization existence and constructs one if exists, while guaranteeing agent privacy on private variables, shared variables and agent identities.

1 Introduction

Some of the essential tasks of a multiagent system (MAS) include distributed probabilistic reasoning, constraint reasoning, and decision making (decision theoretic). Existing frameworks include AEBN [8] and MSBN [10] for probabilistic reasoning, ABT [4], ADOPT [5], DPOP [7], Action-GDL [9], DCTE [1], and MSCN [12] for constraint reasoning, and RMM [2], MAID [3], and CDN [11] for decision making. Frameworks such as AEBN and MAID do not assume specific agent organization. ABT assumes a total order among agents. ADOPT and DPOPS use a pseudo-tree organization. MSBN, MSCN, CDN, Action-GDL, and DCTE all use a junction tree (JT) organization (known as *hypertree* in the first three frameworks), which is the focus of this work.

In JT-based frameworks, the application environment is represented by a set of variables, referred to as the *env*. The env is decomposed into a set of overlapping *subenvs*, each being a subset of env. The subenvs are one-to-one mapped to agents of the MAS. Subenvs (and hence agents) are organized into a JT. A JT is a tree where each node is associated with a set of variables called a *cluster*. The tree is structured such that, the intersection of any two clusters is contained in every cluster on the path between the two (*running intersection*). In a JT agent organization, each cluster corresponds to a subenv and its agent. The organization prescribes communication pathways: an agent can send a message to another, iff they are adjacent in the JT.

JT-based agent organizations support several desirable properties. First, they allow efficient communication. For agents to cooperate by utilizing information available locally at individual agents, it suffices to pass two messages along each link of the JT. Hence, the time complexity of communication is linear in the number of agents. Second,

O. Zaïane and S. Zilles (Eds.): Canadian AI 2013, LNAI 7884, pp. 187–198, 2013.

they support sound inference. Global consistency is guaranteed by local consistency, e.g., probabilistic reasoning in MSBN [10] is exact, constraint reasoning in MSCN [12] is complete, and decision making in CDN [11] is globally optimal.

Third, the above two properties are enabled while message contents involve only shared variables (those in the intersection of subenvs). Hence, JT organizations have the potential capacity to support a high degree of agent privacy. However, such potential has not been fully realized in existing JT-based MAS frameworks.

In Action-GDL [9], the JT is built through a centralized mapping operation from a pseudo-tree where each node is an env variable. The centralized mapping discloses identities of all variables to the mapping agent. The JT in DCTE [1] is constructed according to [6], where clusters (one per agent) of env variables are initially organized into a tree that is generally not a JT. Variable identities are passed along the tree links to transform the tree into a JT, and hence are disclosed beyond the agent initially associated with them.

In MSBN [10], MSCN [12], and CDN [11] frameworks, the JT organization is constructed by a coordinator agent with knowledge of variables shared by agents. Since coordinator knows nothing about private variables (those that are contained in a single subenv), their privacy is ensured. However, shared variables and agent identities are disclosed to the coordinator.

The contribution of this work is a new algorithm suite for JT organization based on the so-called *boundary-set* (see below) of the MAS. Given the boundary-set of a MAS, the algorithm first determines distributively whether a JT organization exists. If it does, then a JT will be constructed. The entire process preserves agent privacy on private variables, shared variables, and agent identities. To the best of our knowledge, no known JT-based MAS frameworks provide the same degree of agent privacy, except an alternative approach for distributed JT construction that we report in a related work [13].

The next section defines the problem that we tackle. The subsequent section presents a necessary and sufficient condition on the existence of JT organization. This insight allows a classification of MAS subenv decompositions relative to JT organization existence. The next section gives another necessary and sufficient condition on JT organization existence, which leads to an agent privacy preserving algorithm suite for JT organization existence recognition and construction, illustrated and then specified in the following two sections.

2 Problem Definition

Consider a MAS populated by a set $\mathscr{A} = \{A_0, ..., A_{\eta-1}\}$ of $\eta > 1$ agents. Let V be the set of env variables, and be decomposed into a collection of *subenvs*, $\Omega = \{V_0, ..., V_{\eta-1}\}$, such that $\cup_{i=0}^{\eta-1} V_i = V$. Each V_i is associated with a unique agent A_i, and vice versa. We refer to the set of shared variables, $I_{ij} = V_i \cap V_j \neq \emptyset$, between A_i and A_j as their *border*. For each agent A_i, we denote the set $W_i = \cup_{j \neq i} I_{ij}$ as its *boundary*, and we refer to $W = \{W_0, ..., W_{\eta-1}\}$ as the *boundary set* of the MAS. Fig. 1 (a) shows a subenv decomposition of a trivial env, with agent boundaries shown in (b). A_i knows A_j and can communicate with A_j, iff they have a border. The message between them can only involve variables included in their border.

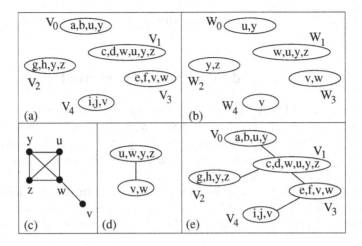

Fig. 1. (a) Subenv decomposition. (b) Agent boundaries. (c) Boundary graph. (d) JT from (c). (e) JT organization for (a).

The general task of JT agent organization is to construct a JT with subenvs as clusters such that running intersection holds.

A variable $x \in V$ is *private* if x is contained in a unique subenv V_i. Agent privacy on private variables is preserved, if no information on any private variable is disclosed to other agents, including its existence, identity, domain (of possible values), associated conditional or marginal probability distribution (in the case of probabilistic reasoning), or constraint (in the cases of constraint reasoning and decision making), and observed or assigned value.

A variable $x \in V$ is *shared* by A_i and A_j, if $x \in I_{ij}$. Agent privacy on shared variables is preserved, if no information on any shared variable is disclosed beyond agents who share it.

An agent is known to another agent, iff they share a border. Privacy on agent identities is preserved, if for every agent, its identity is not disclosed to any non-bordering agent.

The boundary set of a MAS can play an important role in privacy preserving construction of JT organizations, as established by a proposition from [13].

Proposition 1. *Let V be env of a MAS, Ω be its subenv decomposition, W be the boundary set, and T be a JT with boundaries in W as clusters. Let T' be a cluster tree with subenvs of Ω as clusters, such that it is isomorphic to T with each subenv cluster mapped to the corresponding boundary cluster in T. Then T' is a JT.*

Based on Prop. 1, we take the approach to construct a JT organization from the boundary set W, rather than from the subenv decomposition Ω. This approach guarantees agent privacy on private variables, because these variables are excluded from input of the task.

Given a boundary set, a JT made of its boundary clusters may or may not exist. Hence, the problem we tackle in this work is stated as follows. Given the boundary set of a MAS, determine whether a JT agent organization exists and, if so, construct a JT, such that agent privacy on private variables, shared variables, and agent identities are preserved in the process.

3   Boundary Graph Based Condition on Hypertree Existence

Before the existence of JT organization can be determined algorithmically, we analyze conditions of its existence, through an alternative representation of the boundary set. Let W be the boundary set of a MAS. An undirected graph BG is the *boundary graph* of the MAS, if its set of nodes is $N = \cup_{i=0}^{\eta-1} W_i$, and its links are connected so that each W_i is complete (elements are pairwise connected).

Prop. 2 identifies a condition under which a JT can be constructed from a boundary graph such that each JT cluster is a boundary. A set of nodes in a graph is a *clique*, if they are maximally pairwise connected. Two clusters are *comparable*, if one is a subset of the other.

Proposition 2. *Let W be the boundary set of a MAS, and BG be its boundary graph, such that*

1. *BG is chordal, and*
2. *for each clique C of BG, there exists $W_i \in W$ with $C \subseteq W_i$.*

Let T be a JT whose clusters are cliques of BG, and no two clusters in T are comparable. Then for every cluster C of T, there exists a boundary $W_i = C$.

Proof: From subcondition 1, the JT T exists. We prove by contradiction. Suppose there exists a cluster C in T such that $C \neq W_i$ for every $W_i \in W$.

From subcondition 2, there exists W_i such that $C \subseteq W_i$. Since $C \neq W_i$, it follows that $C \subset W_i$. Because BG is a boundary graph, W_i is complete in BG. Therefore, there exists a cluster C_i in T such that $W_i \subseteq C_i$. From $C \subset W_i$ and $W_i \subseteq C_i$, we have $C \subset C_i$. That is, T contains two comparable clusters: a contradiction. Hence, every cluster in T is a boundary.   □

Example 1. *Fig. 1 (a) shows an env decomposition. The set of boundaries is shown in (b). The BG is shown in (c), and it satisfies the two conditions. The JT from the BG is shown in (d), where the two clusters are boundaries W_1 and W_3.*

Utilizing Prop. 2, Theorem 1 establishes a necessary and sufficient condition for the existence of JT organization.

Theorem 1. *Let W be the boundary set of a MAS and BG be its boundary graph. A JT agent organization exists, iff the following hold.*

1. *BG is chordal, and*
2. *for each clique C of BG, there exists $W_i \in W$ such that $C \subseteq W_i$.*

Proof: [Necessity] Suppose a JT H exists, whose clusters are subenvs. For each cluster in H, remove its private variables. The resultant cluster tree T is still a JT, and its corresponding undirected graph is BG. From T being a JT, it follows that BG is chordal. Hence, subcondition 1 holds. The clusters of T are one-to-one mapped to boundaries of agents, from which subcondition 2 follows.

[Sufficiency] Suppose both subconditions hold. We prove by construction.

Since BG is chordal, a JT T exists whose clusters are cliques of BG. Without losing generality, assume that clusters of T are not comparable. By Prop. 2, every cluster in T is a boundary. Hence, for every cluster C such that $C = W_i$ for some i, we can associate C with an agent A_i.

If not every agent is associated with a cluster yet, consider a remaining agent A_i without being associated with any cluster of T yet. Since W_i is complete in BG, there exists a cluster C in T such that $W_i \subseteq C$. Add to T a new cluster W_i, make it adjacent to cluster C, and associate the new cluster with A_i. Repeat this for each remaining agent, until each agent is associated with a cluster in T.

Next, for each agent, add its private variables to its associated cluster in T. The resultant T is a JT agent organization with each cluster being a subenv. □

Theorem 1 provides the following insight. As far as the existence of JT organization is concerned, MAS subenv decompositions can be classified into three types.

Type 1 Boundary graphs are chordal, and their cliques are boundary contained.
Type 2 Boundary graphs are not chordal.
Type 3 Boundary graphs are chordal, but their cliques are not boundary contained.

Example 2. *The boundary graph for the subenv decomposition in Fig. 1 (a) is shown in (c). The subenv decomposition is type 1. The JT of the boundary graph is show in (d). The two clusters are associated with A_1 and A_3. For each of the three remaining agents, a cluster can be added to the JT. The JT organization is shown in (e).*

Example 3. *Fig. 2 (a) shows another subenv decomposition, with agent boundaries in (b) and BG in (c). Since BG is not chordal, the subenv decomposition is type 2. By Theorem 1, it has no JT organization.*

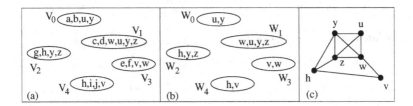

Fig. 2. (a) Subenv decomposition. (b) Agent boundaries. (c) Boundary graph.

Example 4. *For agent boundaries in Fig. 3 (a), the BG is shown in (b). It has two cliques. Since one of them, $\{h, v, w\}$, is not contained in any boundary, the subenv decomposition is type 3. By Theorem 1, it has no JT organization.*

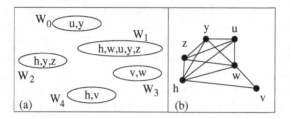

Fig. 3. (a) Agent boundaries. (b) Boundary graph.

4 Boundary Set Based Condition of Hypertree Existence

First, we define an operation to eliminate a boundary from a boundary set W. When a boundary $W_i \in W$ is *eliminated* from W *relative to* boundary $W_j \in W$, where $W_i \cap W_j \neq \emptyset$, it yields a *reduced boundary set* $W' = (W \setminus \{W_i, W_j\}) \cup \{W'_j\}$, where

$$W'_j = \bigcup_{W_k \in W, k \neq i, k \neq j} (W_j \cap W_k).$$

That is, the set W' resultant from eliminating W_i relative to W_j is obtained by deleting W_i and W_j from W, and replacing with W'_j. The W'_j is obtained by the union of borders of A_j, except the border with A_i. In other words, W'_j is the boundary W_j without variables that A_j uniquely shares with A_i. Consider the boundary set for Fig. 1 (b), $W = \{W_0, ..., W_4\}$. After W_0 is eliminated relative to W_1, the *reduced boundary set* is $W' = \{W'_1, W_2, W_3, W_4\}$, where $W'_1 = \{w, y, z\}$.

Without confusion, we refer to each element of W' as a *boundary*, whether or not it is identical to an element of the original boundary set. The elimination operation is well defined on the reduced boundary set, and hence can be performed iteratively.

Example 5. *For the boundary set of Fig. 1, elimination can be performed iteratively as follows.*

$$W = \{W_0 = \{u, y\}, W_1 = \{w, u, y, z\}, W_2 = \{y, z\}, W_3 = \{v, w\}, W_4 = \{v\}\};$$

$$Eliminate\ \{u, y\}\ wrt\ \{w, u, y, z\} : W' = \{\{w, y, z\}, \{y, z\}, \{v, w\}, \{v\}\};$$

$$Eliminate\ \{v\}\ wrt\ \{v, w\} : W' = \{\{w, y, z\}, \{y, z\}, \{w\}\};$$

$$Eliminate\ \{w\}\ wrt\ \{w, y, z\} : W' = \{\{y, z\}, \{y, z\}\};$$

$$Eliminate\ \{y, z\}\ wrt\ \{y, z\} : W' = \{\{y, z\}\}.$$

Note that, each W_i eliminated relative to a W_j has been so chosen to satisfy $W_i \subseteq W_j$. The significance of such a choice will be seen below.

Note also that each reduced boundary set W' (except the final singleton) is a well-defined boundary set, in the sense that each variable is shared by at least two boundaries in W'. Take $W' = \{\{w, y, z\}, \{y, z\}, \{w\}\}$ for example, each of w, y, and z is shared by two boundaries.

Next, we establish another necessary and sufficient condition on hypertree existence, based on boundary elimination.

Theorem 2. *A MAS with the boundary set W has a JT agent organization, iff W can be eliminated iteratively into a singleton, such that each W_i eliminated relative to a W_j satisfies $W_i \subseteq W_j$.*

Sketch of proof: For necessity, suppose a JT H exists. Remove private variables in each cluster. The resultant cluster tree T is a JT, whose set of clusters is W. A leaf cluster satisfying the condition can be found in T, and eliminated iteratively.

For sufficiency, suppose W can be eliminated into a singleton. Denote the sequence of reduced boundary sets as $W^\eta, W^{\eta-1}, ..., W^2, W^1$, where $W^\eta = W$, W^1 is the final singleton, and the superscript indicates the number of boundaries in the set. Boundaries in each W^x, for $x = 2, ..., \eta$, can be organized into a JT. $\square$

5 Distributed Recognition of Hypertree Existence

The condition $W_i \subseteq W_j$ in Theorem 2 is equivalent to $W_i = I_{ij}$. Hence, Theorem 2 suggests a privacy preserving, distributed computation to identify JT organization existence. Agents are self-eliminated one by one as long as possible. An agent A_i can be eliminated if its boundary is equal to the border with another remaining agent A_j. After A_i is eliminated relative to A_j, A_j removes from its boundary the variables that it shares uniquely with A_i. If all agents are eliminated except one, then a JT organization exists for the MAS. Otherwise, the JT does not exist.

We assume that a token is passed between bordering agents, according to depth-first-traversal. The first round of traversal starts at an arbitrary agent, who possesses the token tok^1. If an agent A_i who holds the token has its boundary equal to the border with another agent A_j, then A_i signifies to each bordering agent that it is eliminated, and passes a new token tok^2 to A_j.

A_j then starts the second round of traversal among remaining agents, using tok^2. If an agent starts a new round of traversal, and finds that it has no unelimited bordering agent, then it announces existence of a JT organization.

On the other hand, suppose an agent A_j starts a new round of traversal, with at least another unelimited bordering agent. After the token has traversed every unelimited agent, and comes back to A_j, if A_j still has unelimited bordering agents, then A_j announces non-existence of JT organization.

Example 6. *Consider agents in Fig. 4 (a) with their boundaries shown in ovals, where each link shows a bordering relation. Note that the subenv decomposition is type 1.*

Suppose A_0 starts first round with tok^1. It announces its elimination and passes the token to A_1. In response, A_1 reduces its boundary, as in (b), and starts second round with tok^2. It passes tok^2 to A_2. A_2 announces its elimination and passes tok^3 to A_1. In response, A_1 reduces its boundary again, as in (c), and starts third round. It announces its elimination and passes tok^4 to A_3. In response, A_3 reduces its boundary, as in (d), and starts fourth round. It announces its elimination and passes tok^5 to A_4. Finally, A_4 announces existence of a JT organization.

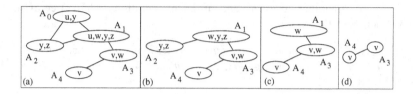

Fig. 4. Distributed recognition of JT organization with type 1 subenv decomposition

Example 7. *Consider agents and their boundaries in Fig. 5 (a). Note that the subenv decomposition is type 2.*

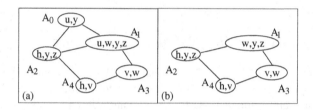

Fig. 5. Recognition of non-existence of JT organization with type 2 subenv decomposition

Suppose A_0 starts first round with tok^1, announces its elimination, and passes tok^2 to A_1. A_1 reduces its boundary, as in (b), starts second round, and passes tok^2 to A_2. A_2 passes tok^2 to A_4, who in turn passes tok^2 to A_3. A_3 has no unvisited agent to pass the token to, and returns tok^2 to A_4. A_4 returns tok^2 to A_2, who in turn returns to A_1. Finally, A_1 announces non-existence of JT organization.

Example 8. *Consider agents and their boundaries in Fig. 6 (a). Note that the subenv decomposition is type 3.*

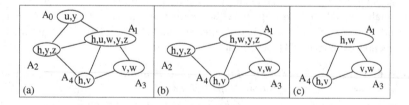

Fig. 6. Recognition of non-existence of JT organization with type 3 subenv decomposition

Suppose A_0 starts first round with tok^1, and passes tok^2 to A_1. A_1 reduces its boundary, as in (b), starts second round, and passes tok^2 to A_2. A_2 announces its elimination and passes tok^3 to A_1. A_1 further reduces its boundary, as in (c), starts third round, and passes tok^3 to A_3. A_3 passes tok^3 to A_4. Eventually, tok^3 is returned to A_1, who announces non-existence of JT organization.

Note that during the traversal, although the *active boundary* for a remaining agent may be reduced, the border between any pair of agents never changes.

6 Algorithm for Hypertree Existence Recognition and Construction

Next, we specify a distributed algorithm suite, that agents execute to implement the computation described intuitively in the previous section. Each agent's activities are driven by responding to the following messages.

- A *StartNewDFT*(*tok*) request that calls the receiver to start a new round of depth-first-traversal with the given token *tok*;
- An *Eliminated* notification sent by an agent who has been self-eliminated;
- A *DFT*(*tok*) request that calls the receiver to perform depth-first-traversal with the given token *tok*;
- A *Report* message sent by an agent who has been called to perform *DFT*(*tok*), signifying either the called agent has been visited in the current round, or it has completed DFT and now backtracks to the caller.

We refer to the receiving agent of a message by A_i, who will act in response, and refer to the message sender by A_c. Every agent performs *Init* to initialize local data. Flag *state* $\in \{IN, OUT\}$ indicates whether A_i has been eliminated. Flag $nbsta(A_k) \in \{IN, OUT\}$ indicates the same for a bordering agent. Variable *curtok* keeps a token value after it has visited A_i, and $visited(A_k)$ indicates whether the token has visited the bordering agent. Y_i maintains the active boundary of A_i.

Procedure 1 (Init)
1   state = IN; parent = null;
2   initialize current token to curtok = null;
3   set active boundary $Y_i = W_i$;
4   for each bordering agent A_k,
5       $nbsta(A_k) = IN$;
6       $visited(A_k) = false$;

At the start of each round of traversal, a remaining agent will be be messaged to *StartNewDFT*. In the first round, an arbitrary *leader* agent messages itself. Agent A_i being messaged does the following.

Procedure 2 (StartNewDFT(tok))
1   if A_c is another agent,
2       $nbsta(A_c) = OUT$;
3   if there exists no A_j with $nbsta(A_j) = IN$,
4   announce "hypertree exists";

5   *return;*
6    $Y_i = \emptyset;$
7   *for each bordering* A_k *where* $nbsta(A_k) = IN,$
8        $Y_i = Y_i \cup I_{ik};$
9 $curtok = tok;$ *parent* $= null;$
10 *run DFT;* // A_i *has IN bordering agents*

When *DFT* below is run from *StartNewDFT*(*tok*), A_i has *parent* $= null$, and have at least one remaining bordering agent A_j. If A_i can be eliminated relative to A_j, it will message A_j to *StartNewDFT*. Otherwise, it will send message *DFT*(*tok*) to A_j.

When *DFT* is run from *DFT*(*tok*) (see below), A_i has *parent* pointing to A_c, and may have no unvisited, remaining bordering agent other than A_c. If A_i cannot be eliminated relative to A_c, it must *Report* to A_c.

Procedure 3 (DFT)
1 *if there exists* A_j *with* $nbsta(A_j) = IN$ *and* $Y_i = I_{ij},$ // *self-eliminate*
2   *state* $= OUT;$
3   *for each* $A_k \neq A_j$ *where* $nbsta(A_k) = IN,$ *send Eliminated to* $A_k;$
4   *send StartNewDFT*(*curtok* $+ 1$) *to* $A_j;$
5 *else* // *no IN agent satisfies* $Y_i = I_{ij}$
6   *for each* $A_k \neq$ *parent where* $nbsta(A_k) = IN,$ *set visited*$(A_k) = false;$
7   *if there exists* $A_k \neq$ *parent where* $nbsta(A_k) = IN$ *and visited*$(A_k) = false,$
8   *send* A_k *message DFT*(*curtok*);
9   *else send Report to parent;*

When a remaining agent A_i receives from A_c the *Eliminated* message, it responds by setting its $nbsta(A_c) = OUT$. When a remaining agent A_i receives from A_c the *DFT*(*tok*) message, it performs the following.

Procedure 4 (DFT(tok))
1 *if curtok* $= tok,$ // *visited*
2   *send Report to* $A_c;$
3 *else* // A_i *has not seen tok before and has IN bordering agent other than* A_c
4   *curtok* $= tok;$
5   *parent* $= A_c;$ *visited*$(A_c) = true;$
6   *run DFT;*

After A_i has messaged A_j with *DFT*(*tok*), it may receive a *Report* from A_j. In response, A_i performs the following.

Procedure 5 (Respond to Report)
1 *visited*$(A_j) = true;$
2 *if there exists* $A_k \neq$ *parent such that* $nbsta(A_k) = IN$ *and visited*$(A_k) = false,$
3   *select* A_k *to send message DFT*(*curtok*) *to it;*
4 *else* // *no unvisited bordering agent*
5   *if parent* $= null,$ *announce "no hypertree exists";* // *DFT starter*
6   *else send Report to parent;*

The algorithm suite, we refer to as HTBS, terminates when a remaining agent announces "hypertree exists" or otherwise. Its soundness and completeness is established below, which follows from Theorem 2.

Corollary 1. *A MAS with the boundary set W has a JT agent organization, iff HTBS terminates with announcement "hypertree exists". Otherwise, HTBS terminates with announcement "no hypertree exists".*

An important product of HTBS is the JT organization emerging upon positive announcement. For every agent A_i self-eliminated relative to A_j, A_i is the sender of *StartNewDFT* message and A_j is the receiver. This relation implies that they are adjacent in the JT organization. Readers are encouraged to verify this by comparing Example 6 with Fig. 1 (e). This result is summarized below, whose proof is omitted due to space limitation.

Theorem 3. *If a MAS with the boundary set W has a JT agent organization, then agent adjacency in the JT is defined by StartNewDFT sender-receiver relation during HTBS.*

Let e be the number of pairs of bordering agents. In each round of HTBS, at most $O(e)$ messages are passed. HTBS halts in at most $O(\eta)$ rounds. Hence, its time complexity is $O(e\,\eta)$.

Since HTBS is based on boundary set, agent privacy on private variables is guaranteed. Since HTBS messages contains no information on shared variables, agent privacy on shared variables is guaranteed. Since HTBS messages are passed between bordering agents only, and the message argument is a token only, privacy on agent identity is guaranteed.

7   Conclusion

The contributions of this research include the following. We proved two necessary and sufficient conditions for the existence of a JT organization given a MAS subenv decomposition. One of them provides insight and classification of subenv decompositions, and the other suggests a distributed reorganization of JT organization existence. Based on the second condition, we have presented an algorithm suite that recognize JT organization existence and construct it if exists. The algorithm guarantees agent privacy on private variables, shared variables, as well as agent identity. To the best of our knowledge, no existing JT-based MAS frameworks provide the same degree of agent privacy, except a related work based on distributed maximum spanning tree construction which we report in [13].

Our algorithm identifies correctly when no JT organization exists for a given subenv decomposition. Further research is needed for distributed revision of the subenv decomposition under the condition where no JT exists, while preserving agent privacy as much as possible.

Acknowledgement. We thank anonymous reviewers for their helpful comments. Financial support through Discovery Grant from NSERC, Canada is acknowledged.

References

1. Brito, I., Meseguer, P.: Cluster tree elimination for distributed constraint optimization with quality guarantees. Fundamenta Informaticae 102(3-4), 263–286 (2010)
2. Gmytrasiewicz, P., Durfee, E.: Rational communication in multi-agent environments. Auto. Agents and Multi-Agent Systems 4(3), 233–272 (2001)
3. Koller, D., Milch, B.: Multi-agent influence diagrams for representing and solving games. In: Proc. 17th Inter. Joint Conf. on Artificial Intelligence, pp. 1027–1034 (2001)
4. Maestre, A., Bessiere, C.: Improving asynchronous backtracking for dealing with complex local problems. In: Proc. 16th European Conf. on Artificial Intelligence, pp. 206–210 (2004)
5. Modi, P., Shen, W., Tambe, M., Yokoo, M.: Adopt: asynchronous distributed constraint optimization with quality guarantees. Artificial Intelligences 161(1-2), 149–180 (2005)
6. Paskin, M., Guestrin, C., McFadden, J.: A robust architecture for distributed inference in sensor networks. In: Proc. Information Processing in Sensor Networks, pp. 55–62 (2005)
7. Petcu, A., Faltings, B.: A scalable method for multiagent constraint optimization. In: Proc. 19th Inter. Joint Conf. on Artificial Intelligence, pp. 266–271 (2005)
8. Valtorta, M., Kim, Y., Vomlel, J.: Soft evidential update for probabilistic multiagent systems. Int. J. Approximate Reasoning 29(1), 71–106 (2002)
9. Vinyals, M., Rodriguez-Aguilar, J., Cerquides, J.: Constructing a unifying theory of dynamic programming DCOP algorithms via the generalized distributive law. J. Autonomous Agents and Multi-Agent Systems 22(3), 439–464 (2010)
10. Xiang, Y.: Probabilistic Reasoning in Multiagent Systems: A Graphical Models Approach. Cambridge University Press, Cambridge (2002)
11. Xiang, Y., Hanshar, F.: Multiagent expedition with graphical models. Inter. J. Uncertainty, Fuzziness and Knowledge-Based Systems 19(6), 939–976 (2011)
12. Xiang, Y., Mohamed, Y., Zhang, W.: Distributed constraint satisfaction with multiply sectioned constraint networks. accepted to appear in International J. Information and Decision Sciences (2013)
13. Xiang, Y., Srinivasan, K.: Construction of privacy preserving hypertree agent organization as distributed maximum spanning tree. In: Zaïane, O., Zilles, S. (eds.) Canadian AI 2013. LNCS (LNAI), vol. 7884, pp. 199–210. Springer, Heidelberg (2013)

Construction of Privacy Preserving Hypertree Agent Organization as Distributed Maximum Spanning Tree

Yang Xiang and Kamala Srinivasan

School of Computer Science, University of Guelph, Canada

Abstract. Decentralized probabilistic reasoning, constraint reasoning, and decision theoretic reasoning are some of the essential tasks of a multiagent system (MAS). Many frameworks exist for these tasks, and a number of them organize agents into a junction tree (JT). Although these frameworks all reap benefits of communication efficiency and inferential soundness from the JT organization, their potential capacity on agent privacy has not been realized fully. The contribution of this work is a general approach to construct the JT organization through a maximum spanning tree (MST), and a new distributed MST algorithm, that preserve agent privacy on private variables, shared variables and agent identities.

1 Introduction

Decentralized probabilistic reasoning, constraint reasoning, and decision theoretic reasoning (referred to below as decision making) are some of the essential tasks of a multiagent system (MAS). Many frameworks exist for these tasks, e.g., AEBN [19] and MSBN [22] for multiagent probabilistic reasoning, ABT [12], ADOPT [13], DPOP [16], Action-GDL [20], DCTE [2], and MSCN [25] for multiagent constraint reasoning, and RMM [7], MAID [10], and CDN [24] for multiagent decision making. Some frameworks do not assume specific agent organizational structure, e.g., AEBN and MAID. Some assume a total order among agents, e.g., ABT. Some use a pseudo-tree organization, e.g., ADOPT and DPOP. Some depend on a junction tree (JT) organization, e.g., Action-GDL, DCTE, MSBN, MSCN, and CDN. The organization is known as *hypertree* in the literature on MSBN, MSCN and CDN, and we refer to JT and hypertree interchangeably.

This work focuses on JT-based agent organizations. Potential advantages of JT organizations include communication efficiency, inferential soundness, and agent privacy. As MAS research progresses, agent privacy has received more attention in recent years [18,4]. However, few studies are known on the privacy issue related to JT-based agent organization (see [22,25] for example). Although existing JT-based MAS frameworks all reap benefits of communication efficiency and inferential soundness, the potential capacity of JT-based organization on agent privacy has not been fully realized.

A framework component critical to agent privacy is construction of the JT-based organization. Some frameworks paid no attention to agent privacy at all during the JT organization construction, and others preserved agent privacy to a limited degree. The main contribution of this work is a new algorithm suite that constructs JT-based agent organizations distributively while preserving agent privacy. To the best of our knowledge, no known JT-based MAS frameworks provide the same degree of agent privacy

O. Zaïane and S. Zilles (Eds.): Canadian AI 2013, LNAI 7884, pp. 199–210, 2013.
© Springer-Verlag Berlin Heidelberg 2013

enabled by the proposed algorithm, except an alternative approach that we report in a related work [26].

In the next section, we introduce background on JT-based organization and the related privacy issue. In the subsequent section, we describe our approach to JT construction as distributed construction of a maximum spanning tree (MST). After reviewing relevant work on distributed MSTs, we present our new algorithm suite, followed by an analysis of its soundness, complexity, and privacy.

2   JT-Based Multiagent Organization and Agent Privacy

JT-based organizations are used in a number of multiagent frameworks, e.g., [21,23,27,20,2]. Under these frameworks, the application environment is represented by a set of variables, referred to as the *env*. The env is decomposed into a set of overlapping *subenvs*, each being a subset of env. The subenvs are one-to-one mapped to agents of the MAS. Fig. 1 (a) shows the subenv decomposition of a trivial env, where subenv V_i is mapped to agent A_i.

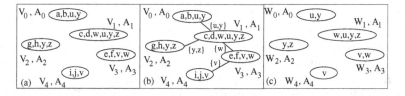

Fig. 1. (a) Subenvs with variables shown in each oval, and their agent association; (b) JT organization with links labeled by shared variables; (c) Agent boundaries.

The subenvs (and hence the agents) are organized into a JT. A JT is a tree where each node is associated with a set called a *cluster*. The tree is so structured that, for any two clusters, their intersection is contained in every cluster on the path between the two (*running intersection*). In a JT-based organization, each cluster corresponds to a subenv and hence an agent. The organization prescribes direct communication pathways between agents. That is, an agent can send a message to another agent, iff they are adjacent in the JT organization.

For each pair of adjacent agents in the JT organization, the intersection of their subenvs is non-empty, and it is used to label the link between corresponding clusters. We refer to the intersection as the set of variables *shared* by the two agents. Fig. 1 (b) shows a JT organization.

Messages between adjacent agents are restricted to be about shared variables only (which is sufficient). In multiagent probabilistic reasoning, a message is the sending agent's belief over shared variables, e.g., in MSBN [21,22]. In multiagent constraint reasoning, a message contains partial solutions over shared variables, e.g., in MSCN [27,25]. In multiagent decision making, a message is either an expected utility function over shared variables, or a partial action plan over them, e.g., in CDN [23,24].

MAS frameworks using JT-based organizations can be grouped based on whether subenvs are assumed *simple* or *complex*. Subenvs are assumed simple under a framework, if each subenv is treated as a subset of variables without internal structure. Subenvs are assumed complex under a framework, if dependence structure within each subenv is explicitly represented and manipulated during multiagent probabilistic reasoning, constraint reasoning, or decision making. Under this criterion, subenvs in Action-GDL [20] and DCTE [2] are simple, while subenvs in MSBN, MSCN, and CDN are complex. In particular, each subenv in a MSBN is modeled as a Bayesian subnet, each subenv in a MSCN is encoded as a constraint subnet, and each subenv in a CDN is represented as a decision subnet.

JT-based agent organizations enable a number of advantages.

Efficient Communication. For agents to cooperate using relevant information local in other agents, it is sufficient to pass two messages along each link of the JT. Hence, time complexity of communication is linear in the number of agents, which is derived from the tree topology of JT.

Soundness of Inference. Global consistency is guaranteed by local consistency, which is derived from running intersection of JT. For instance, multiagent probabilistic reasoning in MSBN [22] is exact. Multiagent decision making in CDN [24] is globally optimal.

Agent Privacy. Most information about individual agents can be kept private while the MAS is fully functioning. This is derived from the restriction of message content in JT-based communication.

Agent privacy above is desirable, when each agent represents an independent principal. For instance, agents collaborating in industrial design with a CDN may represent independent manufacturers in a supply chain [23]. The subenv associated with an agent, modeled as a decision subnet, contains proprietary technical know-hows of a manufacturer, whose non-disclosure is desirable.

More precisely, in a JT-based organization, certain information at individual agents does not need to be exchanged during normal inference, and thus can potentially be kept private. Such information includes the following.

Information Related to a Private Variable. An env variable $x \in V$ is *private*, if it is contained in a single subenv V_i, and hence is associated with a single agent A_i. The information includes its existence, identity, domain (of possible values), associated conditional or marginal probability distribution (in the case of probabilistic reasoning), or constraint (in the cases of constraint reasoning and decision making), and observed or assigned value. In Fig. 1 (a), h is a private variable of A_2.

Information Related to a Shared Variable. An env variable $x \in V$ is *shared*, if it is contained in two or more subenvs, and hence associated with two or more agents. The information about x, as listed above, does not need to be released beyond the agents who share x. In Fig. 1 (a), y is a variable shared by A_0, A_1 and A_2.

Existence and Identity of a Non-bordering Agent. Two agents are *non-bordering* if their subenvs have no common variables. They will not be adjacent in any JT organization, and need not communicate during inference. Hence, they do not need to know the existence or identity of each other. In Fig. 1 (b), A_0 and A_4 are non-bordering.

In summary, agent privacy in a JT-based organization involves at least three types: privacy of private variables, privacy of shared variables, and privacy of agent identity. Although agent privacy is relevant to both MAS frameworks over complex subenvs and those over simple subenvs, it is particularly important for frameworks with complex subenvs, as a simple subenv may contain only a few variables while a complex subenv may include tens or hundreds or more private variables.

For any given JT-based MAS framework, a critical component related to privacy is the construction of JT organization. This is because, once the JT organization is constructed and functioning, it is relatively easy to maintain privacy during normal inference operations, due to message content restriction. It is through the JT construction component, JT-based MAS frameworks demonstrate different degrees of attention to the privacy issue.

In Action-GDL [20], the JT is built from a pseudo-tree, where each node corresponds to one env variable, through a centralized mapping operation. The centralized mapping operation necessarily discloses identities of all variables to the agent who performs the mapping.

The JT used by DCTE [2] is constructed by a method from [15], where agents are initially organized into a tree topology. Each agent starts with a cluster of variables determined by application-based conditions. Hence, the cluster tree generally does not satisfy running intersection property and is not a JT. The cluster tree is transformed into a JT through a message passing process, during which each agent communicates, to each adjacent agent, its local variables as well as variables reachable from other adjacent agents. The message passing thus propagates identities of variables well beyond the agent initially associated with them.

For MSBN, MSCN and CDN frameworks, the JT-based organization is constructed by a centralized coordinator agent [22,25], who has the knowledge of variables shared by any pair of agents, but does not know any private variable. Hence, privacy of private variables is ensured. However, shared variables and identities of non-bordering agents are disclosed to the coordinator, and the corresponding types of privacy are compromised. In this work, we develop a new distributed JT algorithm that respects all three types of privacy.

3 Distributed JT Construction as MST

Consider a MAS consisting of a set $\mathscr{A} = \{A_0, ..., A_{\eta-1}\}$ of $\eta > 1$ agents. Let V be the set of env variables, and be decomposed into a collection of *subenvs*, $\Omega = \{V_0, ..., V_{\eta-1}\}$, such that $\cup_{i=0}^{\eta-1} V_i = V$. Agents in $\mathscr{A}$ are one-to-one mapped into subenvs in Ω. If A_i and A_j have shared variables, i.e., $V_i \cap V_j \neq \emptyset$, we refer to the set of shared variables, $I_{ij} = V_i \cap V_j$ ($i \neq j$), as their *border*. A JT-based agent organization is a cluster tree, where each cluster is a subenv and each link is labeled by a border, such that running intersection property holds. Fig. 1 (b) shows a JT organization, where the border between A_1 and A_2 is $I_{12} = \{y, z\}$.

In [22], it is observed that all private variables of a subenv in the above representation can be congregated into a single private variable. If a JT can be constructed using the congregated subenvs, by replacing the congregated private variable with the original private variables, the new cluster tree is still a valid JT.

In this work, we go one step further and observe that the congregated private variable is not needed. For each agent A_i, we denote the set $W_i = \cup_{j \neq i} I_{ij}$ as its *boundary*. We refer to $W = \{W_0, ..., W_{n-1}\}$ as the *boundary set* of the MAS. Fig. 1 (c) illustrates agent boundaries. The following Proposition establishes the fact that JT-based organization can be investigated without reference to private variables, whose proof is straightforward.

Proposition 1. *Let V be env of a MAS, Ω be its subenv decomposition, W be the boundary set, and T be a JT with boundaries in W as clusters. Let T' be a cluster tree with subenvs of Ω as clusters, such that it is isomorphic to T with each subenv cluster mapped to the corresponding boundary cluster in T. Then T' is a JT.*

Based on Prop. 1, we can construct a JT-based organization based on the boundary set. Note that this task formulation on JT organization construction immediately guarantees privacy of private variables, as they have been excluded from the specification of the task.

Given the boundary set of an MAS, a JT may or may not exist. Figure 2 (a) shows a boundary set whose elements cannot be organized into a JT.

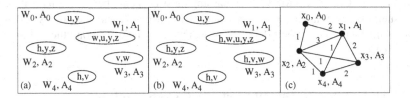

Fig. 2. (a) A boundary set that does not admit JT organization; (b) A boundary set that has a JT organization; (c) Weighted graph defined from the boundary set in (b)

In this work, we assume that there exists a JT for the given boundary set, and we focus on how to distributively construct a JT organization from the boundary set while preserving agent privacy. In a related work [26], we consider how to identify the existence of a JT organization distributively given a boundary set without disclosing agent privacy.

The task of distributed construction of JT-based agent organization can now be stated as follows.

- A set $\mathscr{A} = \{A_0, ..., A_{n-1}\}$ of agents is associated with the boundary set $W = \{W_0, ..., W_{n-1}\}$, such that a JT exists, with elements of W as its clusters.
- Agents A_i and A_j know the identity of each other and can exchange messages, iff they have a border $I_{ij} = W_i \cap W_j \neq \emptyset$.

– Agent A_i knows only $W_i = \cup_{j \neq i} I_{ij}$, and nothing about variables in other boundaries beyond W_i.

The task of agents is to compute a JT organization with boundaries as clusters, such that each agent knows its adjacent agents in the JT, and the process does not disclose information on agent boundaries beyond the initial knowledge state. To solve this problem, we explore the relation between JT and maximum spanning tree (MST) as follows.

– From a boundary set W, define a weighted graph Ψ. For each $W_i \in W$, create a node x_i. Add a link $\langle x_i, x_j \rangle$, iff there is a border between W_i and W_j, and the weight of the link is $w(x_i, x_j) = |I_{ij}|$.
– Let Ψ' be any MST of Ψ. Define a cluster tree T from Ψ', such that for each link $\langle x_i, x_j \rangle$ of Ψ', W_i and W_j are adjacent in T.
– Then T is a JT, iff a JT with elements of W as its clusters exists [8,22].

Fig. 2 (c) illustrates Ψ defined from a boundary set in (b). Note that the above formulation of JT organization as a MST immediately guarantees privacy of shared variables, since the input to MST computation contains only the number of shared variables between each pair of bordering agents, with other information about these variables excluded.

From the relation between JT and MST, the task of distributed construction of JT organization can be cast as follows. Let the weighted graph Ψ be defined distributively, namely, each agent A_i is associated with node x_i, and knows A_j and $w(x_i, x_j)$ iff $\langle x_i, x_j \rangle \in \Psi$. Compute a MST Ψ' by passing messages only between adjacent agents in Ψ, such that each agent knows its adjacent agent in Ψ', and the process does not disclose information on agent identity and weight association beyond the initial agent knowledge state.

4   Work Related to Distributed MST Construction

Before presenting our privacy preserving distributed MST algorithm, we review relevant literature. Since minimum or maximum spanning trees differ only in the comparison operator (*Min* versus *Max*), without confusion, we refer to all of them by MST.

In the pioneering work by Gallager et al. [5], MST fragments (each initially made of a single node) are combined into larger ones according to a level control, until a single fragment is formed. It has a time complexity $O(\eta \log \eta)$. The basic algorithm assumes distinct link weights, which generally does not hold in our application. To accommodate nondistinct link weights, the modified algorithm either appends node identities to link weights or identifies fragments by node identities. Since link weights and fragment identifies are propagated through messages, node identities will be disclosed beyond node adjacency.

Awerbuch [1] proposed a three-stage algorithm, which was later improved by [3], with time complexity $O(\eta)$. It starts with a counting stage to get η. Then the algorithm in [5] is run to grow each fragment to an $\Omega(\eta/\log \eta)$ size. A variant of [5] follows, with a more accurate level updating to speed up computation. Non-distinct link weights are handled using the same technique as [5], appending node identities to link weights. Hence, the method suffers from the same node identity disclosure as [5].

An improved algorithm with time complexity $O(d + \eta^{0.613} \log * \eta)$ is proposed in [6], where d is diameter (maximum length of a simple path) of the weighted graph. It first uses a variant of [5] to produce multiple fragments of small diameters, and then combines them into a MST by a rooted operation. Its limitation on node identity disclosure is identical to [5] and [1].

In [11], a two-part algorithm is developed. In the first part, a $\sqrt{\eta}$-dominating set D of size at most $\sqrt{\eta}$ is computed, as well as a partition of the weighted graph into fragments one per node in D. The second part combines these fragments into a MST by the same rooted operation as [6]. Since the first part employs a simplified version of [5], its limitation on node identity disclosure is identical to the above algorithms.

An approximate MST algorithm is presented in [9]. Due to the necessary and sufficient condition between JT and MST, an approximate MST cannot yield a JT organization, and hence the method is not applicable to our task.

Recently, an algorithm for computing a set of MSTs, one for each component of a disconnected graph, was proposed [14]. As a parallel algorithm, access of the entire graph by each processor is assumed, and hence it is applicable only when privacy is not a concern at all.

In summary, existing work on distributed MST construction has largely ignored the issue of node identity disclosure.

5 Distributed MST Construction with Privacy of Agent Identity

In this section, we present a new distributed algorithm for MST construction that does not disclose node identity to non-adjacent nodes. In the context of JT organization construction, this translates into non-disclosure of agent identity to non-bordering agents. For this purpose of privacy preservation, we take a different approach than [5] and its extensions [1,3,6,11]. Rather than growing multiple fragments simultaneously, we extend Prim's algorithm [17] distributively and grow a MST through a rooted control. As the result, our algorithm does not assume distinct link weights and needs not to append node identities to link weights either. As will be shown below, our algorithm ensures that no node identity is disclosed beyond adjacency.

Precisely stated, the task is as follows. Given a distributed representation of a connected, weighted graph Ψ of η nodes, construct a MST T by distributed computation. As each node is associated with an agent, without confusion, we refer to node and agent interchangeably. Distributed Ψ representation means that each node has the initial knowledge about each adjacent node (called *neighbor* or *nb*) and the link weight. It knows nothing about the existence of other nodes nor the links and weights between them. For any node v and a nb x of v, the weight of link $\langle v, x \rangle$ is $w(v, x)$.

The algorithm initializes T with an arbitrary node, referred to as the *root* and builds T up as a *directed*, single-rooted tree in $\eta - 1$ rounds. An *outgoing* link of T is a link of Ψ with only one end in T. In each round, a best outgoing link $\langle p, c \rangle$ is selected, where p is in T, and c is added to T. We refer to p as the *tree-parent* of c, and c as a *tree-child* of p. For any node in T, we refer to its tree-parent or a tree-child as its *tree-nb*.

In addition to the initial knowledge, each node v maintains the following local data structure.

1. The state of v is indicated by variable $state \in \{OUT, IN, DONE\}$. Value OUT means that v is not yet in T. IN means that v is in T, but not yet finished its computation. $DONE$ means that v is in T and has finished its computation.
2. Knowledge of v on the state of each nb x is maintained by variable $nbstate(x) \in \{OUT, IN, DONE\}$.
3. The *tree-parent* of v in T is indicated by a pointer so named.
4. A *best outgoing weight table* (BOWT) is maintained. Each row is indexed by a nb x of v, that may lead to outgoing links, and contains the best weight of these links known to v, denoted by $bw(x)$.

During MST computation, nbs of Ψ exchange four types of messages.

Announce Sender announces to each nb, that the former is in MST.

Expand A tree-parent instructs a tree-child to expand current MST, by finding a new node to add.

Notify A tree-leaf notifies a nb that the latter is added to current MST.

Report A tree-child sends to its tree-parent, either to report its termination, or to report the best outgoing weight via the tree-child, through an argument.

We assume that transmission of each message takes at most one time unit.

When the algorithm starts, every node v in Ψ runs *Init* to initialize local data structure.

Procedure 1 (Init)

1   state = OUT, tree-parent pointer = null;
2   for each nb x, nbstate(x) = OUT;
3   create BOWT with one row per nb;
4   for each row of BOWT indexed by x, bw(x) = w(v,x);

An arbitrary node is elected as the *root*. It starts the MST computation by executing *Start*. It first adds itself to T, and then runs *Expand* to expand T.

Procedure 2 (Start)

1   state = IN;
2   send Announce message to each nb;
3   run Expand;

Proc. *Expand* can either be called (as in *Start*), or run in response to an *Expand* message. The node selects a nb y that leads to a best outgoing link, adds y to T if y is OUT, otherwise asks y to expand T.

Procedure 3 (Expand)

1   select nb y = arg max$_x$ bw(x) from BOWT table, breaking ties randomly;
2   if nbstate(y) = OUT,
3   send Notify message to y;
4   nbstate(y) = IN; record y as a tree-child;
5   else send Expand message to y; // y must be IN

When node v receives *Notify* message from nb p, it is in T. It runs Proc. 4 in response. In the process, it announces its status in T to its nbs. At the end of the process, v replies to p with a *Report* message, which contains one of two possible arguments, the *state* value of v, or the best outgoing weight *bow*.

Procedure 4 (Response to Notify)
1 $nbstate(p) = IN$; point tree-parent pointer to p;
2 delete the row indexed by p from *BOWT*;
3 $state = IN$;
4 for each nb $x \neq p$, send Announce message to x;
5 run Inform;

A node v runs *Inform* to send *Report* message to its tree-parent p.

Procedure 5 (Inform)
1 if no nb y with $nbstate(y) = OUT$
         and each tree-child c has $nbstate(c) = DONE$,
2      $state = DONE$;
3   if v is not root, send p message $Report(state = DONE)$;
4 else if v is not root,
5   compute $maxbow = \max_x bw(x)$ from *BOWT*;
6   send p message $Report(bow = maxbow)$;

When a node v receives *Announce* message from a nb x, it performs the following. Note that root cannot receive *Announce* from its tree-child, but can receive from its non-child tree-descendent.

Procedure 6 (Response to Announce)
1 $nbstate(x) = IN$; delete the row indexed by x from *BOWT*;
2 if $state \neq OUT$, run Inform;

When a tree-parent v receives *Report* from a tree-child c, it performs Proc. 7. The report allows v to know whether c has terminated and, if not, to update its knowledge on the best outgoing link weight through c.

Procedure 7 (Response to Report)
1 if message argument is $(state = DONE)$,
2      $nbstate(c) = DONE$; delete the row indexed by c from *BOWT*;
3   if no nb y with $nbstate(y) = OUT$
             and each tree-child c has $nbstate(c) = DONE$,
4          $state = DONE$;
5   if v is not root, send tree-parent p message $Report(state = DONE)$;
6   else return; // root termination
7 else if $maxbow \neq bw(c)$ in *BOWT*, // argument is $(bow = maxbow)$
8      $bw(c) = maxbow$;
9 if v is not root,
10   compute $maxbow' = \max_x bw(x)$ from *BOWT*;
11   send tree-parent p message $Report(bow = maxbow')$;
12 else if no pending Report messages, run Expand; // v is root

When Proc. 7 returns from line 6 at root, the algorithm suite halts. Consider the example in Fig. 2 (b). Suppose x_0 is the root, whose $BOWT$ at start is $(x_1 : 2; x_2 : 1)$. It sends $Announce$ to x_2 and $Notify$ to x_1. Node x_1 removes x_0 from its $BOWT$, sends $Announce$ to x_2, x_3 and x_4, and $Report(bow = 3)$ to x_0. Based on the report, x_0 revises its $BOWT$ to $(x_1 : 3; x_2 : 1)$, and sends $Expand$ to x_1.

$BOWT$ at x_1 is $(x_2 : 3; x_3 : 2; x_4 : 1)$. Hence, x_1 sends $Notify$ to x_2, which replies with $Report(bow = 1)$. Based on the report, x_1 sends $Report(bow = 2)$ to x_0. After x_0 has processed $Announce$ from x_2 and $Report(bow = 2)$ from x_1, its $BOWT$ is $(x_1 : 2)$. Hence, x_0 sends $Expand$ to x_1, which in turn sends $Notify$ to x_3.

Eventually, x_4 is notified by x_3 and sets its state to $DONE$. When x_2 receives $Announce$ from x_4, it sets its state to $DONE$ as well. Node x_3 sets $state = DONE$ when it receives $Report$ from x_4, and x_1 does so upon receiving $Report$ from x_3. In the end, x_0 receives $Report$ from x_1 and terminates the computation.

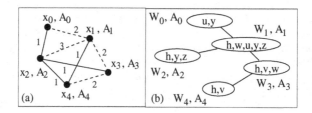

Fig. 3. (a) The MST (dashed links) from Fig. 2 (c); (b) The JT organization.

The resultant MST is shown in Fig. 3 (a) by dashed links, and the corresponding JT organization is in (b). Throughout the computation, no agent identity is communicated.

6   Soundness and Complexity

We refer to the algorithm suite as DPMST, whose soundness is established below.

Proposition 2. *Given a connected, weighted graph Ψ, DPMST computes a MST T of Ψ that is specified distributively, such that each node knows its tree-nbs.*

Proof: DPMST will compute a MST because it extends Prim's algorithm distributively. The recursive executions of Proc. 7 let the root know where an outgoing link with the best weight is located, and recursive executions of Proc. 3 add the other end of the link to T.

When a node v is added to T, it knows its notifier as its tree-parent, and its notifier knows v as its tree-child. Hence, when DSMSTC halts, each node knows its tree-nbs in T.   □

We analyze the communication cost and time complexity below. Let d denote the *diameter* of Ψ, e denote the number of links, and r denote the maximum degree of nodes.

Communication cost: Each node is added to T with at most d $Notify/Expand$ messages: a subtotal of $O(d\,\eta)$ messages. Each link of Ψ passes two $Announce$ messages, one for each end when it is added to T: a subtotal of $O(e)$ messages. After a node is added to T, *Report* messages are propagated to the root from the node (Proc. 4) as well as its nbs (Proc. 6): $O(r\ d)$ messages. This yields a subtotal of $O(r\ d\ \eta)$ messages. Hence, the total number of messages is $O(r\ d\ \eta + e)$.

Time complexity: The $O(d\ \eta)$ *Notify/Expand* messages are sequential, and take $O(d\ \eta)$ time. *Announce* messages by the same sender take $O(r)$ time. The $O(2e)$ *Announce* messages take $O(r\ \eta)$ time. The $O(r\ d)$ *Report* messages due to one node addition form r parallel sequences and take $O(d)$ time. The $O(r\ d\ \eta)$ *Report* messages take $O(d\ \eta)$ time. Hence, time complexity is $O((d+r)\ \eta)$.

Agent privacy: By using the boundary set of a MAS, privacy of private variables is preserved. By using the weighted graph defined from the boundary set, privacy of shared variables is preserved. Since our distributed MST algorithm does not disclose node identity, privacy of agent identity is preserved.

7 Conclusion

Our contribution is a general approach for JT organization construction based on MST, and an algorithm suite for MST construction. Combination of our approach and algorithm suite guarantees agent privacy on private variables, shared variables, as well as agent identity. To the best of our knowledge, no existing JT-based MAS frameworks enable agent privacy at such a degree, except a related work based on boundary set elimination which we report in [26].

The method proposed here assumes that a JT organization exists for the given env decomposition. Whether the condition (JT existence) holds is not dealt with in the current work, and is detected distributively in [26].

Our distributed MST algorithm is efficient, but not as efficient as the most efficient existing algorithms, although they do not allow preservation of agent identity. An open question is whether it is possible for a distributed MST algorithm to be as efficient as these algorithms while preserving agent identity.

Acknowledgement. We thank anonymous reviewers for their helpful comments. Financial support through Discovery Grant from NSERC, Canada is acknowledged.

References

1. Awerbuch, B.: Proc. 19th ACM Symp. Theory of Computing, pp. 230–240 (1987)
2. Brito, I., Meseguer, P.: Cluster tree elimination for distributed constraint optimization with quality guarantees. Fundamenta Informaticae 102(3-4), 263–286 (2010)
3. Faloutsos, M., Molle, M.: Optimal distributed algorithm for minimum spanning trees revisited. In: Proc. 14th Annual ACM Symp. Principles of Distributed Computing, pp. 231–237 (1995)
4. Faltings, B., Leaute, T., Petcu, A.: Privacy guarantees through distributed constraint satisfaction. In: Proc. IEEE/WIC/ACM Intelligent Agent Technology, pp. 350–358 (2008)
5. Gallager, R., Humblet, P., Spira, P.: A distributed algorithm for minimum-weight spanning trees. ACM Trans. Programming Languages and Systems 5(1), 66–77 (1983)

6. Garay, J., Kutten, S., Peleg, D.: A sublinear time distributed algorithm for minimum-weight spanning trees. SIAM J. Comput. 27(1), 302–316 (1998)
7. Gmytrasiewicz, P., Durfee, E.: Rational communication in multi-agent environments. Auto. Agents and Multi-Agent Systems 4(3), 233–272 (2001)
8. Jensen, F.: Junction tree and decomposable hypergraphs. Tech. rep., JUDEX, Aalborg, Denmark (February 1988)
9. Khan, M., Pandurangan, G.: A fast distributed approximation algorithm for minimum spanning trees. Distributed Computing 20(6), 391–402 (2008)
10. Koller, D., Milch, B.: Multi-agent influence diagrams for representing and solving games. In: Proc. 17th Inter. Joint Conf. on Artificial Intelligence, pp. 1027–1034 (2001)
11. Kutten, S., Peleg, D.: Fast distributed construction of smallk-dominating sets and applications. J. Algorithms 28(1), 40–66 (1998)
12. Maestre, A., Bessiere, C.: Improving asynchronous backtracking for dealing with complex local problems. In: Proc. 16th European Conf. on Artificial Intelligence, pp. 206–210 (2004)
13. Modi, P., Shen, W., Tambe, M., Yokoo, M.: Adopt: asynchronous distributed constraint optimization with quality guarantees. Artificial Intelligences 161(1-2), 149–180 (2005)
14. Nobari, S., Cao, T., Karras, P., Bressan, S.: Scalable parallel minimum spanning forest computation. In: Proc. 17th ACM SIGPLAN Symp. Principles and Practice of Parallel Programming, pp. 205–214 (2012)
15. Paskin, M., Guestrin, C., McFadden, J.: A robust architecture for distributed inference in sensor networks. In: Proc. Information Processing in Sensor Networks, pp. 55–62 (2005)
16. Petcu, A., Faltings, B.: A scalable method for multiagent constraint optimization. In: Proc. 19th Inter. Joint Conf. on Artificial Intelligence, pp. 266–271 (2005)
17. Prim, R.: Shortest connection networks and some generalizations. Bell Syst. Tech. J. (36), 1389–1401 (1957)
18. Silaghi, M., Abhyankar, A., Zanker, M., Bartak, R.: Desk-mates (stable matching) with privacy of preferences, and a new distributed CSP framework. In: Proc. Inter. Florida Artificial Intelligence Research Society Conf., pp. 83–96 (2005)
19. Valtorta, M., Kim, Y., Vomlel, J.: Soft evidential update for probabilistic multiagent systems. Int. J. Approximate Reasoning 29(1), 71–106 (2002)
20. Vinyals, M., Rodriguez-Aguilar, J., Cerquides, J.: Constructing a unifying theory of dynamic programming DCOP algorithms via the generalized distributive law. J. Autonomous Agents and Multi-Agent Systems 22(3), 439–464 (2010)
21. Xiang, Y.: A probabilistic framework for cooperative multi-agent distributed interpretation and optimization of communication. Artificial Intelligence 87(1-2), 295–342 (1996)
22. Xiang, Y.: Probabilistic Reasoning in Multiagent Systems: A Graphical Models Approach. Cambridge University Press, Cambridge (2002)
23. Xiang, Y., Chen, J., Deshmukht, A.: A decision-theoretic graphical model for collaborative design on supply chains. In: Tawfik, A.Y., Goodwin, S.D. (eds.) Canadian AI 2004. LNCS (LNAI), vol. 3060, pp. 355–369. Springer, Heidelberg (2004)
24. Xiang, Y., Hanshar, F.: Multiagent expedition with graphical models. Inter. J. Uncertainty, Fuzziness and Knowledge-Based Systems 19(6), 939–976 (2011)
25. Xiang, Y., Mohamed, Y., Zhang, W.: Distributed constraint satisfaction with multiply sectioned constraint networks. Accepted to appear in International J. Information and Decision Sciences (2013)
26. Xiang, Y., Srinivasan, K.: Boundary set based existence recognition and construction of hypertree agent organization. In: Zaïane, O., Zilles, S. (eds.) Canadian AI 2013. LNCS (LNAI), vol. 7884, pp. 187–198. Springer, Heidelberg (2013)
27. Xiang, Y., Zhang, W.: Multiagent constraint satisfaction with multiply sectioned constraint networks. In: Kobti, Z., Wu, D. (eds.) Canadian AI 2007. LNCS (LNAI), vol. 4509, pp. 228–240. Springer, Heidelberg (2007)

The K-Modes Method under Possibilistic Framework

Asma Ammar[1], Zied Elouedi[1], and Pawan Lingras[2]

[1] LARODEC, Institut Supérieur de Gestion de Tunis, Université de Tunis
41 Avenue de la Liberté, 2000 Le Bardo, Tunisie
asma.ammar@voila.fr, zied.elouedi@gmx.fr
[2] Department of Mathematics and Computing Science, Saint Mary's University
Halifax, Nova Scotia, B3H 3C3, Canada
pawan@cs.smu.ca

Abstract. In this paper, we develop a new clustering method combining the possibility theory with the standard k-modes method (SKM). The proposed method is called KM-PF to express the fact that it is a modification of k-modes algorithm under possibilistic framework. KM-PM incorporates possibilistic theory in two distinct stages in application of the SKM combining the possibilistic k-modes (PKM) and the k-modes using possibilistic membership (KM-PM). First, it deals with uncertain attribute values of instances using possibilistic distributions. Then, it computes the possibilistic membership degrees of each object to all clusters. Experimental results show that the proposed method compares favourably to the SKM, PKM and KM-PM.

Keywords: Clustering, possibility theory, uncertainty, k-modes method, categorical data, possibilitic membership degree.

1 Introduction

The k-means algorithm [9] is one of the most known clustering algorithm that deals with numeric data sets. The standard k-modes method (SKM) [5], [6] is a modification of k-means algorithm defined for large data sets with categorical attributes. The k-modes forms the basis for the approach proposed in this paper. The SKM is also called a hard (or crisp) clustering method. It assigns an object to only one cluster. This assignment isolates the object from other clusters even if it shares some characteristics with them. To overcome this limitation, soft clustering approaches [2], [3] have been proposed. For each object, they define the degree of belonging to several clusters. This soft assignment allows the detection of objects that do not belong to exactly one cluster. Based on the SKM, different soft methods have been developed to handle certain databases.

Real-world applications are pervaded by imperfection. Uncertainty in data mining field can be essentially located at two levels. It can be in the attributes' values, e.g. measurement of blood pressure, temperature or humidity levels. Uncertainty may also exist in the belonging of objects to different clusters. Taking

O. Zaïane and S. Zilles (Eds.): Canadian AI 2013, LNAI 7884, pp. 211–217, 2013.

these aspects of uncertainty into account can improve the real-world applications of clustering (e.g. medicine, banking, pattern recognition, data mining). It will lead to a better decision making process. The possibility theory is an uncertainty theory that has been successfully combined with clustering approaches using uncertain data sets from data mining area. For example, the possibilistic c-means [8], where uncertainty is dealt in the clusters, the possibilistic k-modes (PKM) [1], which handles uncertain attributes' values of training instances, and the k-modes using possibilistic membership (KM-PM) [2] that deals with uncertainty in the belonging of objects to different clusters.

Our aim in this work is to adapt the SKM to an uncertain framework and to improve its results. To this end, we will introduce modifications at two levels with the help of possibility theory. The proposed approach will deal with possibilistic attribute values and obtain a soft clustering using possibilistic membership.

2   Possibility Theory

The possibility distribution is a fundamental concept in possibility theory [4], [11]. It is defined by the function π which represents the state of knowledge. It is defined as the mapping from the universe of discourse $\Omega = \{\varpi_1, \varpi_2, ..., \varpi_n\}$ to the interval $[0, 1]$ which represents the possibilistic scale L. ϖ_i presents an uncertain state of knowledge to which a degree of uncertainty illustrated by a possibility degree is defined by $\pi(\varpi) = 0$ if ϖ is impossible and totally excluded and $\pi(\varpi) = 1$ when ϖ is fully plausible.

A possibility distribution is considered as normalized when $\max_i \{\pi(\varpi_i)\} = 1$. Besides, based on π, the complete knowledge ($\exists \varpi_0, \pi(\varpi_0) = 1$ and $\pi(\varpi) = 0$ otherwise) and the total ignorance ($\forall \varpi \in \Omega, \pi(\varpi) = 1$) can be obtained. Furthermore, from the well-known possibilistic similarity measures used to compare two normalized possibility distributions and to compute their degree of similarity, we can mention the information affinity [7] (Equation (1)).

$$IA(\pi_1, \pi_2) = 1 - 0.5 [D(\pi_1, \pi_2) + Inc(\pi_1, \pi_2)]. \tag{1}$$

where $D(\pi_1, \pi_2) = \frac{1}{n} \sum_{i=1}^{n} |\pi_1(\varpi_i) - \pi_2(\varpi_i)|$, $Inc(\pi_1, \pi_2) = 1 - \max(\pi_1(\varpi) Conj \pi_2(\varpi))$ and $\forall \omega \in \Omega, \Pi_{Conj}(\omega) = \min(\Pi_1(\omega), \Pi_2(\omega))$.

3   The K-Modes Method and Its Extensions

3.1   The SKM

The standard k-modes method (SKM), proposed in [5], [6], uses a categorical data set, the simple matching measure to compute the dissimilarity between the modes and the objects and the frequency based function to update the modes.

The simple matching measure computing the dissimilarity between two objects $X_1 = (x_{11}, x_{12}, ..., x_{1m})$ and $Y_1 = (y_{11}, y_{12}, ..., y_{1m})$ with categorical values are defined by $d(X_1, Y_1) = \sum_{t=1}^{m} \delta(x_{1t}, y_{1t})$, where m is the number of attributes and $\delta(x_{1t}, y_{1t})$ is equal to 0 if $x_{1t} = y_{1t}$, and equal to 1 when $x_{1t} \neq y_{1t}$.

Generally, if our aim is to cluster n categorical objects $S = \{X_1, X_2, ..., X_n\}$ into k clusters $C = \{C_1, C_2, ..., C_k\}$ with k modes $Q = \{Q_1, Q_2, ..., Q_k\}$ ($k \leq n$), we have to $minimize\ D\,(W, Q) = \sum_{j=1}^{k} \sum_{i=1}^{n} \omega_{ij} d\,(X_i, Q_j)$, where W is a k-by-n matrix, $\omega_{i,j} \in \{0, 1\}$ is the membership degree of X_i in C_j, $1 \leq j \leq k$ and $1 \leq i \leq n$.

Despite its interesting results, the SKM are effective only in certain case. As there are many aspects of uncertainty in real-world situations, in this paper we propose to adapt the possibility theory to the SKM in order to cluster objects with uncertain values that belong to several clusters.

3.2   The PKM

The PKM approach [1] is a hard and uncertain clustering method. It is based on the SKM and uses the possibility theory to handle uncertain attribute values of objects. It uses a training set containing uncertain attribute values presented through possibility degrees from $[0, 1]$. The similarity measure $InfoAff$ applied in the PKM is based on the information affinity measure [7].

$$InfoAff_{PKM}(X_1, X_2) = \frac{\sum_{j=1}^{m} InfoAff(\pi_{1j}, \pi_{2j})}{m}. \tag{2}$$

Besides, the PKM uses the mean operator $\forall \omega \in A_j, \pi_{jC}(\omega) = \frac{\sum_{i=1}^{p} \pi_{ij}(\omega)}{|C|}$ to update the modes, where $A_j = (a_{j1}, a_{j2}, ..., a_{jt})$ is the set of t values, $1 \leq j \leq m$, and π_{ij} is the possibility distribution defined for A_j related to the object x_i.

3.3   The KM-PM

The KM-PM [2] is a soft approach that clusters each instance of the training set to k clusters using possibilistic membership. It uses a certain training set and the simple matching dissimilarity measure. Besides, it defines a possibilistic membership degree $\omega_{ij} \in [0, 1]$ by computing the similarity degree between the object i and the cluster j then, dividing it by the total number of attributes. The update of modes is obtained by determining ω_{ij}. Then, we sum the ω_{ijvt} relative to each value v of the attribute t corresponding to the object i by $\forall j \in k, t \in A, Mode_{jt} = \max_v \sum_{i=1}^{n} \omega_{ijtv}$, with $\forall i \in n, \max_j (\omega_{ij}) = 1$. Finally, we set the value v that achieves the maximum of the summation as a new value in the mode.

4   The K-Modes Method under Possibilistic Framework

The KM-PF is a combination of possibility theory and SKM. It is characterized by uncertainty in its attribute values and in the belonging of objects to k clusters. Thus, the KM-PF modifies the parameters of the PKM and KM-PM as follows to conserve their performances and overcome their limitations. The KM-PF uses:

1. An uncertain training set where the attribute values can be certain and/or uncertain. The values are presented through possibility degrees from $[0, 1]$ where a possibility distribution is defined for each attribute relative to each object. This possibility value describes the degree of uncertainty. Note that the KM-PF uses the same structure of the training set as the PKM [1].
2. The possibilistic similarity measure applied is already used by the PKM to determine the similarity between the modes and the objects (Equation (2)).
3. The possibilistic membership degree ω_{ij} is obtained by computing the possibilistic similarity measure (Equation (2)). It indicates the degree of belonging of an object i to the cluster j.
4. The update of the cluster mode takes into account the degree of belonging of each object to different clusters. It also depends on the possibilistic degrees of the training instances. To compute the new modes' values, we set the weight w as a new parameter. We follow these steps:
   (a) Step 1: We compute the number of objects NO in each cluster having the maximum of ω_{ij} using $NO_j = count_j(\max_i \omega_{ij})$.
   (b) Step 2: We compute the weight of the cluster j denoted by w_j as follows:

   $$w_j = \begin{cases} \frac{NO_j}{total\ number\ of\ objects} & \text{if } NO_j \neq 0, \\ \frac{1}{total\ number\ of\ objects\ +1} & \text{otherwise.} \end{cases} \quad (3)$$

   (c) Step 3: In order to obtain the new mode $(Mode'_j)$, we multiply the attributes' values of the mode $(Mode_j)$ relative to the cluster j by w_j:

   $$\forall j \in k, Mode'_j = w_j \times Mode_j. \quad (4)$$

---

1. *Randomly, select the k initial modes, one mode for each cluster.*
2. *Allocate each instance to the k clusters based on possibilistic membership degrees ω_{ij} after computing the possibilistic similarity measure using Equation (2).*
3. *Compute the weight w_j for each cluster j using Equation (3) then, update the cluster mode using Equation (4).*
4. *Retest the similarity between objects and modes. Reallocate objects to clusters using possibilistic membership degrees then update the modes.*
5. *Repeat (4) until all clusters are stable.*

Fig. 1. Algorithm KM-PF

5   Experiments

5.1   The Framework

For the experiments, we used eight real-world databases from UCI: Machine Learning Repository [10]. They consist of Shuttle Landing Control (SL), Balloons (Bal), Post-Operative Patient (PO), Congressional Voting Records (CV), Balance Scale (BS), Tic-Tac-Toe Endgame (TE), Solar-Flare (SF) and Car Evaluation (CE). Note that the number of classes corresponds in our case to the k.

5.2   Artificial Creation of Uncertain Data Sets

1. Case of certain attribute values: it expresses the case of complete knowledge in possibility theory. Thus, it is obtained by assigning to the true values the possibility degree 1 and the other values 0.
2. Case of uncertain attribute values (detailed in [1]): we assign to the real values the degree of 1 and to the remaining values a degree between $(0, 1]$.

5.3   Evaluation Criteria

First, the accuracy [5] $AC = \frac{\sum_{j=1}^{k} a_j}{n}$ is used, where $k \leq n$, n is the total number of objects and a_j is the number of correctly classified objects. Then, the iteration number (IN) and the execution time (ET) are computed. Note, there are some other criteria that can be used for the evaluation e.g. the F-measure using the Recall, the Rand index, and the cluster purity.

5.4   Experimental Results

We cross validate by dividing observations randomly into a training set and a test set. The obtained results are as follows.

1. Certain Case: For both the SKM and the KM-PM, we use databases from UCI without any modifications. For the PKM and the KM-PF, each attribute value is transformed to a possibilistic degree corresponding to the case of the complete knowledge. Table 1 shows the performance gain from the KM-PF versus the SKM, the PKM and the KM-PM in terms of AC, IN and ET.

Table 1. The KM-PF vs the SKM, the PKM and the KM-PM

	Data sets	SL	Bal	PO	CV	BS	TE	SF	CE
	AC	0.61	0.52	0.684	0.825	0.785	0.513	0.87	0.795
SKM	IN	8	9	11	12	13	12	14	11
	ET/s	12.431	14.551	17.238	29.662	37.819	128.989	2661.634	3248.613
	AC	0.69	0.694	0.72	0.896	0.789	0.564	0.876	0.84
PKM	IN	2	2	6	3	2	5	6	3
	ET/s	0.017	0.41	0.72	3.51	8.82	37.118	51.527	75.32
	AC	0.63	0.65	0.74	0.79	0.82	0.59	0.91	0.87
KM-PM	IN	4	4	8	6	6	2	10	12	12
	ET/s	10.28	12.56	15.23	28.09	31.41	60.87	87.39	197.63
	AC	0.71	0.74	0.749	0.91	0.834	0.625	0.932	0.897
KM-PF	IN	2	3	6	3	2	3	6	4
	ET/s	2.3	0.9	1.4	6.7	8.51	40.63	55.39	89.63

Looking at Table 1, we remark that KM-PF has improved the quality of the clustering task. This improvement is especially obvious for the first evaluation criterion i.e the accuracy. For the Solar-Flare database for example, the accuracy reaches 0.932 and six iterations (corresponding to 55.39 seconds) are needed for the KM-PF to obtain the final results.

Generally, the KM-PF is more accurate than the other methods. Moreover, the number of iterations of the KM-PF is low and close to the IN of the PKM. For the execution time, it is lower than the ET of both the SKM and the KM-PM, but slightly higher than the ET of the PKM.

2. Uncertain Case: Only uncertain training sets can be used in this case. Hence, we compare the KM-PF to the PKM since both of the SKM and the KM-PM are applied on certain databases. For the KM-PF and the PKM, we introduce uncertainty in the training set by replacing the values of attributes by possibilistic degrees belonging to $(0, 1]$. Table 2 (with A the percentage of uncertain attributes in the training set and d the possibility degree of values) and Table 3 detail the results of the uncertain approaches.

Table 2. The AC of the KM-PF vs the PKM

	Data sets	SL	Bal	PO	CV	BS	TE	SF	CE
$A < 50\%$ and	PKM	0.652	0.632	0.713	0.851	0.783	0.566	0.864	0.821
$0 < d < 0.5$	KM-PF	0.659	0.64	0.73	0.88	0.79	0.589	0.875	0.87
$A < 50\%$ and	PKM	0.638	0.634	0.69	0.843	0.785	0.569	0.868	0.791
$0.5 \leq d \leq 1$	KM-PF	0.647	0.651	0.71	0.85	0.8	0.58	0.89	0.83
$A \geq 50\%$ and	PKM	0.713	0.774	0.758	0.896	0.795	0.572	0.877	0.875
$0 < d < 0.5$	KM-PF	0.735	0.811	0.784	0.921	0.87	0.63	0.912	0.9
$A \geq 50\%$ and	PKM	0.698	0.693	0.737	0.843	0.78	0.558	0.864	0.859
$0.5 \leq d \leq 1$	KM-PF	0.71	0.73	0.72	0.85	0.8	0.62	0.87	0.88

From Table 2, we notice that the KM-PF has the highest accuracy for the different values of the parameters A and d especially when the training set contains more than 50% of attributes with uncertain values (between $(0, 0.5)$).

This result proves that the possibilistic approaches, and especially the KM-PF, provide good accuracy when they deal with uncertain databases. Thus, the KM-PF shows again that it presents a reasonable approach for handling different aspects of uncertainty.

Table 3. The number of iterations and the execution timeof the KM-PF vs the PKM

	Data sets	SL	Bal	PO	CV	BS	TE	SF	CE
PKM	IN	3	2	8	4	2	8	6	5
	ET/s	0.02	0.437	0.794	3.881	8.914	36.219	51.766	76.221
KM-PF	IN	3	3	8	4	2	6	8	4
	ET/s	2.02	0.672	0.95	6.97	9.653	41.31	56.781	90.3

Looking to Table 3, we can remark that the IN and the ET of both the tested methods (i.e. the PKM and the KM-PF) are very close. For example, for Balance Scale and Shuttle Landing Control databases, the two approaches need the same IN. Furthermore, the difference between the PKM and KM-PF especially in terms of execution time is due to the soft clustering of the KM-PF. The new approach needs a little more time to specify the degree of membership of each instance in the training set to the k clusters.

6   Conclusion

In this paper we have highlighted the problem of uncertainty that can be found in the training set or when performing the clustering task. We have solved this issue by the development of the k-modes method under possibilistic framework (KM-PF) which is based on two of our previous works [1] and [2]. The uncertainty is handled by using the possibility theory at two levels. The first application of possibility theory describes the uncertain values of attributes. Its second use specifies the uncertainty in the belonging of an object to several clusters. To evaluate the KM-PF, we used databases from UCI: Machine Learning Repository [10]. The experimental results show the effectiveness of the proposed method.

References

1. Ammar, A., Elouedi, Z.: A New Possibilistic Clustering Method: The Possibilistic K-Modes. In: Pirrone, R., Sorbello, F. (eds.) AI*IA 2011. LNCS (LNAI), vol. 6934, pp. 413–419. Springer, Heidelberg (2011)
2. Ammar, A., Elouedi, Z., Lingras, P.: K-modes clustering using possibilistic membership. In: Greco, S., Bouchon-Meunier, B., Coletti, G., Fedrizzi, M., Matarazzo, B., Yager, R.R. (eds.) IPMU 2012, Part III. CCIS, vol. 299, pp. 596–605. Springer, Heidelberg (2012)
3. Ammar, A., Elouedi, Z., Lingras, P.: RPKM: The Rough Possibilistic K-Modes. In: Chen, L., Felfernig, A., Liu, J., Raś, Z.W. (eds.) ISMIS 2012. LNCS, vol. 7661, pp. 81–86. Springer, Heidelberg (2012)
4. Dubois, D., Prade, H.: Possibility theory: An approach to computerized processing of uncertainty. Plenum Press (1988)
5. Huang, Z.: Extensions to the k-means algorithm for clustering large data sets with categorical values. Data Mining and Knowledge Discovery 2, 283–304 (1998)
6. Huang, Z., Ng, M.K.: A note on k-modes clustering. Journal of Classification 20, 257–261 (2003)
7. Jenhani, I., Ben Amor, N., Elouedi, Z., Benferhat, S., Mellouli, K.: Information affinity: A new similarity measure for possibilistic uncertain information. In: Mellouli, K. (ed.) ECSQARU 2007. LNCS (LNAI), vol. 4724, pp. 840–852. Springer, Heidelberg (2007)
8. Krishnapuram, R., Keller, J.M.: A possibilistic approach to clustering. IEEE Trans. Fuzzy System 1, 98–110 (1993)
9. MacQueen, J.B.: Some methods for classification and analysis of multivariate observations. In: Proceeding of the Fifth Berkeley Symposium on Math, Stat and Prob., pp. 281–296 (1967)
10. Murphy, M.P, Aha, D.W.: Uci repository databases (1996), http://www.ics.uci.edu/mlearn
11. Zadeh, L.A.: Fuzzy sets. Information And Control 8, 338–353 (1965)

Quantitative Aspects
of Behaviour Network Verification

Christopher Armbrust, Thorsten Ropertz, Lisa Kiekbusch, and Karsten Berns

Robotics Research Lab, Department of Computer Science,
University of Kaiserslautern, P.O. Box 3049, 67653 Kaiserslautern, Germany
{armbrust,ropertz,kiekbusch,berns}@cs.uni-kl.de
http://rrlab.cs.uni-kl.de/

Abstract. This paper presents quantitative aspects of an approach for
the modelling and verification of behaviour networks published previously and describes the application of said modelling technique to a complex coordinating behaviour. In order to decrease the number of intercon-
nection failures in behaviour networks, verification techniques focusing
on behaviour interaction can be applied. In previous work, the authors
have introduced a novel approach for modelling behaviour networks as
networks of finite-state automata, to which model checking can be applied as verification technique. This paper presents how the approach
can be used to model complex behaviours and provides calculations of
the numbers of states, transitions, and state variables in the resulting
automata.

Keywords: Behaviour-based System, Behaviour Network, Behaviour
Modelling, Behaviour Network Verification, Quantitative Aspects.

1   Introduction

Behaviour-based robot control systems (BBS) have shown to provide several advantages: In contrast to traditional approaches, complex tasks are decomposed
into simple sub-goals pursued by individual behaviours. Due to the behaviours'
limited complexity, they are easier to develop, implement, and maintain. The
restricted scope also facilitates the reusability, such that common functionalities
can be realised based on well-tested modules.

Behaviours interact via signal exchange in order to provide a suitable overall
behaviour. With a growing demand for numerous functionalities, the behaviour
network complexity increases and the necessity of sophisticated analysis methods
and tools to ensure a certain level of quality arises. Especially regarding safety
critical systems, experience has shown that proving the systems' correctness
is obligatory. A common method for proving correctness is formal verification,
which has to be adapted to the properties of BBS in order to become feasible.

This paper continues the research on behaviour network verification conducted
at the Robotics Research Lab. In [1] and [2], a novel approach for modelling
behaviour-based networks as networks of finite-state automata has been presented. The purpose of this paper is to provide information about how to model

O. Zaïane and S. Zilles (Eds.): Canadian AI 2013, LNAI 7884, pp. 218–225, 2013.

more complex behaviours and about the complexity of the created models. This is done by explaining how a special coordinating behaviour can be properly modelled using the presented approach and by providing information about the number of locations, edges, and variables in the created models.

2 Related Work

Model checking is a common technique for ensuring the correctness of concurrent finite-state systems. Thereby, an abstract model of the system is created and its compliance with a given specification is checked by applying a verification algorithm (see [4] for an overview). The authors of [13] describe the use of model checking for verifying the control system of an unmanned aircraft. In [8] the application and extension of model checking techniques for verifying spacecraft control software is presented. Another example of the use of model checking is given in [7], which applies model checking to verify a distributed coordination algorithm in the field of swarm robotics.

In general, different types of models are possible depending on the model checking method. In [5], X-machines (a computational machine resembling an FSM extended by memory) are proposed for modelling an agent's behaviour. Thereby, each agent is represented by a single X-machine, which leads to large monolithic elements and thus complicates the modelling. The authors of [10] used the synchronous programming language Quartz, which allows for an automatic derivation of a fine-grained model, for implementing single behaviours and to prove their correctness using model checking. UPPAAL is a model checking tool that requires the system to be modelled by automata. In [12] model checking with UPPAAL in combination with fault-tree analysis is applied to prove the absence of critical events in the system. The correctness and completeness of fault-trees can also be verified using model checking as described in [6].

3 Behaviour Network Modelling and Verification

This section introduces the behaviour-based architecture used for the work at hand (the iB2C) and describes how iB2C networks can be modelled and verified.

The presented work uses the behaviour-based architecture iB2C, which has been implemented using the software frameworks MCA2-KL[1] and FINROC[2]. The iB2C is described extensively in [9], while only its essential aspects are presented here. An iB2C behaviour is defined as $B = (f_a, f_r, F)$, with f_a calculating its *activity vector* a, f_r calculating its *target rating* r, and $F : u = F(e, \iota)$ transferring its input vector e and activation ι into the output vector u. The activation is a combination of a behaviour's *stimulating* (s) and *inhibiting* $i = \|i\|_\infty$ inputs and is calculated as $\iota = s \cdot (1 - i)$. A behaviour indicates the amount of influence

---

[1] MCA2-KL: Modular Controller Architecture Version 2 - Kaiserslautern Branch.

[2] FINROC is the downward compatible successor of MCA2-KL (see [11]).

it intends to have in a network using its *activity a*. According to an iB2C principle, a behaviour's activation limits its activity, i.e. $a \leq \iota = s \cdot (1 - i)$. The target rating r describes how satisfied a behaviour is with the current situation. s, i, a, and r are limited to $[0, 1]$. The typical connection between two behaviours is that one behaviour stimulates or inhibits another with its activity. Furthermore, fusion behaviours (FB) can be used to coordinate competing behaviours. Figure 6 depicts two iB2C networks.

iB2C networks are modelled as networks of finite-state automata, on which model checking is performed using the UPPAAL[3] toolbox. In UPPAAL, a network of *automata* (parametrised instantiations of *templates*) is called *system*. An automaton consists of *locations* and interconnecting *edges*. Whether an edge can be taken is restricted by *guards* (side-effect free Boolean expressions) and synchronisations. The latter are realised via so-called *channels*. Edges can be labelled with a channel name followed by "!" for a sending or "?" for a receiving channel. Furthermore, so-called *updates* (assignments) can be added to edges. Figure 3 shows an automaton with the described elements. When modelling an iB2C network, each behaviour is represented by an instantiation of each of five basic templates, namely `StimulationInterface`, `InhibitionInterface`, `ActivationCalculation`, `ActivityCalculation`, and `TargetRatingCalculation`. Three of the five basic templates are shown in Figs. 1 to 3. Due to reasons of complexity, the value range of the behaviours is limited to the set of $\{0, 1\}$. Experience has shown that this does not pose a problem to the effectiveness of the presented approach.

As `InhibitionInterface` depends on the number of inhibiting behaviours, different versions for different numbers of inhibiting behaviours are created by the modelling algorithm, while the other basic templates are just instantiated and connected using the appropriate channels. The structure of the different versions of `InhibitionInterface` is a hypercube that has the number of inhibiting behaviours as dimension. To verify the proper operation of a BBS, queries are given to UPPAAL's model checker, which evaluates them to true or false.

4   Modelling Complex Behaviour Nodes

As an example of how to model the activity function of a complex behaviour node in a more precise way than depicted in Fig. 2, the modelling of a maximum fusion behaviour shall be described here. For an FB B_{Fusion} with n_c connected behaviours B_{Input_d}, $a_{\text{Fusion}} = 1$ if $(\exists d : (1 \leq d \leq n_c) \wedge (a_{\text{Input}_d} = 1))$ and $(\iota_{\text{Fusion}} = 1)$, otherwise $a_{\text{Fusion}} = 0$.

A naive implementation of a template for calculating f_a of an FB would result in a basic structure resembling that of a hypercube with dimension $n_c + 1$—one for each connected input behaviour and one for monitoring ι_{Fusion} (Version 1). In order to reduce the number of locations and transitions (see Sec. 5), the modelling of f_a of an FB has been split up into different automata: For each connected behaviour B_{Input_d}, an instance of `FBIBActivityChanged` (see Fig. 4) is created. Furthermore, there is one single instance of `FBActivityCalculation` (Version

---

[3] See `http://www.uppaal.org/` and [3].

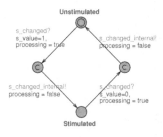

Fig. 1. `StimulationInterface` of a single behaviour

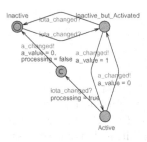

Fig. 2. `ActivityCalculation` (calculating the activity)

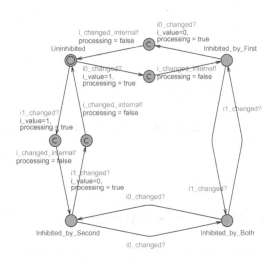

Fig. 3. `InhibitionInterface` for a behaviour that is inhibited by two others (double circle: initial location; "C" in circle: committed location; purple: name of location; turquoise: channel; blue: update)

2) that combines the signals of the individual automata. Each instance of `FBIB-ActivityChanged` monitors $\iota_{\text{Fusion}} = 1$ and the activity of the corresponding input behaviour. If both equal 1, it transitions to location `Active` and sends a signal to the combining automaton. In contrast to `FBIBActivityChanged`, `FBActivityCalculation` (Version 2) depends on n_c as it has to process signals from n_c different automata. Figure 5 depicts this automaton for the case of $n_c = 2$. In the initial location, it is assumed that B_{Fusion} is inactive, i.e. $a_{\text{Fusion}} = 0$. If one of the instances of `FBIBActivityChanged` signals that the corresponding input behaviour is active *and* $\iota_{\text{Fusion}} = 1$, then `fb_a_value` is set to 1 and this change is signalled via `fb_a_changed`. If now the second instance of `FBIBActivityChanged` also indicates that its corresponding input behaviour is active, another change of location takes place—but no change of `fb_a_value` and no signalling is performed. In case of an instance signalling a change now, `FBActivityCalculation` does not signal a change or update a variable. But if then the other instance of `FBIBActivityChanged` also signals a change, `FBActivityCalculation` goes back to its initial location, setting `fb_a_value` to 0 again and signalling the change via `fb_a_changed`. The structure of `FBActivityCalculation` for n_c connected input behaviours is a hypercube with dimension n_c. Just like the different versions of `InhibitionInterface`, it is automatically created by the modelling algorithm.

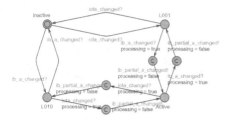

Fig. 4. FBIBActivityChanged of a fusion behaviour

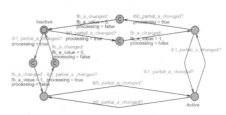

Fig. 5. FBActivityCalculation (Version 2) for a fusion behaviour with two connected input behaviours

5   Quantitative Aspects

The number of locations and edges in a model for a given behaviour network provides an estimate of the complexity of the model. For the basic templates, the locations and edges can simply be counted (see Tab. 1). As already mentioned in Sec. 3, InhibitionInterface for n_i inhibiting behaviours has the basic structure of a hypercube of dimension n_i. The number of vertices in such a hypercube is 2^{n_i}. It has $2^{n_i-1} \cdot n_i$ edges. As the edges in InhibitionInterface are bidirectional, this has to be multiplied by 2, resulting in $2^{n_i} \cdot n_i$. For each dimension (i.e. for each inhibiting behaviour), there are two additional committed locations with two additional edges. Summing that up yields $2^{n_i} + 2 \cdot n_i$ locations and $n_i \cdot (2^{n_i} + 2)$ edges. The same numbers can be calculated for FBActivityCalculation (Version 2) and FBTargetRatingCalculation. As mentioned in Sec. 4, the basic structure of FBActivityCalculation (Version 1) is a hypercube with dimension $n_c + 1$. Additional committed locations are added for the following cases: (1) ι_{Fusion} changes from 0 to 1 or vice versa *and* at least one of the competing behaviours is active ($2 \cdot (2^{n_c} - 1)$ cases). (2) The activity of one competing behaviour changes from 0 to 1 or vice versa *and* $\iota_{\text{Fusion}} = 1$ and *none* of the other competing behaviours is active ($2 \cdot n_c$ cases). This results in a total of $2^{(n_c+2)} + 2 \cdot n_c - 2$ locations. Taking into account that the edges of FBActivityCalculation (Version 1) are bidirectional and for every additional committed location one edge is added yields as total number of edges $2^{(n_c+1)} \cdot (n_c + 2) + 2 \cdot (n_c - 1)$. For $n_c \leq 3$, FBActivityCalculation (Version 1) needs less locations and for $n_c \leq 2$, it also needs less edges than FBActivityCalculation (Version 2). However, in the current implementation, Version 2 is always used.

Additionally, five variables of type **bool** (**processing** flags) and four of type **int[0,1]** (for values of behaviour signals) are needed for each behaviour along with different channels. There are three binary channels (for internal signalling of changes of s, i, and ι) and two broadcast channels (for signalling changes of a and r). No channels are needed for signalling changes of s or i from other behaviours as these are the channels signalling a change of a of the respective

Table 1. The number of locations and edges of each template (n_i: number of inhibiting behaviours; n_c: number of competing behaviours)

Template	#Locations	#Edges
StimulationInterface	4	4
InhibitionInterface	$2^{n_i} + 2 \cdot n_i$	$n_i \cdot (2^{n_i} + 2)$
ActivationCalculation	4	7
ActivityCalculation	4	6
TargetRatingCalculation	2	2
FBActivityCalculation (V. 1)	$2^{(n_c+2)} + 2 \cdot n_c - 2$	$2^{n_c+1} \cdot (n_c + 2) + 2 \cdot (n_c - 1)$
FBIBActivityChanged	8	12
FBActivityCalculation (V. 2)	$2^{n_c} + 2 \cdot n_c$	$n_c \cdot (2^{n_c} + 2)$
FBTargetRatingCalculation	$2^{n_c} + 2 \cdot n_c$	$n_c \cdot (2^{n_c} + 2)$

stimulating or inhibiting behaviour. For an FB with n_c competing behaviours, n_c binary channels for signalling between an instance of FBIBActivityChanged and FBActivityCalculation as well as n_c processing flags for the n_c instances of FBIBActivityChanged have to be added.

6   Application Example

In this section, the modelling and verification of the behaviour networks presented in [1] (see Figs.6a and 6b) shall be investigated further using the results of Secs. 4 and 5. *(G) Drive Control* with the sub-network *(G) Mediator* shall control the motion of an autonomous off-road vehicle based on different navigation approaches. Due to page restrictions, the reader has to be referred to [1] for details. What is relevant here is that the BBS shall fulfil four requirements. That this is the case was proven using model checking.

(G) Drive Control (excluding the contents of *(G) Mediator*) consists of a standard behaviour, three FBs with two input behaviours, and two FBs with no input behaviours that are taken into account for this analysis. Three of the FBs are inhibited by another behaviour. In total, this results in 166 locations, 230 edges, 40 variables of type bool, 24 of type int[0,1], 28 binary channels, and 12 broadcast channels. *(G) Mediator* alone consists of five normal behaviours (*(G) Local Path Planner* is considered as normal behaviour here.) and two FBs with two input behaviours each. Two of the normal behaviours are inhibited by another behaviour. This yields 158 locations, 221 edges, 39 variables of type bool, 28 of type int[0,1], 25 binary channels, and 14 broadcast channels. Due to a number of optimisations in the modelling algorithm, the actual numbers are partly slightly lower. However, these numbers show that even simple behaviour networks can result in rather large networks of automata.

Several queries have been verified with UPPAAL to check the fulfilment of the aforementioned requirements. For each requirement, verifyta of the UPPAAL suite was called with its default values and passed the model of the system and the query file corresponding to the requirement. Table 2 shows CPU and memory usage on an Intel® Core™ i7 920 CPU @ 2.67 GHz with 12 GB RAM. The

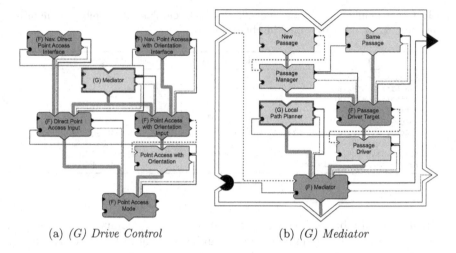

(a) *(G) Drive Control*   (b) *(G) Mediator*

Fig. 6. The two networks (grey node: simple behaviour; blue: maximum fusion behaviour; double-bordered grey: behavioural group; dashed green edge: stimulation; red: inhibition; blue: activity transfer; dotted brown: target rating; bold grey: data; filled stimulation input: always stimulated, i.e. $s = 1$)

Table 2. CPU and memory usage for verifying the requirements

Req.	#Queries	CPU User Time (Sum)	Virtual Memory (Max)
1	5	5.950 ms	86.944 KiB
2	3	5.880 ms	85.624 KiB
3	1	1.890 ms	84.300 KiB
4	2	2.170 ms	84.964 KiB
Sum	11	15.890 ms	-

figures show that even the solving of rather simple queries leads to a considerable memory consumption. Further experiments with larger networks (see [2]) have indicated that this is currently the main drawback of the presented approach and will have to be dealt with in the context of future work.

7   Conclusion and Future Work

This paper has demonstrated how a novel approach for modelling behaviour networks with the aim of verification can be applied to complex behaviours. Quantitative aspects of said approach have been presented and illustrated using the verification of a behaviour network from the control system of a real off-road vehicle. Future work will deal with optimising the modelling algorithm in order to reduce CPU and memory requirements of the verification process. Furthermore, the approach shall be used to verify more complex networks, taking into account aspects of the robot's environment.

Acknowledgements. The authors gratefully acknowledge Prof. Roland Meyer from the Concurrency Theory Group[4] of the University of Kaiserslautern for his helpful comments and suggestions. Furthermore, the authors acknowledge the technical work done by the student Maryla Rittmann. The research leading to these results has received funding from the European Union Seventh Framework Programme (FP7/2007-2013) under grant agreement number 285417.

References

1. Armbrust, C., Kiekbusch, L., Ropertz, T., Berns, K.: Verification of behaviour networks using finite-state automata. In: Glimm, B., Krüger, A. (eds.) KI 2012. LNCS, vol. 7526, pp. 1–12. Springer, Heidelberg (2012)
2. Armbrust, C., Kiekbusch, L., Ropertz, T., Berns, K.: Tool-assisted verification of behaviour networks. In: ICRA 2013, Karlsruhe, Germany, May 6-10 (2013)
3. Behrmann, G., David, A., Larsen, K.G.: A tutorial on UPPAAL. In: Bernardo, M., Corradini, F. (eds.) SFM-RT 2004. LNCS, vol. 3185, pp. 200–236. Springer, Heidelberg (2004)
4. Clarke, E.M., Grumberg, O., Peled, D.A.: Model Checking. MIT Press (1999)
5. Eleftherakis, G., Kefalas, P., Sotiriadou, A., Kehris, E.: Modeling biology inspired reactive agents using x-machines. In: Okatan, A. (ed.) International Conference on Computational Intelligence 2004 (ICCI 2004), December 17-19, pp. 93–96. International Computational Intelligence Society, Istanbul (2004)
6. Faber, J.: Fault tree analysis with Moby/FT. Tech. rep., Department for Computing Science, University of Oldenburg (2005),
   http://csd.informatik.uni-oldenburg.de/
   ˜jfaber/dl/ToolPresentationMobyFT.pdf
7. Juurik, S., Vain, J.: Model checking of emergent behaviour properties of robot swarms. Proceedings of the Estonian Academy of Sciences 60(1), 48–54 (2011)
8. Lowry, M., Havelund, K., Penix, J.: Verification and validation of AI systems that control deep-space spacecraft. In: Raś, Z.W., Skowron, A. (eds.) ISMIS 1997. LNCS, vol. 1325, pp. 35–47. Springer, Heidelberg (1997)
9. Proetzsch, M.: Development Process for Complex Behavior-Based Robot Control Systems. RRLab Dissertations, Verlag Dr. Hut (2010)
10. Proetzsch, M., Berns, K., Schuele, T., Schneider, K.: Formal verification of safety behaviours of the outdoor robot ravon. In: Zaytoon, J., Ferrier, J.-L., Andrade-Cetto, J., Filipe, J. (eds.) ICINCO 2007, pp. 157–164. INSTICC Press (May 2007)
11. Reichardt, M., Föhst, T., Berns, K.: On software quality-motivated design of a real-time framework for complex robot control systems. In: Proceedings of the 7th International Workshop on Software Quality and Maintainability (SQM), in conjunction with the 17th European Conference on Software Maintenance and Reengineering (CSMR) (March 5, 2013)
12. Schäfer, A.: Combining real-time model-checking and fault tree analysis. In: Araki, K., Gnesi, S., Mandrioli, D. (eds.) FME 2003. LNCS, vol. 2805, pp. 522–541. Springer, Heidelberg (2003)
13. Webster, M., Fisher, M., Cameron, N., Jump, M.: Model checking and the certification of autonomous unmanned aircraft systems. Tech. Rep. ULCS-11-001, University of Liverpool Department of Computer Science (2011)

---

[4] http://concurrency.cs.uni-kl.de/

A Causal Approach
for Mining Interesting Anomalies

Sakshi Babbar, Didi Surian, and Sanjay Chawla

School of Information Technologies
University of Sydney, Australia
sakshib@it.usyd.edu.au, dsur5833@uni.sydney.edu.au,
sanjay.chawla@sydney.edu.au

Abstract. We propose a novel approach which combines the use of Bayesian network and probabilistic association rules to discover and explain anomalies in data. The Bayesian network allows us to organize information in order to capture both correlation and causality in the feature space, while the probabilistic association rules have a structure similar to association mining rules. In particular, we focus on two types of rules: (i) *low support & high confidence* and, (ii) *high support & low confidence*. New data points which satisfy either one of the two rules conditioned on the Bayesian network are the candidate anomalies. We perform extensive experiments on well-known benchmark data sets and demonstrate that our approach is able to identify anomalies in high precision and recall. Moreover, our approach can be used to discover contextual information from the mined anomalies, which other techniques often fail to do so.

Keywords: Bayesian network, anomaly, causality, probabilistic association rules.

1 Introduction

In this paper, we propose an approach based on Bayesian network (or BN) to capture knowledge about inter-relationship among features with the objective of mining interesting anomalies. A Bayesian network is a kind of probabilistic graphical model to capture causal relationships among a set of variables using a graph in which variables are nodes and causation is indicated by arrows. In Bayesian terminology, a node is a parent of a child, if there is an arc from the former to the latter. Assuming discrete variables, the strength of the relationship between variables is quantified by conditional probability distributions associated with each node. An excellent introduction and theory of Bayesian networks can be found in [8].

The motivations of using BN for anomaly detection are three-fold. *First*, BN encodes knowledge on cause-effect relationships that can be utilized to mine real and meaningful anomalies. *Second*, BN could answer complex probabilistic queries which are an advantage in mining low probability events that exist in the observations. *Third*, BN has a capacity to model the joint probability

O. Zaïane and S. Zilles (Eds.): Canadian AI 2013, LNAI 7884, pp. 226–232, 2013.

distributions (or JPD) compactly, which means that we can study each causal interaction encoded in the model independently. The JPD over all variables X_1, $X_2...,X_{|X|}$ is represented using chain rule as shown in Equation 1, where notation $P(X_i \mid Pa(X_i))$ denotes probability of X_i given a set of its parent nodes denoted by $Pa((X_i))$. This special feature of BN is very useful to explain the reasons for the unusual behavior of anomalies. Throughout this paper, we address each causal interaction, i.e., $P(X_i \mid Pa(X_i))$ as *causal subspaces*.

$$P(X_1, X_2....,X_{|X|}) = \prod_{i=1}^{|X|} P(X_i|Pa(X_i)) \qquad (1)$$

Bayesian network has also been used for mining outliers in the classification settings. However, there exist several variants where Bayesian network has been used in an unsupervised setting [11] [10]. In this present paper, we used Bayesian network to discover the relationship among attributes in the analyzed problem in order to mine the outliers in an unsupervised environment. With this objective, we propose use of two robust probabilistic association rules which are based on two measures namely, *support* (unconditional probability) and *confidence* (conditional probability) on a Bayesian network to discover low probability events. However, unlike traditional association rule mining we are interested in mining infrequent patterns whose occurrence suggests the presence of uncommon and exceptional situations. Application of rules on BN extracts anomalous patterns to which we address as domain specific anomalous patterns (or **DSAPs**). In order to test if a particular test case is an anomaly for a given domain, we check if it carries "any" pattern from the discovered set of DSAPs. We call our method as a causal outlier mining (or **COM**) approach.

Our contributions are as follows:

1. We propose a novel approach that combines the use of Bayesian network and probabilistic association rules to discover anomalies in data. We focus on the *causality* effect that describes why an observation is anomalous.
2. Our proposed approach is designed specifically to give contextual information of an anomaly, which can also be used to enrich our knowledge about the anomaly.
3. We perform extensive experiments and show that our proposed approach gives results in high precision and recall.

The remainder of the paper is organized as follows. In Section 2, we present our detailed methodology. Experiments and results are discussed in Section 3. And, finally we give a conclusion in Section 4.

2   COM Methodology for Anomaly Detection

In this section, we explain two probabilistic rules which we call as **R₁** and **R₂** to mine interesting low probability patterns from a given domain whose knowledge is captured by a Bayesian network. Before we proceed, we would like to make following clarification on theory of causal subspaces and rules.

- Rules are applied on each causal subspace encoded in the Bayesian network in order to reveal low probability patterns.
- A parent node in one causal subspace could appear as a child node in another causal subspace and vice-versa.
- In any causal subspace there could exist more than one parent of a child node, but more than one child node is not possible.

With this clarification, we now define $\mathbf{R}_1$ and $\mathbf{R}_2$ as follows:

1. $\mathbf{R}_1$: *In every causal subspace, select that state in child node which have a high confidence conditioned on all its parents in low support.*
2. $\mathbf{R}_2$: *In every causal subspace, select those state(s) in child node which have a low confidence conditioned on all its parents in high support.*

Both of the these rules work on principle of two measures namely *support* and *confidence*. The definitions of support and confidence of a variable in BN are defined using Equation 2 and 3 respectively. Support of variable X is like a prior probability in some state of x_i. In contrast, confidence is a conditional probability of variable X in some state x_i given set of observations Y.

$$support(X = x_i) = P(X = x_i) \tag{2}$$

$$confidence(X = x_i) = Pa(X = x_i | Y) \tag{3}$$

In our work, we use the concept of support for all parent nodes in each causal subspace structured in the Bayesian network whereas, confidence is computed for each child node encoded in the causal subspace. This implies, Equation 2 and Equation 3 can be rewritten as Equation 4 and Equation 5 respectively for each causal subspace (CS) encoded in the Bayesian model.

$$support(X = x_i)_{X \in CS_j} = P(X = x_i)_{X \in CS_j} \tag{4}$$

$$confidence(X = x_i)_{X \in CS_j} = Pa(X = x_i \mid Pa(X))_{X, Pa(X) \in CS_j} \tag{5}$$

Intuitively, rules $\mathbf{R}_1$ and $\mathbf{R}_2$ mine those suspicious patterns which do not provide enough evidence to accept them as the usual theory of the domain, but actually are an indicator of an alternative theory not favored by the domain. The $\mathbf{R}_1$ focuses on the extraction of the *"low support & high confidence"* patterns, which refers to the patterns whose "cause" appears with low probability but, interestingly the impact on the "effect" is strong. On the other hand, the rule $\mathbf{R}_2$ aims for the *"high support & low confidence"* patterns, which means that $\mathbf{R}_2$ mines those patterns whose "cause" appears with high probability, but has low impact on the respective "effect". We exclude *"low support & low confidence"* patterns because the causal relationship that showing low conditional probability conditioned on low prior is more like a *noise* rather than an anomaly.

We refer the low support, high support, low confidence, and high confidence as *minsupp, maxsupp, minconf,* and *maxconf* respectively. The first two are Bayesian specific, while the last two are parameters defined by a user. Equations 6 and 7 define the mathematical definitions for *minsupp* and *maxsupp* respectively.

$$minsupp(X = x_i)_{X \in CS_j} = \min_i \left(P(X = x_i) \right)_{X \in CS_j} \qquad (6)$$

$$maxsupp(X = x_i)_{X \in CS_j} = \max_i \left(P(X = x_i) \right)_{X \in CS_j} \qquad (7)$$

Application of these rules in each causal subspace of BN results in mining DSAPs which has an implication expression of the form:

$$X[x_i] \rightarrow C[c_j] \qquad (8)$$

where the left hand side of the arrow represents parent nodes and the right hand side of the arrow is their respective child node. Information enclosed in the square brackets represents states satisfying rules taken by parent and child nodes respectively. Equations 9 and 10 present the formal definitions of these rules.

Rule 1 (R_1)
$$\forall X \in Pa(C) \in CS_j \; s.t. \; (P(X = x_i) = minsupp) \wedge (P(C = c_k|X) > maxconf) \qquad (9)$$

Rule 2 (R_2)
$$\forall X \in Pa(C) \in CS_j \; s.t. P(X = x_i) = maxsupp) \wedge (P(C = c_k|X) < minconf) \qquad (10)$$

Our assumption that considers each DSAP as an indicator of anomalous event may lead to high false positive rate because of multiple hypothesis testing problem especially in the case when total number of DSAPs (denoted by notation |DSAPs|) from a BN is large. In order to control false positive rate, we propose to rank extracted DSAPs on how interesting they are from the Bayesian perspective only if condition: $|DSAPs| > 2 * |X|$ is satisfied. We apply the concept of *sensitivity analysis* in Bayesian networks, which is a measure of how sensitive is the conclusion to the findings for ranking discovered DSAPs. Sensitivity analysis in BN is performed by entering the known observations and studying sensitivity incurred in variable of interest. If the findings give negligible impact on a node under study, then the findings are considered sufficiently influential. On the other hand, if the impact on a node under study is significant, then those observations are considered least interesting for the investigated node. To score every extracted DSAP on a sensitivity measure, we use the DSAP notation (Equation 8) where the observations in variables are on the left hand side of the arrow and the computed sensitivity in variable present on the right hand side of arrow. We then sort DSAPs in an ascending order and consider the top τ patterns with the lowest scores as the most interesting unlikely patterns. We present the causal outlier mining in Bayesian network (**COMBN**) algorithm in Algorithm 1.

The computational complexity of the algorithm COMBN is governed by factors such as, Bayesian graphical structure (number of nodes, links and total number of unconditional and conditional probability entries) and Bayesian inference. In our approach, we are dealing with every causal subspaces present in the BN which helps reducing complexity of application of our proposed rules

---

Algorithm 1. COMBN

Input: BN, parameters *minconf, maxconf*, |X|, τ and a test set
Output: DSAPs, anomalies
1. Compute *minsupp* and *maxsupp* for every parent node in BN
2. For all causal subspace in BN, repeat:
      2.1. Apply $\mathbf{R}_1$ and $\mathbf{R}_2$ using equations 9 and 10 to discover DSAP
      2.2. Compute sensitivity of discovered DSAP in BN
3. If (|DSAPs| > 2 * |X|) then,
      3.1 Sort DSAPs
      3.2 Output top (τ * |DSAPs|) low scored DSAPs
   else
      Output all the DSAPs extracted
4. Output test cases with DSAPs within as the anomalies

---

on sparse Bayesian networks. Both Bayesian inference and sensitivity analysis are known to be NP-hard problem [8]. However, we are only performing simple queries of the form $P(X=x_i \mid Pa(X))$ which are always tractable as compared to complex queries, $P(X=x_i \mid Y)$ where, Y belongs to set of descendent nodes of X in BN which may require operations such as, marginalization over irrelevant variables for computing such probability of interest.

3   Experiments

We performed experiments over three alternatives for anomaly detection besides COMBN algorithm namely Latent Dirichlet Allocation (LDA) [9], k^{th} nearest neighbour (k^{th}-NN) [6] and Local Outlier Factor (LOF) [7]. For each anomaly detection technique, we followed a training & testing environment. We invite interested readers to investigate [5] for more details on experimental set up we followed for these techniques. We performed experiments on six real data sets taken from UCI repository [1]. The column 1 of Table 1 lists data set names. Column 2 and 3 of the same table presents the summary of results achieved using COMBN, LDA, k^{th}-NN and LOF anomaly detection techniques. We set parameter τ=50% in the COMBN algorithm. For a reasonable comparison between our approach and LDA, we took the top n low probability patterns mined by LDA where, n was equal to (τ * |DSAPs|) set in COMBN algorithm. In k^{th}-NN approach, we set k=5 for experiments. We obtained encouraging results by our algorithm with precision and recall more than 70% for almost every data set.

The LDA result for KDD Cup data set are not shown because this data set contained imbalance proportion of instances belonging to each class (normal and 22 different attack types). In order to mine the patterns of low probability for each class type, it was important to train LDA model on these classes. However, we formed a ten randomized data set from original KDD Cup data set addressed as KDD Cup* especially for LDA in order to investigate its performance on a real network intrusion detection data set. Due to the space constraint, we present the performance of LDA over KDD Cup* data set in [5].

Table 1. Summary of results achieved using COMBN, LDA, kth-NN and LOF anomaly detection techniques

Data set	Precision (COMBN, LDA, kNN, LOF)	Recall (COMBN, LDA, kNN, LOF)
Zoo	(.91, .69, .62, .56)	(.99, 1, .62, .52)
House vote	(.91, .59, .54, .45)	(.95, .94, .64, .48)
Lymphography	(.72, .50, .69, .57)	(.83, 1, .69, .66)
Statlog	(.86, .49, .52, .49)	(.77, .49, .49, .45)
Mushroom	(.62, .56, .52, .66)	(.71, 1, .62, .61)
KDD Cup	(.96, -, .72, .41)	(.99, -, .66, .44)

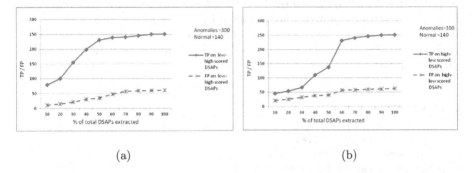

(a)   (b)

Fig. 1. (a) For Statlog test set pattern of TP and FP achieved on parameter τ scaled from 10% to 100% when DSAPs sorted in a ascending order (b) Same as (a) but DSAPs were sorted in a descending order

Recall from Section 2 that if the number of DSAPs extracted is large for a data set of low dimensionality then, it may lead to high false positive rate because of multiple hypothesis problem. In order to show the relation among the number of attributes, numbers of DSAPs extracted and false positive rate, first refer Fig. 1(a). It shows an increase in TP and FP with an increase in percentage of number of DSAPs extracted for Statog data set. We set the parameter τ from 10% to 100% and, calculated TP and FP at every scale of parameter τ. The number of normal and anomalous data points was 140 and 300 respectively in the test set of this data set. TP is represented by a thick line, whereas FP is shown by a dashed line. The graph clearly shows an increase in TP with an increase in percentage of total DSAPs until τ=100%. After this point, there is not much change in TP. However, there is always a constant increase in FP till τ=100%. This explains the fact that if we consider all DSAPs extracted to discover anomalies then, we may end up having good recall but poor precision. On the other hand, if we consider only top few low scored interesting DSAPs then; we can get both good precision and recall. We also show in Fig. 1(b) that the patterns of TP and FP for the same data set on similar scale of τ but, this time DSAPs were ranked in a descending order. Here the trend of

TP until $\tau=50\%$ is increasing at a very low pace. This indicates that top high scored DSAPs were least interesting from the anomaly discovery perspective. Interestingly, there is an increase in TP right after $\tau=50\%$ which clearly shows the contribution of low scored DSAPs in mining true anomalies.

4   Conclusion and Future Research Direction

In this paper we proposed two robust probabilistic association rules which are based on causal knowledge captured by a Bayesian network (BN) to mine anomalous patterns for the domain. We extracted patterns which are examples of either *low support & high confidence* or *high support & low confidence* events. Extracted patterns were then tested on new data points to discover anomalies. We prove the credibility of our approach over existing well known outlier detection techniques by taking well known benchmark data sets. For future directions, we are interested in improving definition of minimum and maximum support presented in this paper. Also, we plan to optimize of our approach to work on much higher dimensional data sets.

References

1. http://archive.ics.uci.edu/ml/
2. http://www.norsys.com/
3. http://b-course.cs.helsinki.fi/obc/
4. https://sites.google.com/site/bayesianoutlier
5. http://sydney.edu.au/engineering/it/~sakshib/
6. Ramaswamy, S., Rastogi, R., Shim, K.: Efficient Algorithms for Mining Outliers from Large Data Sets. In: Proceedings of International Conference on Management of Data, pp. 427–438 (2000)
7. Breunig, M.M., Kriegel, H., Ng, R.T., Sander, J.: LOF: Identifying Density-Based Local Outliers. In: Proceedings of the ACM SIGMOD International Conference on Management of Data, pp. 93–104 (2000)
8. Koller, D., Friedman, N.: Probabilistic Graphical Models: Principles and Techniques. MIT Press (2009)
9. Blei, D.M., Ng, A.Y., Jordan, M.I.: Latent Dirichlet Allocation. Journal of Machine Learning Research 3, 993–1022 (2003)
10. Wong, W.K., Moore, A., Cooper, G., Wagner, M.: Bayesian Network Anomaly Pattern Detection for Disease Outbreaks. In: Proceedings of International Conference on Machine Learning, pp. 808–815 (2003)
11. Babbar, S., Chawla, S.: On Bayesian Network and Outlier Detection. In: Proceedings of International Conference on Management of Data (2010)

Pathfinding by Demand Sensitive Map Abstraction

Sourodeep Bhattacharjee and Scott D. Goodwin

School of Computer Science
University of Windsor
Windsor ON N9B 3P4, CA
{bhattac1,sgoodwin}@uwindsor.ca

Abstract. This paper deals with the problem of pathfinding in real-time strategy games. We have introduced a new algorithm: Demand Sensitive Map Abstraction (DSMA) to overcome some of the challenges faced by the benchmark hierarchical pathfinding algorithm: Hierarchical Pathfinding A* (HPA*). DSMA is a type of hierarchical pathfinding algorithm in which we vary the granularity of the abstract map based on pathfinding request demand associated with various regions in the abstract map and the time taken by DSMA to find the previous path. Results from experiments show that dynamically varying the granularity of abstraction helps in maintaining a balance between path quality and search time.

Keywords: Pathfinding, Hierarchical, Abstract.

1 Introduction

Pathfinding is defined as the problem of finding a route of desired quality from a given start to a given goal location in a game map. This paper deals with pathfinding in game maps using hierarchical pathfinding techniques. Hierarchical pathfinding involves the use of a high level abstract map created from the low level, game map. This is employed to address the high search time of A* search [10]. We assume that the map is known a priori. All our maps use a grid representation at their lowest level and the grids have octile navigation which means a mobile agent can move in all eight directions. In this paper we present a new algorithm which we call Demand Sensitive Map Abstraction (DSMA), where we vary the granularity of the abstract map dynamically depending on the demand of pathfinding in a particular section and the last pathfinding time.We claim that, by varying the granularity of abstraction dynamically we can make better use of resources (CPU time and memory space) to find a suitable path, as opposed to keeping the granularity constant throughout the game-play.

The experimental results support our claim, since the performance metric curves of DSMA lie between the curves of the two constant granularity variations. Our work is significant because it varies the granularity of abstraction associated with specific regions instead of adding more hierarchical levels to suit the resources available.

O. Zaïane and S. Zilles (Eds.): Canadian AI 2013, LNAI 7884, pp. 233–240, 2013.

2   Related Work

The topic of Hierarchical Pathfinding applied to game world maps was suggested in the paper Near Optimal Hierarchical Pathfinding [2]. Their paper addressed the problem of pathfinding on "large" maps where limited CPU and memory resources create severe bottlenecks. They achieved this by employing hierarchical pathfinding techniques in their algorithm - Hierarchical Pathfinding A* (HPA*). The authors refer to [1] as related previous work. The authors state that A* is slightly better than HPA* when the solution length is small. The authors explain the difference in performance by stating that the overhead of inserting start and goal states into the abstract graph and other such techniques becomes an unnecessary expense when the map is mostly empty or the path is possibly a straight line through the grid. However, given a real game scenario with standard number of obstacles and mobile agents, the authors claim that HPA* is up to 10 times faster than a highly optimized A*.

3   Demand Sensitive Map Abstraction

3.1   Motivation

HPA* effectively solves the problem of pathfinding in large search spaces in a reasonable amount of time. The technique, however, has its own limitations which prevent it from being used in commercial games. The expensive pre-caching of intra edges step saves a lot time on scenarios where most of the map is being used for pathfinding. On the other hand, the same feature becomes undesirable when only a part of the map is traversable and pre-caching intra edges of unused sectors of the map is entirely unnecessary [3]. A corollary of the same situation occurs when the map is dynamic and a change in traversability in a portion of the map calls for an expensive update of the abstract map. The motivation for our algorithm sprouts from the idea of tessellating terrain using triangular bin-trees [4] [5] and the fact that it is reasonable to put a cap on the time required to find a path from start to goal. An ideal time range to find an initial path would be 1 millisecond to 3 milliseconds (ms) [7] [8] given the current technology standards.

3.2   Triangle Bin-Trees

A triangle Binary tree [9] (triangle bin-tree) is a spatial data structure. Triangle bin-trees are binary trees with space representational properties of Quad Trees. A triangle bin-tree is comprised solely of right isosceles triangles and hence never develops cracks or T-junctions. Triangle bin-trees are mainly used for tessellating terrain [5]. Let us consider figure 1. The triangle bin-tree consists of a triangle (I) and two possible children- a left child (III) and a right child (II). When triangle I is decomposed, we obtain II and III. Similarly, when triangle III is decomposed, triangles IV and V are produced. Conversely, it is also possible to compose or merge two neighbor triangles to reduce granularity. Henceforth, the term "triangle bin-tree" and "triplet" will be used interchangeably, referring to any arbitrary triangle in the abstract map.

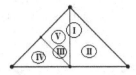

Fig. 1. Illustration of Triangle Binary Tree

3.3   Measuring Demand

We vary the granularity of abstraction dynamically according to the demand associated with triplets. Hence, in addition to level we add another parameter- Demand, to triplets. A value associated with demand represents how many times a triplet was explored in subsequent hierarchical A* searches. Demand is increased by one, for triplets containing start and goal nodes and for every triplet that gets expanded (explored) in hierarchical A* search. We also decrease demand by one for every triplet that does not get expanded in the high level A* search. When the last execution time of A* search rises above three milliseconds, we decompose the highest demand triplet (breaking ties arbitrarily) thereby forcing the low level A* search to expand fewer nodes so that the total search time is reduced. This step is expected to return less optimal path for lower execution time. Conversely, if the last execution time of A* search falls below one millisecond, we merge two of the lowest (collective) demand neighbor triplets (breaking ties arbitrarily).

3.4   Decomposition and Composition Operations

Decomposition. Consider the game map on the left hand side in figure 2; the dots represent the paths taken by games agents. In other words, the dots are the foot prints of the agents. As shared previously, we keep track of these footprints and use it to dynamically vary the demand of the triplets. In addition to that, we also keep track of the running time in A* search. If the last running time of A* search went beyond of 3 milliseconds, we apply the decomposition operation. When the decomposition operation is applied, we extract the highest demand

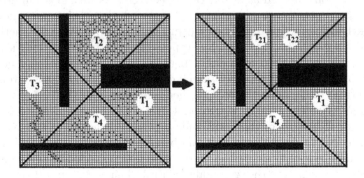

Fig. 2. Decomposition

triplet from the decomposition queue (T_2 for the map shown above) and we decompose it; thereby replacing T_2 by its children T_{21} and T_{22} as shown on the right hand side of figure 2.

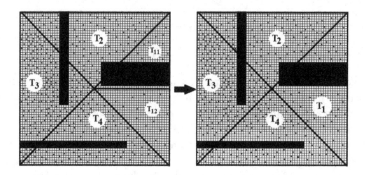

Fig. 3. Composition

Composition. The reverse of the decomposition operation is the composition operation; which we apply when the last running time of an A* search went below 1 millisecond. In this operation, we pick a pair of triangles/triplets from the composition queue with lowest collective demand. Usually this is the first element of the queue since we sort the composition in ascending order of collective demand at regular intervals. For the game world depicted in figure 3, if the last A* search time went below 1 millisecond,we would compose/merge T_{11} and T_{12} resulting in the game world shown on the right side of figure 3.

3.5   Pathfinding by DSMA

In this section we explain how DSMA finds path between two points; given a certain map and a given instance of an abstract high level map. Let us consider the map in figure 4, with a fixed high level abstract map and with the start and goal positions marked as S and G, respectively: As we can see in figure the start and the goal cells are ambiguously placed such that their centers lie on the border of the triplets. We resolve such conflicts arbitrarily. With the method we have used, the high level A* search can possibly take two directions as shown in figure 4. Let us proceed with case I because it poses a new challenge which we will explore. The high level A* search considers an abstract path from start to the centroid of T_1 as shown with dashed line. The high level A* search does not consider obstacles, unless a triplet is over 90% full or an obstacle covers the centroid. If it is more than 90% full, a penalty is added to the triplet's f-value during high level A* search. If an obstacle is covering a centroid, the low level A* search attempts to connect the present way-point to the next way-point (centroid), after skipping the way-point/centroid being covered by an obstacle.

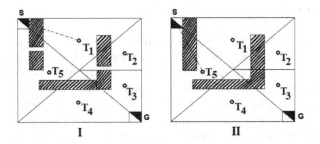

Fig. 4. Two possible abstract paths

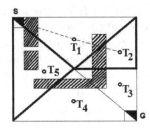

Fig. 5. Choosing the Adjacent Triplets

The high level A* search considers two of its adjacent triplets shown in figure 5 and selects T_2 assuming it has a lower f-value. The tentative abstract path from T_1 to T_2 is shown in the dashed line in figure 5. Now, let us assume that the goal is in T_3. The only adjacent triplet to T_2 is T_3 and T_3 contains the goal. Hence the high level search is successful and the complete abstract path is shown in figure 6.

Thus the high level A* search has discovered way-points which the low level A* search must now connect to get the actual path. The challenge of crossing the obstacle, we mentioned above has been resolved by the low level A* search, as shown in the right side of figure 6.

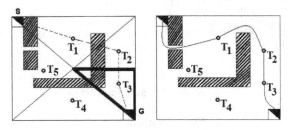

Fig. 6. Complete Abstract and Actual Path

4 Experiments

We have used ten different maps inspired from commercial RTS games for the purpose of experimentation. Each of the maps are scaled to three different pixel sizes: 256 by 256, 512 by 512 and 1024 by 1024. All maps are grid worlds with octile navigational freedom. In addition to hard coded obstacles in the maps, we also introduce random obstacles into the map (without modifying the map when a certain instance of path planning is in progress). The random obstacles are varied in density from 20% to 40%; always ensuring that the start and goal points are connected. All obstacles are made to fit to cells. In order to compare the algorithms we generate 500 random start and goal locations for every map. We compare DSMA to two versions of a generic hierarchical A* search (not same as HA* [1]). The first version-Sparse HA*, has a sparse and constant (single level) abstract map, containing eight triplets. The second version (Dense HA*) has a denser and constant (single level) abstract map, containing 64 triplets. The sparse map has 8 triplets because that is the minimum number of triplets DSMA is allowed to have and the dense has 64 because that is the maximum number of triplets we allow DSMA to create.

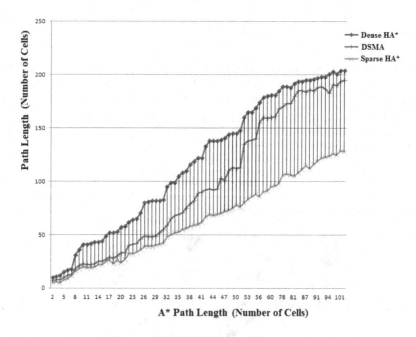

Fig. 7. Path Length

5   Results and Discussion

Firstly, let us consider the graph in figure 7 depicting relative path lengths of Dense HA*, DSMA and Sparse HA* against the path length given by A* in the x axis. This graph is taken from the maps of size 1024 by 1024 pixels. The points are average path lengths returned for a given A* path length. The tables are sorted in ascending order by A* path lengths, before the averages are determined. We observe that the DSMA algorithm balances the path length according to the time taken by the respective last A* search.

The graph in figure 8 shows the relative number of nodes expanded by Dense HA*, DSMA and Sparse HA* against the optimal path length for map size 1024 by 1024 pixels. It is evident that DSMA can keep the nodes expanded between those given by the dense and sparse configurations.

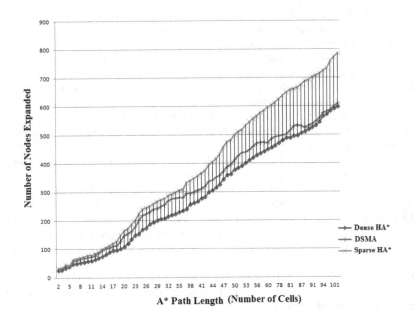

Fig. 8. Nodes Expanded

6   Conclusion and Future Work

In this paper we presented a new hierarchical pathfinding algorithm- Demand Sensitive Map abstraction in which we varied the granularity of abstraction dynamically depending on the pathfinding demand associated with various regions of the high level map and the time taken by DSMA to find the previous path. DSMA is an alternative to the HPA*, as HPA* pre-caches all intra-edges and DSMA does not have intra or inter edges. Instead DSMA uses a hybrid abstract edge that is computed on the fly. The results we derived are promising as DSMA is *successful in balancing the path quality and search time and continuously evolves the abstract map to keep the balance.* In future we plan to implement forced composition and decomposition. A forced composition is one where the triplets being composed do not come from the same parent while in forced decomposition we have to perform an additional decomposition in order to decompose a particular triplet.

References

1. Holte, R.C., Perez, M.B., Zimmer, R.M., MacDonald, A.J.: Hierarchical A*: searching abstraction hierarchies effiiently. In: Proceedings of the Thirteenth National Conference on Artificial Intelligence, vol. 1, pp. 530–535. AAAI Press (1996)
2. Botea, A., Muller, M., Schaeffer, J.: Near optimal hierarchical pathfinding. Journal of Game Development 1(1), 7–28 (2004)
3. Jansen, M.R., Buro, M.: HPA* enhancements. In: Proceedings of the Third Artificial Intelligence and Interactive Digital Entertainment Conference, Stanford, California, USA, pp. 84–87 (2007)
4. Samet, H.: The design and analysis of spatial data structures, vol. 85, p. 87. Addison-Wesley, Reading (1990)
5. Duchaineau, M., Wolinsky, M., Sigeti, D.E., Miller, M.C., Aldrich, C., Mineev-Weinstein, M.B.: ROAMing terrain: real-time optimally adapting meshes. In: Proceedings of the IEEE Visualization 1997, pp. 81–88 (1997)
6. Demyen, D.J., Buro, M.: Efficient triangulation-based pathfinding. Masters Abstracts International 45(03) (2006)
7. Dalmau, D.S.C.: Core techniques and algorithms in game programming. New Riders Pub. (2004)
8. Bulitko, V., Sturtevant, N., Lu, J., Yau, T.: Graph abstraction in real-time heuristic search. JAIR 30, 51–100 (2007)
9. Lindstrom, P., Koller, D., Ribarsky, W., Hodges, L.F., Faust, N., Turner, G.A.: Real-time, continuous level of detail rendering of height fields. In: Proceedings of the 23rd Annual Conference on Computer Graphics and Interactive Techniques, pp. 109–118. ACM (1996)
10. Hart, P.E., Nilsson, N.J., Raphael, B.: A formal basis for the heuristic determination of minimum cost paths. IEEE Transactions on Systems Science and Cybernetics 4(2), 100–107 (1968)

Detecting and Categorizing Indices in Lecture Video Using Supervised Machine Learning

Christopher Brooks, G. Scott Johnston, Craig Thompson, and Jim Greer

University of Saskatchewan, Department of Computer Science
110 Science Place, Saskatoon, SK
cab938@mail.usask.ca, {g.scott.j,craig.thompson,jim.greer}@usask.ca

Abstract. This work reports on the evaluation of detecting scene transitions in lecture video through supervised machine learning. It expands on previous work by gathering training data from multiple human raters. We include a robust evaluation that compares predictions against the entire set of expert classifications in disagreement. Finally, we explore some of the issues around constructing training data from multiple human experts, specifically emphasizing that evaluation strategies should be carefully considered when using aggregated training data.

1 Introduction

Computer-based lecture capture and media systems, such as Opencast Matterhorn[1] and echo360[2], provide the ability to record video of classroom lectures, including projector content such as PowerPoint slides or other desktop activity. Our interest in these technologies is to enable fast, accurate navigation through content by way of thumbnails that represent the start of new segments in a lecture. These segments can be thought of as roughly corresponding to new slides in a PowerPoint presentation, though our intent is to work with broader forms of presentation and not rely on any particular technology or lecture paradigm. Unlike previous work which has used static algorithms [1] or learning algorithms trained on data from a single human rater [2] for determining these kinds of indices, we provide a method to compare algorithms based on multiple raters that are in disagreement. The result is an increase in the quality of indexing compared to non-trained algorithms, and a method that considers training data from multiple raters.

One contribution of this work is an exploration of the effects of considering multiple raters when annotating data for supervised machine learning. In previous work [2], algorithms trained with data from a single rater were shown to produce better results than static algorithms. We go further and demonstrate *a*) the challenges in achieving agreement between multiple raters in this domain, and *b*) how aggregates for training can be formed on these conflicting ratings, and how these aggregates compare with the current state-of-the-practice.

---

[1] http://www.opencastproject.org
[2] http://echo360.com/

O. Zaïane and S. Zilles (Eds.): Canadian AI 2013, LNAI 7884, pp. 241–247, 2013.

This paper is organized as follows: Section 2 presents a case study, the kind of data we are interested in, and how human raters perform the categorization task. Section 3 compares our approach to other popular methods. Finally, the paper closes in Section 4 with a description of ongoing work in how supervised learning can be used to classify segments of video and other avenues of further research.

2   Human Indexing in Lecture Video: A Case Study

We performed a case study to collect training data for our supervised approach and to measure how well different people agree on indexing in educational video. In this study we have a set of videos captured from the data projector for nine lectures from a single undergraduate course in Computer Science. These videos are roughly 80 minutes each, and are broken into still images at one frame per second, resulting in a total of 43,770 video frames. These video frames were shown to six study participants who were selected to control for gender (three males and three females) and educational experience (they all had taken a university level course within the last two years). In contrast to our previous work [2], we explicitly sought out participants who were neither graduate students nor instructors, as we wanted to examine how non-pedagogues would perform educational video content indexing. While a group of six participants may be too small to make generalizations about how the general population indexes video, it is large enough to illustrate some of the problems encountered and to compare with unsupervised algorithms.

Study participants were instructed to mark index points in each video using a tool that allowed them to navigate through the video on a frame-by-frame basis. A purpose-based goal was used as a motivator for this task: learners were to "...mark all transitions as if [they] were building the left hand navigation window for a lecture video player," and were shown an image of a production lecture capture system to help better understand the task. Based on previous experiments, learners were dissuaded from semantically analyzing the content, and were asked to "mark transitions based on visual changes that [they thought] would be helpful for this lecture and other similar lectures." Finally, subjects were asked to limit their index choices to between fifteen and thirty indices per lecture video. Participants were able to perform this task at their own pace, and were free to navigate back and forth through the video, and select/unselect index points as they saw fit.

Fleiss' κ [3] is a measure of interrater reliability[3] useful for determining the agreement between pairs of people or within a larger group. We use it here to examine the relationship between our human participants; and later to evaluate the accuracy of our models. It ranges from -1.0 to 1.0. κ is chance-corrected:

---

[3] We use two vocabularies to refer to our study subjects: participants and raters. This is because the κ literature refers to raters that provide disagreeing ratings of instances; when discussing κ in general, we default to a rater/rating vocabulary to describe participants and their classifications of instances.

thus a score of 0 indicates only chance agreement, and positive scores indicate agreement beyond what would be expected by chance alone.[4] An issue with using κ in groups is that as the number of total raters increases, the amount of change any single rater can make to the overall level of agreement within the group decreases. To avoid ignoring differences between human raters, it is therefore useful to measure all pairwise κs (for our study, Figure 1a) to understand the significance of each individual's ratings and to observe any outliers.

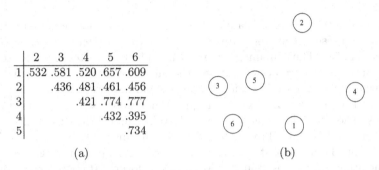

	2	3	4	5	6
1	.532	.581	.520	.657	.609
2		.436	.481	.461	.456
3			.421	.774	.777
4				.432	.395
5					.734

(a)   (b)

Fig. 1. (a) contains pairwise κ between raters. These were calculated with the standard formula for κ; the input data is a series of yes/no classifications of each video image by two participants. Two simultaneous yeses or noes is an agreement. (b) is an agreement map for 6 raters where distances between nodes are approximately inversely proportional to κ (raters closer together have higher agreement).

Using terminology from Landis and Koch [4], the κs between pairs of our participants (Figure 1a) range from "fair" ($\kappa = 0.395$) to "substantial" ($\kappa = 0.777$). These values vary from the group κ of the participants; in fact, their group κ is 0.577, and the mean of their pairwise κ is 0.551. This would indicate that the whole of their agreement is better than the sum of the parts of their agreement. Figure 1b visualizes the agreement between raters using a spring model. Each node represents an individual rater, and the Euclidian distance between nodes approximates the level of agreement between two raters. In the figure, raters two and four have the largest distance between themselves and other raters, indicating that their ratings of indices shared less is common with the other raters. Raters three, five, and six form a more tightly knit group, a reflection of their high level of agreement relative to the others ($\kappa \geq 0.734$ from Figure 1a).

After the first phase of the study, we had a consensus building activity with the participants that included a verbal discussion of the strategies each individual used to generated indices. We found no evidence that the human experts could predict who they were most similar to or different from, further illuminating

---

[4] κ is chance-corrected, as it is a ratio of the agreement achieved in a group, $\bar{P} - \bar{P}_e$, to the agreement that would have been achieved by chance alone, $1 - \bar{P}_e$. The details of its calculation are omitted for space.

the complexity inherent in ill-defined domains. Beyond the collection of human rater data, the lack of agreement between subjects is an important takeaway for our machine indexing task. This suggests that forming training data based on multiple raters will lead to more robust and less over-trained results. This is explored in the next section, with a focus on quantifying the effects of using multiple raters in disagreement.

3   Training Strategies from Ratings in Disagreement

As our goal is to mark significant slide transitions, our approach used attributes that each represent some measure of the difference between two images that are adjacent in time. Thus, an image at time t has a feature vector of attributes, each based on the difference between it and the image at time $t - 1$. For tractability, the images were taken from videos at one second intervals.

In total, nine different high level approaches were used to generate our attributes. The attribute groups are mostly taken unmodified from [2], or, were either minimally modified from [2] or first used in this work. In total, our dataset consisted of 134 different attributes for each sequential pair of images. For example, one attribute first used in this work was a "form" attribute, where our goal was to replicate a preceding phase that determined a semantic tag for images. To represent this attribute, we hand-coded each image as one of four such tags, and used a decision tree's classifications as an attribute in our main classification task. While some of our attributes are novel and worth reporting on, we leave the specifics of attribute formation and evaluation for future work, and instead focus on the overall process of gathering training data from multiple human raters and appropriately evaluating results from this type of data.

We generated models for classifying pairs of images as scene transitions, using our attributes and the J48 decision tree algorithm, as implemented in the WEKA Machine Learning toolkit.[5] As this algorithm requires a single classification per training instance, and because there was a lack of consensus from our human raters, we trained six J48 decision trees ($T_1, \ldots, T_6$) using six different aggregations of the participants' ratings. To aggregate the training data for each T_i, i is the minimum number of experts who must agree in order for an instance to be marked as a transition in the training set. Thus, the aggregate for T_1 required only one individual to indicate an index for a given instance, and T_6 required all six to agree; the T_1 aggregate therefore had the most positive classifications of indices, and T_6 had the fewest (approximately 30). All algorithms were evaluated using ten-fold cross-validation.

For comparison, we also evaluated the accuracy of three non-learning algorithms. The first of these, *Time*, is a naïvealgorithm that would select an index every 180 seconds into the video regardless of the content of video frames. The second, *Opencast*, was the method used in production code by the Opencast Matterhorn project. This algorithm uses differences in RGB intensities between frames to detect changes. The final algorithm we considered, *Dickson*, from [1],

---

[5] http://www.cs.waikato.ac.nz/ml/weka/

is a multi-pass image processing function that examines both pixel and block characteristics of video to determine stable events. All of these algorithms have seen real-world deployment in lecture capture systems, though few studies have been done as to the quality of these methods.

It is important to note that our usage of κ in this section differs from its typical usage in the supervised machine learning literature. Often, κ is used to compare the agreement between the model that results from a machine learning algorithm, and a test dataset; in general, κ is used to measure agreement between one algorithm and one training set, or within a group of humans. However, our goal is not to evaluate the ability of the algorithm to develop an accurate model of the aggregated training data, as we are using existing algorithms that have been demonstrated to be capable. Instead, our goal is to evaluate the ability of the resulting model itself (not the algorithm that generated a model). Specifically, we are evaluating a model's ability to replicate the pattern of decision making that our human raters used when they classified our training data. Thus our goal in using κ is to evaluate the ability of a trained model to match the unaggregated human raters' opinions. Future work to evaluate the suitability of the algorithm should also consider other metrics of evaluation (such as accuracy and precision/recall curves).

As κ can evaluate the agreement between multiple raters simultaneously, we can use κ to evaluate how well our model's predictions agree with the set of all six human raters at the same time, rather than an aggregation. This allows us to compare our six aggregation strategies to each other more fairly, by consistently measuring the agreement of T_1 through T_6 with all six human raters, rather than comparing each one to its own training aggregate. Recall that we are interested in finding the aggregation strategy that produces a model that best agrees with the human raters, not the model that best agrees with its own training aggregate. This distinction is subtle, but important.

We therefore believe it is more appropriate to use κ to compare the learned model with the entire group of expert raters, rather than comparing the learned model to the training set that was aggregated from the group of expert raters. The results presented in this section thus represent the ability of a model (T_1 through T_6, *Time, Opencast, Dickson*) to agree with the opinions of human raters, rather than the model's ability to agree with its training data.

However, using κ for groups instead of pairs requires a different method of interpreting the values of κ. This is because the group of human raters is large in size compared with the addition of one new rater, and because the group of human raters are often largely in agreement. A group κ includes the human raters' agreement, and is only somewhat affected by the addition of a rater. To more clearly see the effect of any particular model, we should contextualize the κs with how it changes based on one rater's choices. To do this, we calculate an upper and lower bound on the possible κ by computing κ when an artificial rater is added to the group. Since it is more likely that a set of raters will disagree partially on any given instance than that they will agree unanimously, the maximum agreement an added artificial set of ratings could achieve is determined by adding a rater who always agrees with the majority; the same is true for the minimum level of agreement, obtained by adding a rater who disagrees with the majority. We can construct these two maximally agreeing and minimally agreeing ratings for any group of raters, and their group κs with the original group of raters, are the upper and lower bounds of possible κ values for the given group and any additional rater.

Table 1 presents these lower and upper bounds on κ given our group of human raters, and shows that any given algorithm can at best raise the group κ to 0.626, or at worst lower the group κ to -0.15. The results of adding ratings by the different comparison algorithms (*Time, Opencast, Dickson*) to the human rater values, as well as the results of our trained algorithms (T_1 through T_6) are also shown in Table 1.

Table 1. Group κ between raters and algorithms. *upper* and *lower* are the max and min values any algorithm could provide for κ. Recall that κ between the six expert raters without an algorithm was 0.577.

κ Bounds		Comparison Algorithms			Our Trained Algorithms					
upper	*lower*	*Time*	*Opencast*	*Dickson*	T_1	T_2	T_3	T_4	T_5	T_6
0.626	-0.15	0.391	0.370	0.448	0.574	0.565	0.565	0.537	0.530	0.487

To interpret the κs of a system's trained models versus comparison algorithms, we can compare the κs with the group κ of the six participants together but without an algorithm (0.577). We note that each comparison algorithm lowers κ more than any of our trained algorithms. We can interpret this as saying that our training strategies provide better indexing results when compared with human experts than the automated algorithms.

In the case of a study that does not have comparison algorithms, because of the expense of reimplementing other works or for other reasons, the value of the κs of the trained algorithms can still be compared versus the calculated bounds. While Landis and Koch [4]'s subjective categorizations would indicate that κs near 0.5 show only "moderate agreement," the κs achieved here are reasonably close to the maximum possible once you compare with the entire group of human raters. Further, our model T_1 provided nearly the same level of agreement as the group of expert raters alone ($\kappa = 0.574$ versus $\kappa = 0.577$); this suggests that our model T_1 finds index points as reliably as any one of our human experts.

It is also interesting to note that as we require more consensus from our training set (e.g. T_6 instead of T_1), κ decreases. We believe the increasing sparseness of positive instances decreases the models' accuracy. This further casts doubt on generic aggregation strategies in the case of significant class imbalance.

4   Conclusions and Future Work

This paper advances the state of the art in generating indices from video using supervised machine learning. This is a subjective and ill-defined task, requiring human raters to mark points within a video to be used for navigation. The task results in a subjective dataset that contains disagreement between raters that should not be dismissed as noise. In particular, we have provided two contributions to the fields of machine learning and user modelling.

1. Often, in the field of machine learning, predictive accuracy of a model is quantified by comparing the model's predictions with true values with a given evaluation metric. We disagree with this strategy in the case of gathering human participants' opinions for training an applied system; here, the goal of evaluation should be to judge the model's ability to predict the participants' opinions, not the ability of an algorithm to create a model that agrees with its training data. We draw this conclusion directly from our pairwise use of κ in Section 2: although participants given the same task produce reasonable aggregated training data, the process of aggregation removes nuances of their disagreement, which can influence model accuracy. In the future, we hope to continue examining how individual raters' opinions are better predicted by specific attribute types, and how to better aggregate multiple ratings as training data for algorithms that accept only one class attribute.

2. We demonstrated how bounds on κ could be determined so that κ can be used effectively for comparing models with groups of human raters. Although κ can be used to compare a model's predictions with test data, by comparing an algorithm with humans that represent a target audience for a practical system, we can directly evaluate the algorithm's usefulness at emulating opinions humans would find useful. Interpretation of group κs with bounds is important because of how κ is normally reported as significant for values larger than 0.0, as in [4]; but we demonstrate that even naïve algorithms such as *Time* achieve "significant" agreement because of the inherent agreement in the group itself. Instead, we suggest determining whether a κ measured with a group of humans *a*) is close to the upper bound on achievable κ, and *b*) is close to the contribution to κ that each disagreeing human rater makes.

References

1. Dickson, P., Adrion, W., Hanson, A.: Automatic Capture of Significant Points in a Computer Based Presentation. In: Eighth IEEE International Symposium on Multimedia (ISM 2006), pp. 921–926 (2006)
2. Brooks, C., Amundson, K.: Detecting Significant Events in Lecture Video using Supervised Machine Learning. In: 2009 Conference on Artificial Intelligence in Education (2009)
3. Fleiss, J.L.: Measuring nominal scale agreement among many raters. Psychological Bulletin 76(5), 378–382 (1971)
4. Landis, J.R., Koch, G.G.: The Measurement of Observer Agreement for Categorical Data. Biometrics 33(1), 159–174 (1977)

Improvements to Boosting with Data Streams

Erico N. de Souza[1] and Stan Matwin[2,*]

[1] School of Information Technology and Engineering
University of Ottawa
edeso096@uottawa.ca
[2] Faculty of Computer Science,
Dalhousie University
stan@cs.dal.ca

Abstract. Data Streams (DS) pose a challenge for any machine learning algorithm, because of high volume of data - on the order of millions of instances for a typical data set. Various algorithms were proposed, in particular, OzaBoost - a parallel adaptation of AdaBoost - creates various "weak" learners in parallel and updates each of them with new instances during training. At any moment, OzaBoost can stop and output the final model. OzaBoost suffers with memory consumption, which avoids its use for certain types of problems. This work introduces OzaBoost Dynamic, which changes the weight calculation and the number of boosted "weak" learners used by OzaBoost to improve its performance in terms of memory consumption. This work presents the empirical results showing the performance of all algorithms using data sets with 50 and 60 million instances.

1 Introduction

Data Stream mining is the process of extracting knowledge from rapidly produced data records. The data can be from different sources, such as networks or the stock market. Since the data set is typically generated very quickly, it is difficult to store this information for further evaluation, making hard to use traditional Machine Learning (ML) algorithms in data stream environment.

One of the first algorithms adapted to deal with data streams was Decision Trees (DT). Domingos and Hulten [4] proposed Very Fast Decision Trees (VFDT) that modified the traditional node splitting policy used by decision trees, such as InfoGain, and applied the Hoeffding bound as the splitting criterion.

Another option is to use ensemble methods to combine various learners to predict results from data streams. The problem with ensemble methods is that they are slower than DT's, though they do offer better accuracy if the data stream cannot be linearly separated. The two main challenges to adapt the two main ensemble methods, AdaBoost and Bagging, to data streams are: 1) both

---

* The author is also affiliated with the Institute of Computer Science, Polish Academy of Sciences, Poland.

O. Zaïane and S. Zilles (Eds.): Canadian AI 2013, LNAI 7884, pp. 248–255, 2013.

algorithms depend on multiple passes on the data set, and 2) they depend on previous knowledge of the size of the data set.

Oza and Russell introduce OzaBoost [6], which is a parallel boosting strategy that follows AdaBoost, with the exception of weight calculation, since AdaBoost depends on previous knowledge of the number of instances available for training - impossible in a data stream setting. In order to solve this issue, Oza and Russell used a Poisson distribution to calculate weights. The disadvantages of OzaBoost are related to model size and evaluation time of the algorithm. Both can grow exponentially, making this approach not suitable for some data sets.

This work adapts the weight calculation and the variation of the learner as used by [3] to the OzaBoost algorithm proposed by [6]. This new algorithm is called OzaBoost Dynamic (OzaDyn), and its main advantage is not related to accuracy, since it performs like OzaBoost in this respect, but to improvements in model size and evaluation time. In addition, an extra modification to OzaDyn was to allow it boost two "weak" learners, instead of one single learner. This idea comes from the works introduced by de Souza and Matwin [2,3] who presented a version of AdaBoost using various learners during the training process. In the context of data streams, learner variation is also affecting model size and memory consumption, which are reduced, because a complex model (VFDT) and a simple model (Naive Bayes Multinomial, or MBN) are combined. Each learner is executed in an alternating fashion - one algorithm in one iteration and the other in following step.

This paper is organized as follows: Section 2 presents OzaBoost; Section 3 discusses On-line Dynamic Boosting, its origins, and how the weight calculation was inserted; Section 4 describes the results and the evaluation procedure used; and Section 5 presents the conclusions and suggestions for future work.

2   OzaBoost

In [6], Oza and Russell present their On-line Boosting algorithm known as the OzaBoost algorithm, a kind of parallel boosting algorithm. The algorithm adapts AdaBoost.M1[5] to process data streams, and updates the weights with a Poisson distribution.

The algorithm presented in Table 1 performs as follows: the first step initializes the array h_t indexing all the learners, and the variables that will store the number of correct classifications (λ_t^{sc}) and incorrect classifications (λ_t^{sw}). Then for all training examples, the algorithm updates a variable k with the Poisson value (line 5), then uses this value to increase or decrease the weight of the instance for each h_t (line 6). After this, the algorithm checks if the new h_t correctly classifies the instance (line 7). If it does, λ_t^{sc} is updated with the value of λ_d (line 8), and λ_d is updated so its value is equal to $\lambda_d \frac{N}{2\lambda_t^{sc}}$ (line 9). When it is detected that h_t misclassified the example, the algorithm updates λ_t^{sw} in the same way as in line 8, and λ_d will be equal to $\lambda_d \frac{N}{2\lambda_t^{sw}}$ (lines 11 and 12). At any time, the algorithm can output the final hypothesis $H_{final} = \arg\max_{y \in Y} \sum_{t:h_t(x)=y} \log \frac{1}{\beta_t}$, that is equal to the original AdaBoost.M1, with $\beta_t = \frac{\epsilon_t}{1-\epsilon_t}$, and $\epsilon_t = \frac{\lambda_t^{sw}}{\lambda_t^{sw}+\lambda_t^{sc}}$.

Table 1. Oza and Russell's Online Boosting as implemented by MOA. N is the number of examples seen, and T is the number of models.

01. Initialize base models h_t for all $t \in \{1, 2, ..., T\}$, $\lambda_t^{sc} = 0, \lambda_t^{sw} = 0$
02. **for** *all training examples* **do**
03.   Set "weight" of example $\lambda_d = 1$
04.   **for all** t **do**
05.   Set $k = Poisson(\lambda_d)$
06.   Update h_t with the current example with the weight k
07.   **if** h_t correctly classifies the example **then**
08. $\lambda_t^{sc} \leftarrow \lambda_t^{sc} + \lambda_d$
09. $\lambda_d \leftarrow \lambda_d \frac{N}{2\lambda_t^{sc}}$
10.   **else**
11. $\lambda_t^{sw} \leftarrow \lambda_t^{sw} + \lambda_d$
12. $\lambda_d \leftarrow \lambda_d \frac{N}{2\lambda_t^{sw}}$
13.   **end if**
14.   **end for**
15. **end for**
Anytime output
Calculate the $\epsilon_t = \frac{\lambda_t^{sw}}{\lambda_t^{sw}+\lambda_t^{sc}}$, and $\beta_t = \frac{\epsilon_t}{1-\epsilon_t}$ for all m
Return: $H_{final} = \arg\max_{y \in Y} \sum_{t:h_t(x)=y} \log \frac{1}{\beta_t}$

Massive On-line Analysis (MOA) [1] documentation shows that the authors tried to use the original boosting weight calculation, instead of the Poisson, but the performance was no better than OzaBoost's [1].

3   On-line Dynamic Boosting

Table 2 shows the proposed modifications to OzaBoost. The main changes in relation to OzaBoost (Table 1) are the weight update changes in line 6 and the error calculation in line 13. Line 6 changes the weight calculation that was based on the Poisson distribution to a new weight calculation based on the sigmoid function. Line 13 calculates how many times h_t misclassified the instance. Like OzaBoost, OzaDyn could stop at any time and calculate the final model with $\beta_t = 1 - (\epsilon_t(1 - \epsilon_t))$ applied in AD-AC, and the output will be $H_{final} = \arg\max_{y \in Y} \sum_{t:h_t(x)=y} \beta_t$.

Notice that each β_t, and the final hypothesis (H_{final}), does not use logarithms in their calculations, and this, combined with applying a different learner (VFDT and Multinomial Naive Bayes) in each iteration, helps to make the algorithm significantly faster than the original OzaBoost, while maintaining the accuracy for the data sets tested.

Table 2. OzaDyn algorithm that adapts the weight calculation used in AD-AC algorithm to OzaBoost algorithm. N is the number of examples seen, and T is the number of base models.

01. Initialize base models h_t for all $t \in \{1, 2, ..., T\}$, $\lambda_t^{sc} = 0, \lambda_t^{sw} = 0$
02. **for** *all training examples* **do**
03.   Set "weight" of example $\lambda_d = 1$
04.   Set current error $\varepsilon = 0.99$
05.   **for all** t **do**
06.   Update h_t with current example with weight $\frac{1}{1+e^{-(1-(\varepsilon(1-\varepsilon)))\phi}}$
07.   **if** h_t correctly classifies the example **then**
08. $\lambda_t^{sc} \leftarrow \lambda_t^{sc} + \lambda_d$
09. $\lambda_d \leftarrow \lambda_d \frac{N}{2\lambda_t^{sc}}$
10.   **else**
11. $\lambda_t^{sw} \leftarrow \lambda_t^{sw} + \lambda_d$
12. $\lambda_d \leftarrow \lambda_d \frac{N}{2\lambda_t^{sw}}$
13. $\varepsilon \leftarrow \frac{\lambda_t^{sw}}{\lambda_t^{sw}+\lambda_t^{sc}}$
14.   **end if**
15.   **end for**
16. **end for**
Anytime output
Calculate the $\epsilon_t = \frac{\lambda_t^{sw}}{\lambda_t^{sw}+\lambda_t^{sc}}$, and $\beta_t = 1 - (\epsilon_t(1 - \epsilon_t))$ for all t
Return: $H_{final} = \arg\max_{y \in Y} \sum_{t:h_t(x)=y} \beta_t$

4   Algorithm Evaluation

MOA implements various data stream generators. Those used in this work are the Random Tree Generator(RTG) and Function Generator (also known as Agrawal data stream). We chose these particular two data streams generators because, unlike the others, they are well documented in MOA. Furthermore, MOA allows the addition of artificial drifts to any type of data stream with support of a noise generator based on sigmoid function. The only variable changed was the angle of the drift, which was set up to 50 degrees, for the Agrawal data set.

Tests were conducted as follows: First RTG and Agrawal data stream were executed without any artificial drift, and then the Agrawal data stream was executed with addition of concept drift. The number of instances used for the RTG was 60 million, and 50 million for the Agrawal data stream. The difference is due to the fact that OzaBoost crashes once the Agrawal data set reaches 50 million instances.

All simulations were executed a machine with 4 Core Intel(R) Xeon(R) CPU 2.67GHz 64 bits, and 8GB RAM. Operating system was Linux, with Kernel 2.6.32.26-175. The Java heap size used was 3GB for all algorithms, with exception of OzaBoost that required the memory to be extended until 7GB.

Cross-validation is not an alternative for data streams, because on certain data sets with high number of drifts, the separation between training and testing data may hide these drifts, and give wrong results. In MOA, the Prequential approach is a kind of stream adjusted leave-one-out approach in which each individual example is used to test the model before it is used for training, and the accuracy is incrementally updated. No holdout set is used for testing, and cumulative nature of the Prequential scheme ensures a smooth plot of accuracy over time, as each individual become increasingly less significant over time. The Prequential approach was the chosen evaluation procedure, because the accuracy, processing time and memory occupation measures are reliable[1].

OzaDyn was tested in two configurations. The first considered the algorithm with one classifier (VFDT), and the second considered the combination of results of two different classifiers (VFDT and NBM) by alternating their executions. OzaBoost used VFDT as its base classifier. All OzaDyn and OzaBoost algorithms considered only ten iterations, and used their default configurations.

Figure 1(a) presents the accuracy results using NBM, VFDT, OzaBoost with VFDT, OzaDyn with VFDT, and OzaDyn with both VFDT and NBM, applied on the RTG data set. Figure 1(b) shows the accuracy results for the same classifiers with an Agrawal data set. Both data sets were without concept drift. All algorithms had accuracy close to 100%, with exception of NBM.

Figures 1(c) and 1(d) present the processing time in seconds it took to execute the model evaluation for the same algorithms. Figure 1(c) shows that all algorithms had linear performances relative to the number of instances. However, OzaBoost used more time to evaluate the models (more than four hours to evaluate), OzaDyn with two classifiers had better performance than OzaBoost and OzaDyn with one classifier, since it used NBM, and took approximately 2 hours to finish the task. VFDT and NBM had the best performances. Figure 1(d) shows that OzaBoost used with Agrawal data set took almost 14 hours to evaluate, when dealing with 50 million instances. The main reason for this behaviour can be explained by the fact that OzaBoost's output is a sum of various logarithms, and this operation, with increasing number of instances, is very costly. In comparison, the performance of OzaDyn with one classifier that based on the sums of simple multiplications is exponentially better.

Figures 1(e) and 1(f) show the evolution of the model size (in bytes) in relation to number of instances. Figure 1(e) shows that OzaBoost with VFDT applied to RTG kept an almost constant size level until it reaches 50 million instances. After this, OzaBoost's size leveled off near 350Mb, where it stayed until the end of the simulation. The explanation for this unusual change in size is that the implementation removes the worst classifier when a change (drift) is detected, and a new classifier is added. OzaBoost is sensitive to this variation, because the weights are calculated iteratively, by adapting a new window size, and increasing the evaluation time. OzaDyn with one classifier is less prone to this variation, because the weights are based on the error calculated by each model, and it reduces the weights of instances that did not help improve the model, which makes the algorithm almost linear in relation to model size and evaluation time. Interestingly, OzaDyn combining two classifiers (VFDT and NBM)

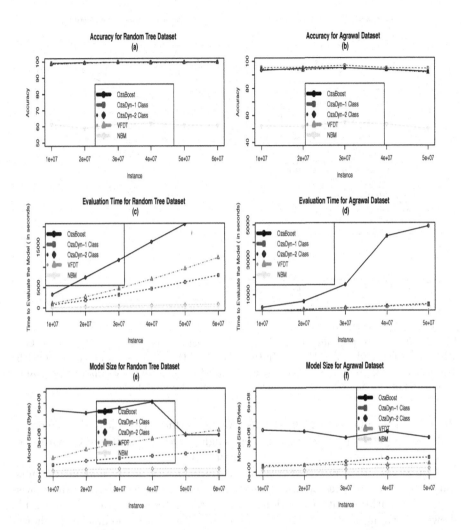

Fig. 1. This figure presents the performance of OzaDyn with one classifier (OzaDyn - 1 Class) and with two classifiers (OzaDyn - 2Class) compared to original OzaBoost, VFDT and Naive Bayes Multinomial (NBM) for the Random Tree and Agrawal data sets. Figures (a) and (b) compare the accuracy of the algorithms, figures (c) and (d) compare the evaluation time, and (e) and (f) present the model size.

had smaller memory consumption than the other algorithms. The reason is that NBM maintains memory below 200Mb, reducing the memory footprint of the entire model. Figure 1(f) shows the memory consumption for the Agrawal data

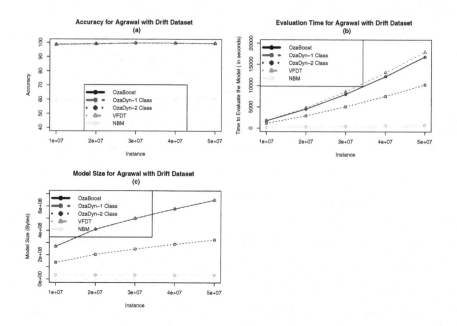

Fig. 2. This figure presents the performance of OzaDyn compared with original Oz-aBoost, VFDT and Naive Bayes Multinomial (NBM) for the Agrawal data set with drifts. Figure (a) compares the accuracy of algorithms, figure (b) compares the evaluation time of the algorithms and figure (c) presents the model size.

set, using the same set of algorithms. All algorithms had linear performance, with only a OzaDyn with two classifiers slightly worse than OzaDyn with one classifier. Both algorithms were better than regular OzaBoost.

Figures 2(a), 2(b) and 2(c) present the performance of the same algorithms using the Agrawal data set, when an artificial drift is added. Figure 2(a) shows that all algorithms, with exception of NBM, reach 100% accuracy. Figure 2(b) shows the evaluation time in seconds for all algorithms, and the first result shows that OzaDyn with one classifier is worse than OzaBoost, while OzaDyn with two classifiers is better than the others. This is because OzaDyn with two classifiers combines a complex classifier (VFDT), and a simple classifier (NBM). The same pattern can be seen in Figure 2(c), which presents the model size (in bytes) for all algorithms. OzaDyn with one classifier and OzaBoost have exactly the same model size, while OzaDyn with two classifiers saves memory space by combining the complex and simpler models. One important observation is that when the data stream presents drifts, models tend to increase their respective sizes and evaluation time. VFDT and NBM also increase both measures.

5   Conclusions

This paper presented a modification of the OzaBoost weight calculation, and an adaptation which enables it to work with more than one classifier. The new algorithm, OzaBoostDynamic, originated in previous work by de Souza and Matwin [2,3] that presented a variation in boosting weight calculation that allowed more than one base classifier to be boosted.

In order to better evaluate the effect of each modification of OzaBoost two versions were implemented and tested: one using a single base classifier, and the other using two base classifiers. This allowed us to check how the algorithm performs with the sole modification of the weight calculation. Another reason was to check how the use of more than one classifier improves different aspects of the OzaBoost algorithm.

These modifications were tested with two very large data sets generated by MOA software. Two tests were conducted with both data sets without concept drift, and one test was done with artificial drift added to the Agrawal data set. Tests on the data sets without concept drift showed that OzaBoostDynamic with either one or two classifiers is a significant improvement of OzaBoost, considering time taken to evaluate the models and the respective model sizes. Considering the data set with concept drift, OzaBoostDynamic with one classifier had inferior performance than OzaBoost in terms of evaluation time, but OzaBoostDynamic with two classifiers had better performance than OzaBoost. The reason is that when there is a drift in the data set, it tends to increase model size and evaluation time, and the combination of a simple model with complex model helps to maintain memory consumption and processing times below that of OzaBoost performance. A valuable direction for future work would be to apply OzaBoost-Dynamic on real data stream problem, eg. in a network performance analysis task.

References

1. Bifet, A., Holmes, G., Kirkby, R., Pfahringer, B.: Data stream mining: a practical approach. Technical report, University of Waikato (May 2011)
2. de Souza, É.N., Matwin, S.: Extending adaBoost to iteratively vary its base classifiers. In: Butz, C., Lingras, P. (eds.) Canadian AI 2011. LNCS (LNAI), vol. 6657, pp. 384–389. Springer, Heidelberg (2011)
3. de Souza, E.N., Matwin, S.: Improvements to adaBoost dynamic. In: Kosseim, L., Inkpen, D. (eds.) Canadian AI 2012. LNCS (LNAI), vol. 7310, pp. 293–298. Springer, Heidelberg (2012)
4. Domingos, P., Hulten, G.: Mining high-speed data streams. In: KDD, pp. 71–80 (2000)
5. Freund, Y., Schapire, R.E.: A decision-theoretic generalization of on-line learning and an application to boosting. Journal of Computer and System Sciences 55(1), 119–139 (1997)
6. Oza, N.C., Russell, S.: Online bagging and boosting. In: Jaakkola, T., Richardson, T. (eds.) 8th International Workshop on Artificial Intelligence and Statistics, Key West, Florida, USA, pp. 105–112. M. Kaufmann (2001)

Revisiting the Epistemics of Protocol Correctness

Aaron Hunter

British Columbia Institute of Technology
Burnaby, BC, Canada
aaron_hunter@bcit.ca

Abstract. Formal logics of knowledge have been employed for the verification of cryptographic protocols. Intuitively, the idea is that a correct run of a protocol should lead the participating agents to "know" some facts, due to the structure of the protocol. While there has been a great deal of work in this area, fundamental questions remain regarding the foundations of the approach. In fact, it is not even clear if any intrinsic properties of knowledge are actually required; in most cases the relevant notion of knowledge reduces directly to universal quantification over protocol runs. Moreover, there is no consensus about the appropriate epistemic domain for protocol participants, the nature of epistemic change, or the significance of initial knowledge. In this paper, we address these fundamental questions by revisiting the epistemic foundations of protocol verification.

1 Introduction

Cryptographic protocols are structured sequences of messages that are used to facilitate secure communication between agents on a hostile network. One well-known approach to the formal verification of cryptographic protocols is based on encoding the protocols in a logic of knowledge, and then proving that particular agents "know" certain facts at the conclusion of the protocol. This approach originated with the pioneering work on BAN logic [2], and it has continued with a variety of different logic-based approaches (e.g. [1,3,5,8,14,15]). Nevertheless, there is still relatively little consensus about the precise nature of the epistemic concepts that are relevant to protocol verification. In particular, there is some ambiguity about the nature of knowledge required for protocol verification, the domain of things that can be known, and the manner in which knowledge changes during a run of a protocol. In this paper, we aim to clarify the role of knowledge in reasoning about protocol correctness.

This is an exploratory paper examining the foundations of knowledge-based protocol verification. We make several contributions to the existing literature on the subject. First, we clarify the epistemic domain required for protocol verification by illustrating the distinct roles played by knowledge of *messages* as opposed to knowledge of *exchanges*. Second, we demonstrate that the notion of *protocol correctness* is implicitly epistemic and it can be formulated as a promotion from an individual's view of a message exchange to a global view. Finally, our work makes it explicit how compromised information can make a secure protocol vulnerable to attack. The purpose of this preliminary work is to lay the foundation for the future analysis of specific protocols in a knowledge-based formalism.

O. Zaïane and S. Zilles (Eds.): Canadian AI 2013, LNAI 7884, pp. 256–262, 2013.

2 Background

2.1 Cryptographic Protocols

Cryptographic protocols are typically described using the notation $A \rightarrow B : m$ to indicate that agent A sends agent B the message m. A complete protocol is then a numbered sequence of message exchanges, as in the following example:

The Needham Schroeder Public Key Protocol
1. $A \rightarrow B : \{N_A, A\}_{K_B}$
2. $B \rightarrow A : \{N_A, N_B\}_{K_A}$
3. $A \rightarrow B : \{N_B\}_{K_B}$

A symbol of the form N_A denotes a random number generated by an agent A, whereas an expression of the form $\{M\}_K$ denotes a message M encrypted with a key K. Each key in this protocol is a public key, and the subscript indicates the owner of the key. The goal of this particular protocol is *mutual authentication*. Hence, at the conclusion of the protocol, we would like for each agent to *know* that they have communicated with the other.

Most formal work on protocol verification is based on the Dolev-Yao intruder model, where a hostile intruder may intercept, read and/or forward any message sent on the network [6]. As such, the recipient of a message is never aware who sent the message.

2.2 Protocol Logics

Protocol verification is the task of proving that a given protocol satisfies some pre-specified goal. Starting with BAN logic [2], formal logics of knowledge and belief have been used for protocol verification. The details of BAN logic are not important for our purposes, and it is not our intention to survey existing protocol logics. The literature in this area is vast, but we are taking a foundational approach that is essentially self-contained. We assume that the reader is familiar with basic the multi-agent systems of [7], which is provably as expressive as the *strand space* model commonly used in security research [10]. The strand space model is, in turn, closely related to the operational semantics of some powerful automated tools for protocol verification (e.g. [4]).

2.3 Motivating Questions

We are motivated by several foundational issues related to knowledge in protocol verification. In existing approaches, it is not always clear if the notion of "knowledge" is being employed as a convenient metaphor or if knowledge is intrinsically relevant to protocol correctness. There are several related foundational questions.

1. What is the appropriate epistemic domain for proving protocol correctness?
2. Does verification require knowledge at the *agent*-level? Or is it sufficient to consider knowledge at the *system*-level?
3. Do fallible *beliefs* play any role in protocol verification?

We address the first two questions in this paper, as we have discussed the significance of beliefs in protocol verification in previous work [11].

3   Protocol Correctness

3.1   A Transition System for Message Passing

We introduce a simple formal model of message passing. In order to simplify the discussion, we do not introduce encrypted messages at this stage. Let $\mathbf{A}$ be a set of *agents* and let $\mathbf{M}$ be a set of *messages*. A *message set* for an agent is a subset of $\mathbf{M}$, intuitively representing the messages held by that agent. An *assignment function* is then a collection of message sets for all agents. More formally, we have the following definition.

Definition 1. *An assignment function is a function* $G : \mathbf{A} \rightarrow 2^{\mathbf{M}}$. *For* $A \in \mathbf{A}$, *the message set for* A *is the set* $G(A)$.

We stress that this definition and those that follow are not intended to be used as representational tools for real protocol verification. Instead, our focus is on making the role of knowledge explicit in a manner that can be applied across a wide range of formalisms.

If $A, B \in \mathbf{A}$ and $M \in \mathbf{M}$, we call $\langle A, B, M \rangle$ a *message exchange*. Informally, a message exchange represents the fact that the message M was sent from agent A to agent B. A sequence of message exchanges will be called a *trace*. We define a natural progression operator on assignment functions with respect to message exchanges.

Definition 2. *For any assignment function* G *and message exchange* $\langle A, B, M \rangle$, *define* $G + \langle A, B, M \rangle = G'$ *where* G' *is the assignment function satisfying:*

1. $G'(B) = G(B) \cup M$.
2. *For all* $C \neq B$, $G'(C) = G(C)$.

Using this definition, we can define a transition system that specifies the effects of message exchanges. The nodes in the transition system are labelled with assignment functions and the edges are labelled with $+\langle A, B, M \rangle$. We define sending actions and receiving actions in terms of sets of message exchanges:

$$send(A, M) = \{\langle A, B, M \rangle \mid B \in \mathbf{A}\}.$$
$$receive(A, M) = \{\langle B, A, M \rangle \mid B \in \mathbf{A}\}.$$

In order to define the semantics of sending and receiving, we extend the progression operator $+$ to apply to sets of messages.

Definition 3. *If* G *is an assignment function and* μ *is a set of message exchanges, then*

$$G + \mu = \{G + M \mid M \in \mu\}.$$

This definition allows us to define a transition system over sending actions and receiving actions. Again, the nodes are labelled with assignment functions. However, we can now label edges with *send* actions and *receive* actions. The effects are non-deterministic, so each send/receive action will be represented with several edges. If $T = M_1, \ldots, M_n$ is a trace, we use the shorthand $G + T$ to denote $G + M_1 + \cdots + M_n$.

It is useful to introduce a precise notion of a single agent's perspective on a trace. Given a trace T and an agent A, let $T \upharpoonright A$ be the trace obtained by removing all message exchanges that do not involve A. Let $T^A = \langle T_1^A, \ldots, T_n^A \rangle$ be the following sequence:

$$T_i^A = \begin{cases} send(A, M) & \text{if the } i^{th} \text{ element of } T \upharpoonright A \text{ is } \langle A, B, M \rangle \text{ for some } B \\ receive(A, M) & \text{otherwise} \end{cases}$$

Thus, T^A is a sequence of *sets of message exchanges*, representing A's view of T.

3.2 Specifying Protocols

Restricting to two agents, the general format of a protocol is the following.

Generic Protocol Description
1. $A \to B : M_1$
2. $B \to A : M_2$
$\ldots$
n. $A \to B : M_n$

Most "attacks" considered in the literature consist of traces that do not represent valid runs of the protocol. As such, we understand the protocol specification to be a set of *action templates*; one action template is specified for each agent in the protocol.

Definition 4. *Let $A \in \mathbf{A}$. An action template for A is a sequence $\langle ACT_1, \ldots, ACT_n \rangle$ where each ACT_i is either a send action for A or a receive action for A.*

An action template specifies the actions of one agent, whereas a trace specifies all exchanges by all agents.

Definition 5. *An instance of a template $\langle ACT_1, \ldots ACT_n \rangle$ is a trace $\langle T_1, \ldots, T_n \rangle$ such that $T_i \in ACT_i$, for each i.*

Hence, an instance of an action template for A consists of a sequence of message exchanges where A is the recipient or sender of each message, as dictated by the template.

A protocol consists of an action template and a set of goals for each communicating agent. We distinguish between two kinds of goals: a *possession goal* is a set of global states, whereas an *exchange goal* is a set of traces.

Definition 6. *A protocol is a function P with $domain(P) \subseteq \mathbf{A}$ such that, for each $A \in domain(P)$, $P(A)$ is a triple $\langle ACT, G_p, G_e \rangle$ where ACT is an action template, G_p is a possession goal, and G_e is an exchange goal.*

A protocol is correct just in case, whenever each honest agent follows the protocol structure, the only possible traces satisfy all protocol goals. To formalize this idea more precisely, we first define a satisfaction relation for protocol goals.

Definition 7. *Let G be an assignment function, let T be a trace, let P be a protocol, and let A be an agent. Then $\langle G, T \rangle \models P(A)$ if and only if $G + T \subseteq G_p$ and $T \in G_e$.*

The satisfaction relation on traces specifies what it means to say that the goals of a protocol are satisfied. Using this relation, we can formally define protocol correctness.

Definition 8. *Let P be a protocol. Then P is correct with respect to a set $\mathcal{G}$ of assignment functions if and only if for every $G \in \mathcal{G}$, every trace T, and every A with $P(A) = \langle ACT, G_p, G_e \rangle$:*

  – If T has a subsequence T' that is an instance of ACT, then $\langle G, T \rangle \models P(A)$

Note that we have defined protocol correctness without any explicit reference to knowledge, or the local state of an agent. However, the notion of an agent's perspective is implicit in our definition of an action template. It is also worth noting that we define correctness with respect to a set of assignment functions, which can be understood to represent an agent's knowledge about message possession.

3.3 Proving Correctness through Knowledge

Two different kinds of knowledge are required for protocol verification: knowledge *about message possession* and knowledge *about message exchanges*.

Definition 9. *An m-knowledge state for an agent A is a set $\mathcal{G}$ of assignment functions such that $G_1, G_2 \in \mathcal{G}$ implies $G_1(A) = G_2(A)$.*

Definition 10. *An e-knowledge state for an agent A is a set of traces $\mathcal{T}$ such that $T_1, T_2 \in \mathcal{T}$ implies $(T_1)^A = (T_2)^A$.*

We remark that action templates are e-knowledge states.

Definition 11. *Let $\langle G, T \rangle$ be a state, and let A be an agent. A knowledge state for A is a pair $\langle \mathcal{G}, \mathcal{T} \rangle$ where*

  – $\mathcal{G}$ is an m-knowledge state such that $G \in \mathcal{G}$.
  – $\mathcal{T}$ is an e-knowledge state such that $T \in \mathcal{T}$.

In general, we are interested in the knowledge of an agent after a successful run of a protocol. As such, we introduce some specialized notation. For any agent A, protocol P and m-knowledge state $\mathcal{G}$, let $knows(A, P, \mathcal{G})$ denote the knowledge state $\langle \mathcal{G}, ACT \rangle$, where ACT is the action template that P assigns to A.

Proposition 1. *Let P be a protocol. Then P is correct for $\mathcal{G}$ if and only if, for every agent A and every $\langle G, T \rangle \in knows(A, P, \mathcal{G})$, it holds that $\langle G, T \rangle \models P(A)$.*

This result illustrates how the definition of correctness can be stated in terms of knowledge states. A protocol is correct if the execution of the protocol causes the participating agents to *know* that all protocol goals are satisfied provided they execute their actions correctly.

4 Discussion

4.1 Knowledge at the Agent Level

In authentication protocols, most attacks involve an intruder that fools an agent into believing that a successful run of the protocol has been completed. In order to prove that such an attack does not exist, one must prove that everything that looks like a successful run to A must in fact be a successful run.

What our main result demonstrates is that the correctness of a protocol must be proved with respect to an m-knowledge state, which is an assignment of initial information to different agents. In practice, we need to prove protocol correctness with respect to a wide range of assignments. The strongest proof of correctness would establish the protocol is correct regardless of the information held by an intruder, but such a proof is unlikely to be established for any real protocol. Knowledge-based protocol verification can be seen as a parametrized proof of correctness, where we can vary the initial ownership of messages and test the correctness of a protocol. In terms of knowledge over messages and knowledge of traces, we have proved that a protocol is correct if it allows an agent to make valid inferences about the set of possible traces given some information about the initial knowledge of messages.

4.2   Comparison with Related Work

The approach suggested in this paper is intended to promote discussion on the nature of knowledge required for protocol verification. Most existing work in the area does not dwell on this problem. The two most common approaches are to simply take knowledge of messages as the relevant notion [3] or to take a modal view of knowledge based on an accessibility relation on states [15].

In [13], a distinction is drawn between different kinds of knowledge in protocol verification. In particular, knowledge of messages is explicitly distinguished from knowledge of *facts*. Facts in this context do not include information about the senders and recipients of messages. Although the goal of this work is quite different, it is similar to our approach in the sense that it proposes that protocol verification requires different kinds of knowledge.

In our own work, we have explored different forms of knowledge in the analysis of protocol attacks [12]. In this work, a distinction is drawn between three kinds of knowledge: knowledge of messages, knowledge of exchanges, and knowledge of facts. However, the goal in that instance was not to re-examine the foundations of the epistemic approach. Instead, the goal was to determine what an intruder was trying to achieve by sending a particular message.

5   Conclusion

In this paper, we have explored the foundations of the epistemic approach to cryptographic protocol verification. This approach has been widely exploited, though there has been little consensus on the basic epistemic concepts required. We have suggested that proving a protocol correct consists in proving that every apparent run of a protocol is actually part of a valid trace. It follows from this observation that protocol verification is inherently agent-based and epistemic, though it does not require explicit epistemic operators or nested knowledge.

As this is a preliminary exploration, we have provided few formal results. Instead, we have focused on presenting an elementary development of an epistemic approach to protocol verification. Our goal is twofold. First, by emphasizing the importance of an agent-level perspective, we hope to validate the importance of epistemic concepts in reasoning about protocol correctness. Our second goal is simply to illustrate that existing work does not provide a unified picture of the kind of knowledge required, and that foundational work is still required. There are several important topics to consider in extensions of this foundational work. In particular, one of the important features of most logics of knowledge is the fact that *nested knowledge* can be represented. It is not obvious, however, if nested knowledge carries formal significance in protocol verification. It would also be interesting to revisit our past work on fallible beliefs in protocol verification, with respect to the epistemic domain presented here.

References

1. Agray, N., van der Hoek, W., de Vink, E.: On BAN logics for industrial security protocols. In: Dunin-Keplicz, B., Nawarecki, E. (eds.) CEEMAS 2001. LNCS (LNAI), vol. 2296, pp. 29–36. Springer, Heidelberg (2002)
2. Burrows, M., Abadi, M., Needham, R.: A logic of authentication. ACM Transactions on Computer Systems 8(1), 18–36 (1990)
3. Carlucci Aiello, L., Massacci, F.: Verifying Security Protocols as Planning in Logic Programming. ACM Transactions on Computational Logic 2(4), 542–580 (2001)
4. Cremers, C.J.F.: The Scyther Tool: Verification, Falsification, and Analysis of Security Protocols. In: Gupta, A., Malik, S. (eds.) CAV 2008. LNCS, vol. 5123, pp. 414–418. Springer, Heidelberg (2008)
5. Dechesne, F., Mousavi, M.R., Orzan, S.: Operational and Epistemic Approaches to Protocol Analysis: Bridging the Gap. In: Dershowitz, N., Voronkov, A. (eds.) LPAR 2007. LNCS (LNAI), vol. 4790, pp. 226–241. Springer, Heidelberg (2007)
6. Dolev, D., Yao, A.C.: On the Security of Public Key Protocols. IEEE Transactions on Information Theory 2(29), 198–208 (1983)
7. Fagin, R., Halpern, J., Moses, Y., Vardi, M.: Reasoning About Knowledge. MIT Press (1995)
8. Gong, L., Needham, R., Yahalom, R.: Reasoning About Belief in Cryptographic Protocols. In: Proceedings of the IEEE Computer Society Symposium on Research in Security and Privacy, pp. 234–248 (1990)
9. Guttman, J., Thayer, J.: Authentication tests. In: Proceedings of the 2000 IEEE Symposium on Security and Privacy (2000)
10. Halpern, J., Pucella, R.: On the relationship between strand spaces and multi-agent systems. ACM Transactions on Information and System Security (TISSEC) 6(1) (2003)
11. Hunter, A., Delgrande, J.P.: Belief Change and Cryptographic Protocol Verification. In: Proceedings of the National Conference on Artificial Intelligence, AAAI 2007 (2007)
12. Hunter, A.: Dissecting the Meaning of an Encrypted Message: An Approach to Discovering the Goals of an Adversary. In: Ortiz-Arroyo, D., Larsen, H.L., Zeng, D.D., Hicks, D., Wagner, G. (eds.) EuroISI 2008. LNCS, vol. 5376, pp. 61–72. Springer, Heidelberg (2008)
13. Kramer, S.: Reducing Provability to Knowledge in Multi-Agent Systems. In: Intuitionistic Modal Logics and Applications Workshop, IMLA 2008 (2008)
14. Syverson, P.: Knowledge, Belief, and Semantics in the Analysis of Cryptographic Protocols. Journal of Compuer Security 1, 317–334 (1992)
15. Syverson, P., Cervesato, I.: The logic of authentication protocols. In: Focardi, R., Gorrieri, R. (eds.) FOSAD 2000. LNCS, vol. 2171, pp. 63–137. Springer, Heidelberg (2001)

Sensory Updates to Combat Path-Integration Drift

Xiang Ji, Shrinu Kushagra, and Jeff Orchard[*]

Cheriton School of Computer Science
University of Waterloo
jorchard@uwaterloo.ca

Abstract. Even without sensory input, an animal can estimate how far it has moved by integrating its velocity, a process called path integration. The entorhinal cortex (EC) and hippocampus seem to be involved in path integration, and in an animal's perceived location in space. However, path integration is highly susceptible to accumulating errors. A real animal avoids this problem by incorporating sensory input (e.g. vision) and updating its perceived position. The best path integration models do not yet incorporate this sensory-updating feature. In this paper, we extend one such model to enable sensory updating, and demonstrate its effectiveness in a series of computer simulations of spiking neural-network models.

1 Introduction

Path integration is the process of tracking one's believed location by integrating velocity. Recent research has tried to explain how animals employ path integration when navigating through their environments. Robotic navigation is a broad topic, in which drift in "dead reckoning" is an issue, but we focus solely on solutions consistent with biological neuroscience.

Biological findings show that certain neural activities are involved in path integration. Some neurons in the hippocampus fire action potentials (spikes) only when the animal occupies a specific location in its environment, suggesting that these so-called *place cells* encode location. Some neurons in the entorhinal cortex (EC) fire bursts of activity at locations that form a hexagonal grid in the environment [1]; these neurons are called *grid cells*. It is now thought that place cells and grid cells are involved in path integration. Also, many neurons in the EC and hippocampus exhibit membrane potential oscillations at a frequency between 4 and 12 Hz, which varies with the animal's velocity and have thus been dubbed *velocity-controlled oscillators*, or *VCOs* [2,3]. It is becoming evident that path integration results from the interaction between the animal's velocity and these oscillators [2,4].

Many biologically-based path integration models use interference between oscillators to explain how neurons generate such spatial activity patterns. As the animal moves around, the relative phases of the oscillators change and result in

---

[*] Corresponding author.

O. Zaïane and S. Zilles (Eds.): Canadian AI 2013, LNAI 7884, pp. 263–270, 2013.

a spatial pattern of constructive and destructive interference. Different combinations of oscillators can be used to create different firing patterns [5,6,7]. For example, [4] proposes a theory of path integration that ties together grid cells, place cells, and theta-phase precession (not discussed here) into a coherent and expandable framework using VCOs.

However, path integration tends to be effective only for short paths, and accumulates error quickly [8]. Scientists have discussed how sensory input might update or correct the path integration system [6,9]. Generally speaking, they propose a theory that place cells are activated by recognition of a location or landmark, and that these place cells feed back to the grid cells to induce the corresponding phase differences.

A spiking-neuron model of path integration incorporating sensory updates was proposed in [10]. However, their model is not consistent with the oscillator-based models and so does not exhibit grid-cells or phase precession. But it demonstrates that sensory update is feasible.

In this paper, we propose an extension to the interference-based model in [4]. Our extension enables updating of the internal state with sensory input, but automatically reverts back to path integration when sensory input is not available.

2   Background

The oscillating frequency θ of a VCO is a linear function of the animal's velocity $\mathbf{v}$, given by $\theta = \mathbf{c} \cdot \mathbf{v} + \theta_0$, where $\mathbf{c}$ is a constant vector specific to the VCO, and θ_0 is a constant [2]. The phase difference at time t between two VCOs with frequencies $\theta_1(\tau) = \mathbf{c}_1 \cdot \mathbf{v}(\tau) + \theta_0$ and $\theta_2(\tau) = \mathbf{c}_2 \cdot \mathbf{v}(\tau) + \theta_0$, can be written as

$$\phi(t) = \int_0^t \left((\mathbf{c}_1 \cdot \mathbf{v}(\tau) + \theta_0) - (\mathbf{c}_2 \cdot \mathbf{v}(\tau) + \theta_0) \right) d\tau = \mathbf{c}_\delta \cdot \int_0^t \mathbf{v}(\tau) d\tau = \mathbf{c}_\delta \cdot \mathbf{x}(t),$$

where $\mathbf{x}(t)$ is the animal's position at time t, and $\mathbf{c}_\delta = \mathbf{c}_1 - \mathbf{c}_2$. In this way, the phase difference between VCOs encode position.

For simplicity, let us consider the case when the animal is restricted to move only in 1-D (e.g. along the x-axis). In this case, both $\mathbf{v}$ and $\mathbf{c}$ would be 1-D vectors or scalars. Let us consider 3 VCOs with the $\mathbf{c}$ values $c_1 = 5$, $c_2 = 7$ and $c_3 = 9$, respectively. The phase difference between the first and the second VCO is then $2 \cdot x(t)$. Thus we can see that it encodes the actual position. Similarly, the phase difference between the second and third VCO is also $2 \cdot x(t)$. One more point to note is that the phase difference between equally spaced VCOs is the same. Thus, if we have an array of uniformly spaced VCOs, then we expect a constant phase increment from one VCO to the next.

At any instant of time, the phases of the VCOs form a straight line (considering 1-D motion) or a plane (considering 2-D). We call this a *phase-ramp*; its slope encodes the animal's position. As the animal moves from one place to another, the phase ramp tilts to track its location. As discussed in [4], these phase ramps can be used to generate various spatial activity patterns like place cells, grid cells, and other spatial maps.

3   Architecture and Methods

3.1   Phase Coupling

If we had perfect VCOs, we could expect the phase to keep its linear trend, even as the rat's movements altered the slope. In reality, however, neural oscillators are not perfect and tend to drift out of phase. With time, the phase along the array loses its ramp-like structure and no longer encodes the animal's position.

To overcome this problem, additional nodes are used to couple adjacent VCOs. These so-called *phase-step* nodes store the slope of the phase-ramp and supply feedback between the VCOs so that they maintain the correct phase difference [4]. Each pair of adjacent VCOs is coupled using a phase-step node, so for an array of D VCOs we have $D - 1$ phase-step nodes. The phase-step nodes work in three steps as follows:

1. Each phase-step node computes the phase difference between its two connected VCOs. The phase-step node gets the unit-length phase vector (x, y) from each of the two VCOs and computes their phase difference using,

$$(c', s') = (x_1, y_1) \overline{(x_2, y_2)} = (x_1 x_2 + y_1 y_2, -x_1 y_2 + y_1 x_2)$$

   where (x_1, y_1) and (x_2, y_2) are the oscillator states and (c', s') the computed phase difference (expressed in the form of a unit-length phase vector).
2. Each phase-step node broadcasts its computed phase difference to the other phase-step nodes, and they arrive at a consensus and store the result.
3. Each phase-step node sends a correction to its afferent VCOs to bring their phase difference closer to the consensus. Given the consensus phase difference (c, s), we rotate (x_1, y_1) clockwise to get (x_2', y_2'), an approximation of (x_2, y_2). Likewise, we rotate (x_2, y_2) counter-clockwise to get (x_1', y_1'). This is done using,

$$(x_1', y_1') = (x_2, y_2) \overline{(c, s)} \quad and \quad (x_2', y_2') = (x_1, y_1) \, (c, s).$$

   We then compute the error using $(\Delta x, \Delta y) = (x', y') - (x, y)$. We only need to compensate for a fraction of that error. Hence the phase of the 1st VCO (x_1, y_1) is adjusted by $0.2 \cdot (\Delta x_1, \Delta y_1)$ and similarly for the other VCO. This process is depicted in Fig. 1.

This design has the effect of maintaining a constant phase difference along the VCOs. The design is shown in Fig. 2 (left). This phase difference, along the array of VCOs, encodes the animal's perceived position.

3.2   Sensory Input

As mentioned, the VCOs are not perfect oscillators. Even though there is phase coupling, the slope of the phase ramp drifts as it accumulates error over time. Hence, the animal's perceived position drifts from its actual position. An actual animal avoids this problem by using sensory feedback. In terms of our model,

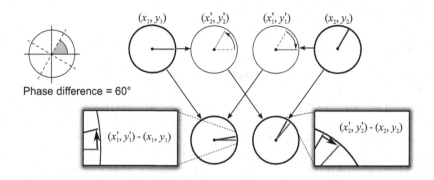

Fig. 1. The operations performed by a phase-step node. The VCOs have phase vectors (x_1, y_1) and (x_2, y_2). In this example, the consensus phase difference is $60°$.

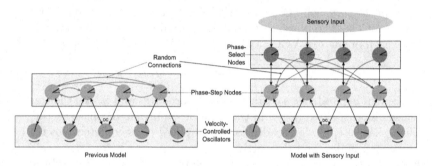

Fig. 2. Architecture of the VCOs, phase-step nodes, and phase-select nodes. The left side shows the architecture from [4], before sensory input was incorporated into the model. The right figure shows the modified architecture to incorporate sensory input.

the sensory input needs to update the slope of the phase ramp (which is stored in the phase-step nodes) to reflect the correct location. This updating then propagates to the VCOs.

This correction is achieved by introducing *phase-select* nodes in the architecture. This is the main contribution of this paper.

From the sensory system, we assume the phase-select nodes receive the correct phase difference that should be observed between adjacent VCOs. There are $D-1$ phase-select nodes, in 1-to-1 correspondence with the phase-step nodes, as shown in Fig. 2 (right).

In the previous model, the phase-step nodes arrived at a consensus phase difference among themselves. However, in our extended model, each phase-step node simply sends its computed phase difference to its corresponding phase-select node, which stores it. These phase-select nodes also receive and store the true phase from the sensory input (if it is available).

The job of each phase-select node is to select one of those two options and send it back into the path-integration system. It sends back the true phase difference if the sensory input is available, and sends back the estimated phase difference otherwise. In our implementation, this selection is controlled by a variable that takes the value of 0 or 1.

The phase differences sent back from each phase-select node is broadcast to a random selection of phase-step nodes, and the phase-step nodes store the average. This round-trip (phase-step → phase-select → phase-step) has the same consensus-building effect as in step 2 of the previous architecture [4].

Here we have described the architecture for motion in 1-D. This method extends trivially to 2-D. In our simulations, we used three such 1-D arrays, called *propellers*, oriented at 0°, 120° and 240°. Each propellor has 9 VCO nodes (300 neurons each), 8 phase-step nodes (500 neurons each), and 8 phase-select nodes (500 neurons each). The model uses the Neural Engineering Framework (NEF) [11], and is implemented using Nengo (nengo.ca). The NEF is a framework for encoding and decoding data using populations of spiking leaky integrate-and-fire neurons. More details about the model's implementation can be found in [4].

4 Experiments and Results

We conducted several experiments in which a virtual rat runs around in a circular pen. Our aim is to examine our model's performance in path integration, both with and without sensory input, and compare that performance to behaviours we would expect from a real animal. Throughout this section, we will use the term "lights on" or "bright environment" to indicate the situation when the sensory input is available, and "lights off" or "dark environment" for the case when the sensory input is not available.

Each propeller encodes the rat's displacement along its length, while all three collectively encode the rat's location in 2-D. The environment is a round platform with a radius of 1. The velocity profile of the virtual rat is generated randomly. Also, the rat changes direction when it reaches the boundary of the platform. The actual position is computed by numerically integrating the velocity, whereas the path integration system sets its position from the phase ramp of the VCOs.

Experiment 1. We start by testing the model's performance in both bright and dark environments. When lights are turned on, the rat should be able to sense the actual position and correct the state of the VCOs. Conversely, in dark environments the rat can only determine its position by path integration.

We ran the model both with and without sensory input, and examined the Euclidean-distance error between the actual and the perceived position. Five trials were run for each of the two conditions, each lasting 30 seconds of simulation time. We computed the average error over the five runs, and filtered the error with an averaging window of 100 ms. The filter removes much of the spike noise, similar to how a synapse would filter the incoming spikes.

The results are shown in Fig. 3. The figure shows that in bright environments, the mean Euclidean error maintains a low and somewhat-constant value of around 0.08 (about 10% of the radius). However, when the lights are off,

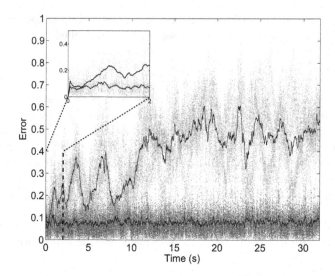

Fig. 3. Euclidean distance error of the perceived location over time. Each dot in the figure is a decoded location, while lines are the filtered average of errors. Black indicates the dark condition, and blue indicates bright.

the error grows to 0.5 over several seconds, indicating that the rat's perception of its position has diverged substantially from its true location. The inset in the figure shows that the error grows a lot even in the first 2 seconds in the dark. The results illustrate that our model effectively uses the actual position information to guide the perceived position.

Experiment 2. In this experiment, we displace the virtual rat from the centre of the platform, but set its path integration system so it thinks it is at the centre. Then we turn the lights on and measure how quickly the perceived position is corrected. We expect a real animal to update its percept quickly.

We set the initial perceived position at $(0, 0)$, and chose different actual positions for sensory input. Sixteen different positions were used: $(\pm 0.2, \pm 0.2)$, $(\pm 0.3, \pm 0.3)$, $(\pm 0.4, \pm 0.4)$, $(\pm 0.5, \pm 0.5)$. Each trial ran for 1 second of simulation time. We averaged the Euclidean distance error across all initial positions with the same displacement, then filtered the result with an averaging window of 10 ms.

We measured how long it took for the error to drop below 0.1. The error typically dropped below 0.1 within 200 ms. We also noted that convergence time increased as the displacement increased. This shows us that even when the perceived position is very different form the actual position, the animal is very quickly (approx. 200 ms) able to adjust its perception to match reality.

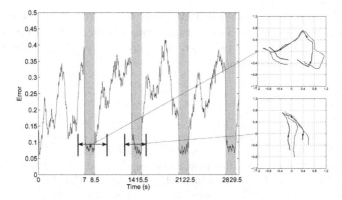

Fig. 4. Position error as the lights are turned on and off. The grey bands indicate when the lights are on. In the sample trajectories on the right, blue indicates when the lights are off, and red indicates when the lights are on. The solid line shows the actual trajectory, while the dotted line shows the perceived trajectory.

Experiment 3. For this experiment, the light is turned on and off repeatedly while we track the rat's perceived and actual position. We expect the perceived position to drift from the actual when the light is off, and then quickly converge back to the true position when the light is turned back on.

We set the lights to be on for 1.5 seconds out of every 7 seconds. Five trials were recorded, each lasting 30 seconds. We took the average error across all trials, and filtered the results with a 100 ms averaging window, as was done in experiment 1.

Figure 4 shows the error versus time. The intervals when the light is turned on is shaded in grey. The figure clearly shows that when the lights are turned on, the error falls almost instantaneously and holds its value. Notice the steep decline in error in Fig. 4 as soon as the lights are turned on (sensory feedback). However, when the lights are off, the error quickly climbs. Figure 4 also shows two examples of the actual trajectory of the rat superimposed on the perceived trajectory.

5   Conclusions and Future Work

Our model extends the EC model presented in [4] by including a mechanism for incorporating sensory corrections into the animal's perceived position. The path integration system, by itself, accumulates error and drifts from the correct position, consistent with the behaviour of animals [8]. However, supplying the correct position from sensory input alters the state of the system and brings it into harmony with the true position.

If our model is correct, then electrophysiological studies of the EC should uncover neurons whose activity is modulated heavily by the presence, or absence, of sensory positional information. When there is no sensory context for location, these neurons should behave differently than when there is.

The model that we extended included only three propellers. We expect that a full biological system would have far more propellers. Adding propellers would probably reduce the drift. We plan to build a more complete version of the model and investigate these questions.

References

1. Hafting, T., Fyhn, M., Molden, S., Moser, M., Moser, E.: Microstructure of a spatial map in the entorhinal cortex. Nature 436(7052), 801–806 (2005)
2. Welday, A.C., Shlifer, I.G., Bloom, M.L., Zhang, K., Blair, H.T.: Cosine Directional Tuning of Theta Cell Burst Frequencies: Evidence for Spatial Coding by Oscillatory Interference. Journal of Neuroscience 31(45), 16157–16176 (2011)
3. Zilli, E.A., Hasselmo, M.E.: Coupled Noisy Spiking Neurons as Velocity-Controlled Oscillators in a Model of Grid Cell Spatial Firing. Journal of Neuroscience 30(41), 13850–13860 (2010)
4. Orchard, J., Yang, H., Ji, X.: Navigation by path integration and the fourier transform: A spiking-neuron model. In: Zaïane, O., Zilles, S. (eds.) Canadian AI 2013. LNCS (LNAI), vol. 7884, pp. 138–149. Springer, Heidelberg (2013)
5. Blair, H., Welday, A.C., Zhang, K.: Scale-Invariant Memory Representations Emerge from Moire Interference between Grid Fields That Produce Theta Oscillations: A Computational Model. Journal of Neuroscience 27(12), 3211–3229 (2007)
6. Burgess, N., Barry, C., O'Keefe, J.: An oscillatory interference model of grid cell firing. Hippocampus 17(9), 801–812 (2007)
7. Hasselmo, M.E., Brandon, M.P.: Linking Cellular Mechanisms to Behavior: Entorhinal Persistent Spiking and Membrane Potential Oscillations May Underlie Path Integration, Grid Cell Firing, and Episodic Memory. Neural Plasticity (2008)
8. Etienne, A.S., Maurer, R., Seguinot, V.: Path Integration in Mammals and its Interaction with Visual Landmarks. Journal of Experimental Biology 199, 201–209 (1996)
9. Williams, J.M., Givens, B.: Stimulation-induced reset of hippocampal theta in the freely performing rat. Hippocampus 13(1), 109–116 (2003)
10. Conklin, J., Eliasmith, C.: A Controlled Attractor Network Model of Path Integration in the Rat. Journal of Computational Neuroscience 18, 183–203 (2005)
11. Eliasmith, C., Anderson, C.H.: Neural engineering: Computation, representation, and dynamics in neurobiological systems. MIT Press, Cambridge (2003)

Quantitatively Evaluating Formula-Variable Relevance by Forgetting

Xin Liang[1], Zuoquan Lin[1], and Jan Van den Bussche[2]

[1] School of Mathematical Sciences, Peking University, Beijing 100871, China
{xliang,lz}@pku.edu.cn
[2] Hasselt University and Transnational University of Limburg, 3590 Diepenbeek,
Belgium
jan.vandenbussche@uhasselt.be

Abstract. Forgetting is a feasible tool for weakening knowledge bases by focusing on the most important issues, and ignoring irrelevant, outdated, or even inconsistent information, in order to improve the efficiency of inference, as well as resolve conflicts in the knowledge base. Also, forgetting has connections with relevance between a variable and a formula. However, in the existing literature, the definition of relevance is "binary" – there are only the concepts of "relevant" and "irrelevant", and no means to evaluate the "degree" of relevance between variables and formulas. This paper presents a method to define the formula-variable relevance in a quantitative way, using the tool of variable forgetting, by evaluating the change of model set of a certain formula after forgetting a certain variable in it. We also discuss properties, examples and one possible application of the definition.

Keywords: knowledge representation, forgetting, relevance, inconsistency.

1 Introduction

Forgetting [1] is an accessible tool for weakening formulas in knowledge bases. Variable forgetting is the most basic form of forgetting. The concept of forgetting has connections with the concept of independence, which is also discussed by Lin and Reiter [1], as well as Lang, Liberatore and Marquis [2]. They have defined the concept of independence in different approaches.

However, current approaches of defining independency only give the judgement of "dependent" or "independent". In other words, these approaches only give a "black or white" answer, and ignore the intermediate conditions (i.e., there is no "greyscale"). For example, consider the two simple formulas: $p \wedge q$ and p. Intuitively, and by the definition presented by Lang, Liberatore and Marquis [2], both formulas are dependent of p. However, by intuition, the latter one is more "dependent" of p than the former one. How to characterize this "degree" of "dependency" (or "relevance"), in order to capture the difference between, say, the two formulas mentioned above, on a certain variable? This paper presents an

O. Zaïane and S. Zilles (Eds.): Canadian AI 2013, LNAI 7884, pp. 271–277, 2013.

approach, which is based on the work of Lin and Reiter [1] and the work of Lang, Liberatore and Marquis [2], extending those ideas to the quantitative analysis of the sets of models of formulas. The key point is to measure the change of the set of models after forgetting. After the definition, we will discuss properties, examples, and one possible application of it.

2   Preliminaries

This paper is based on propositional logic. Let PS be a set of propositional variables (e.g., p, q, r, p_1, p_2, ..., p_n, ..., etc.), and $PROP_{PS}$ denotes the set of all propositional formulas defined on the set PS. Propositional variables occuring on the formulas in $PROP_{PS}$ are all in the set PS, and there exist formulas in $PROP_{PS}$ in which only some (not all) variable(s) in PS occur(s). The language of $PROP_{PS}$ consists of the commonly used logical connectives, such as $\neg$, $\rightarrow$, $\vee$, $\wedge$, $\leftrightarrow$. An interpretation is a truth-value assignment, assigning every variable in PS a truth value from $\{T, F\}$. Here we use a subset of PS to denote an interpretation: If a variable is in the subset, then it is assigned T, otherwise it is assigned F. Given an interpretation ω and a formula φ in $PROP_{PS}$, we can tell whether ω satisfies φ (denoted by $\omega \models \varphi$, stating that φ is true under ω), by the semantics of logical connectives. A model of a formula φ is an interpretation in which φ is true. $\text{Mod}(\varphi)$ denotes the set of models of φ with respect to the variables in the set PS. We say formula ϕ entails formula ψ, denoted by $\phi \models \psi$, if $\text{Mod}(\phi) \subseteq \text{Mod}(\psi)$. We say ϕ and ψ are logically equivalent, denoted by $\phi \equiv \psi$, if $\text{Mod}(\phi) = \text{Mod}(\psi)$.

Variable forgetting on propositional logic was proposed by F. Lin and R. Reiter [1]. Lang, Liberatore and Marquis proposed literal forgetting on propositional logic [2]. Here we only consider the semantic definition of variable forgetting: Let φ be a propositional formula, p be a variable in it, $\text{ForgetVar}(\varphi, p)$ denotes the result of *forgetting p in φ*, whose models can be obtained as follows:

$$\text{Mod}(\text{ForgetVar}(\varphi, p)) = \text{Mod}(\varphi) \cup \{\text{Switch}(\omega, p) \mid \omega \models \varphi\} , \qquad (1)$$

where $\text{Switch}(\omega, p)$ means to change the truth-value assignment of p in the interpretation ω to the opposite one.

3   Formula-Variable Relevance

In this section, we define the method to evaluate formula-variable relevance in a quantitative way. First, let us consider the models of ψ_i and $\text{ForgetVar}(\psi_i, p)$ under the variable set $PS = \{p, q\}$, where i is a variable for the enumeration of propositional formulas, ranging from 1 to 8. The models of ψ_i and $\text{ForgetVar}(\psi_i, p)$ are listed in Table 1. (Just ignore the last column of Table 1, which will be useful later in this section.)

An intuitive idea to define the degree of formula-variable relevance $R_{\text{FV}}(\psi, p)$ is using the "amount" of the *increment* of models (from ψ to $\text{ForgetVar}(\psi, p)$).

Table 1. The models of ψ_i and ForgetVar(ψ_i, p), as well as the value of formula-variable relevance, which is defined later in this section

i	ψ_i	Mod(ψ_i)	Mod$($ForgetVar$(\psi_i, p))$	$R_{\mathrm{FV}}(\psi_i, p)$
1	$p \wedge q$	$\{\{p, q\}\}$	$\{\{p, q\}, \{q\}\}$	$1/2$
2	$p \vee q$	$\{\{p\}, \{q\}, \{p, q\}\}$	$\{\emptyset, \{p\}, \{q\}, \{p, q\}\}$	$1/2$
3	p	$\{\{p\}\}$	$\{\emptyset, \{p\}\}$	1
4	$\neg p$	$\{\emptyset\}$	$\{\emptyset, \{p\}\}$	1
5	$p \rightarrow q$	$\{\emptyset, \{q\}, \{p, q\}\}$	$\{\emptyset, \{p\}, \{q\}, \{p, q\}\}$	$1/2$
6	$q \rightarrow p$	$\{\emptyset, \{p\}, \{p, q\}\}$	$\{\emptyset, \{p\}, \{q\}, \{p, q\}\}$	$1/2$
7	$p \leftrightarrow q$	$\{\emptyset, \{p, q\}\}$	$\{\emptyset, \{p\}, \{q\}, \{p, q\}\}$	1
8	$(p \wedge q) \vee ((\neg p) \wedge q)$	$\{\{q\}, \{p, q\}\}$	$\{\{q\}, \{p, q\}\}$	0

The underlying intuition is: if the model set of ForgetVar(ψ, p) remains unchanged or little changed comparing to the model set of ψ, it means that p is already or almost "forgotten" in ψ, and ψ contains models in which p is *true* and almost the same amount of models in which p is *false* (and these models are almost the same of the former models in the valuation of variables except p). In this sense, p is more irrelevant with ψ under this circumstance, because the operation of "forgetting p in ψ" does not change the model set of ψ too much, which means that the formula ψ *itself* does not have much relevance with the variable p. An extreme condition is that the model set of ψ remain unchanged after forgetting p, i.e. $\psi \equiv$ ForgetVar(ψ, p). This extreme case is equivalent to the "binary" definition of independence which is proposed in the existing literature [2]. If the model set of ForgetVar(ψ, p) changes a lot after the forgetting operation, we can say that the formula ψ is more relevant with the variable p. This is because the operation of forgetting *does* change a lot in the model set of ψ, which could be seen as the result of the fact that ψ is more relevant with p. The more the models change (increase), the more relevant ψ and p are.

By the above intuition, we may give a formal definition of formula-variable relevance.

Definition 1. *Let ψ be a formula in $PROP_{PS}$, p a variable occurring in ψ. The degree of formula-variable relevance $R_{\mathrm{FV}}(\psi, p)$ is defined as:*

$$R_{\mathrm{FV}}(\psi, p) = \frac{|\mathrm{Mod}(\mathrm{ForgetVar}(\psi, p)) \setminus \mathrm{Mod}(\psi)|}{2^{|PS|-1}} . \tag{2}$$

Proposition 1. *For any ψ, and p occurring in ψ,*

$$0 \leq R_{\mathrm{FV}}(\psi, p) \leq 1 , \tag{3}$$

and for any $i \in \{0, 1, 2, \ldots, 2^{|PS|-1}\}$, we can find some ψ and p, such that $R_{\mathrm{FV}}(\psi, p) = i/2^{|PS|-1}$.

Note that if we have a set of variables PS' such that $PS' \supseteq PS$, then for a formula ψ defined on PS, and a variable $p \in PS$, we may get the same formula-variable relevance both under PS and PS'.

Example 1. Now we have some formulas denoted by ψ_i, and let us consider $R_{\text{FV}}(\psi_i, p)$ for each i. The readers may find different ψ_i's and the corresponding results back in Table 1.

Next we show some properties of the relevance defined above.

Proposition 2. *Let p be a variable in PS, and ϕ, ψ two formulas defined on PS.*

1. *If ϕ and ψ are logically equivalent, then $R_{\text{FV}}(\phi, p) = R_{\text{FV}}(\psi, p)$.*
2. *If ϕ is a tautology or a contradiction, then $R_{\text{FV}}(\phi, p) = 0$.*
3. *If $p \models \phi$ and $\neg p \models \phi$, then $R_{\text{FV}}(\phi, p) = 0$.*
4. *If $\phi \models p$ or $\phi \models \neg p$, and ϕ is satisfiable, then $R_{\text{FV}}(\phi, p) > 0$.*

Now let us discuss the issue on computational complexity. We propose a problem (named COUNTREL), which has important connections with the problem of formula-variable relevance. COUNTREL is stated as follows:

- *Input.* A propositional formula ψ, and a variable p occuring in ψ. Let PS be all the variables occurring in ψ.
- *Output.* $|\text{Mod}(\text{ForgetVar}(\psi, p)) \setminus \text{Mod}(\psi)|$.

Theorem 1. *The problem COUNTREL is #P-complete.*

4 Defining Preference among Recoveries with Formula-Variable Relevance

Lang and Marquis [3,4] presented a method to resolve inconsistencies of a given knowledge base via variable forgetting. As we know, it is a common thing that inconsistencies may occur in a knowledge base due to different sources of knowledge or different views of the world for each person. In classical logics, if inconsistency lies in a belief set, then *ex falso quodlibet* occurs, which means that, all the well-formed formulas are theorems of the formal system. Obviously this is meaningless and the formal logic system becomes useless. To avoid this, researchers developed several means for inconsistency tolerance [5].

As one possible approach to resolve inconsistency, Lang and Marquis [3,4] proposed a concept called *recovery*, which is based on the concept of *forgetting vector*, and it plays a key role in the whole process. A forgetting vector is a vector whose items are sets of variables which are to be forgotten in each formula in the belief base, satisfying a certain *forgetting context*. For the notations, a forgetting vector is $V = (V_1, \ldots, V_n)$, where V_i is a set of variables to be forgotten in ϕ_i, and ϕ_i is a formula in belief set B (all the formulas in B could be listed as $\phi_1, \phi_2 \ldots \phi_n$). The set of all forgetting vectors for a knowledge base B given a forgetting context C is denoted as $\mathcal{F}_C(B)$. A *recovery* for a knowledge base under a certain forgetting context is a forgetting vector after "forgetting" which the knowledge base is consistent. For the details, the readers may refer to the original paper by Lang and Marquis [3,4].

Then a problem occurs: For a certain (inconsistent) knowledge base, there may be various recoveries. The authors state that we can define a kind of *preference relation* among recoveries. The preference relation is a kind of binary relations. There are many ways to define the preference relation, reflecting different views of which vector of variable sets to "forget".

Hereby we can apply the formula-variable relevance defined in the previous part of this paper to this circumstance. The idea is to "forget" the vector of variable sets which is least relevant to the formulas in the belief set.

To define the preference relation, we firstly define a ranking function according to formula-variable relevance. The idea is that for each element (variable set) of the forgetting vector, we evaluate the relevance between the corresponding source of knowledge (ϕ_i), and each variable in the variable set, then add them together.

Definition 2.

$$\mathrm{Rank}_B(V) = \sum_{i=1}^{n} \sum_{v \in V_i} R_{\mathrm{FV}}(\phi_i, v) \ , \tag{4}$$

where $B = (\phi_1, \ldots, \phi_n)$, $V = (V_1, \ldots, V_n)$.

Now we have the ranking function, then the definition of preference relation is straightforward.

Definition 3. *The preference relation among forgetting vectors, based on formula-variable relevance, denoted by* $\sqsubseteq_{\mathrm{Rel}}$, *is defined as follows:*

$$V \sqsubseteq_{\mathrm{Rel}} V' \ \textit{if and only if} \ \mathrm{Rank}_B(V) \leq \mathrm{Rank}_B(V') \ . \tag{5}$$

Lang and Marquis [3,4] mentioned a property named *monotonicity*. The preference relation defined above satisfies this property.

Proposition 3. $\sqsubseteq_{\mathrm{Rel}}$ *satisfies monotonicity, i.e. for all* $V, V' \in \mathcal{F}_C(B)$, *if* $V \subseteq_p V'$, *then* $V \sqsubseteq_{\mathrm{Rel}} V'$.

Note: $V \subseteq_p V'$ means that for all i, $V_i \subseteq V_i'$.

Example 2. Let us come to an example which was stated in the papers of Lang and Marquis [3,4]. It is an example of building tennis court and/or swimming pool. The knowledge base (which is inconsistent) and the forgetting context are given as follows [3,4], where "$\oplus$" is the "xor" connective, and formulas with this connective can be easily transformed to an equivalent formula only with the five basic connectives ($\neg, \rightarrow, \vee, \wedge, \leftrightarrow$):

- $\phi_1 = (s \rightarrow (s_r \oplus s_b)) \wedge (\neg s \rightarrow \neg s_r \wedge \neg s_b) \wedge (c_2 \leftrightarrow (s \wedge t)) \wedge (c_1 \leftrightarrow (s \oplus t)) \wedge (c_0 \leftrightarrow (\neg s \wedge \neg t))$;
- $\phi_2 = (c_0 \vee c_1) \wedge (s \rightarrow s_r)$;
- $\phi_3 = (s \vee t) \wedge (s \rightarrow s_b)$;
- $\phi_4 = s$;
- $\phi_5 = s \wedge t$;

– The forgetting context $\mathcal{C}$ is the conjunction of the following formulas (where $forget(x, i)$ means that atom x may be forgotten in ϕ_i, and $Var(\phi)$ is all the variables occuring in ϕ):

1. $\bigwedge_{x \in Var(\phi_1)} \neg forget(x, 1)$;
2. $\bigwedge_{i=2}^{5}(forget(s, i) \rightarrow forget(s_r, i)) \wedge (forget(s_r, i) \leftrightarrow forget(s_b, i)) \wedge (forget(c_0, i) \leftrightarrow forget(c_1, i)) \wedge (forget(c_1, i) \leftrightarrow forget(c_2, i))$.

In this example, in order to restore consistency, the authors gave 9 possible forgetting vectors as recoveries, which are listed in Table 2. Actually, we can choose any of them in order to resolve inconsistency. However, the preference defined by formula-variable relevance can give a criteria for choosing recovery. For the knowledge base B and the recoveries V^1 to V^9, we may compute $\text{Rank}_B(V^i)$ for each i, which is listed in Table 2.

Table 2. Different recoveries and the corresponding ranking values

i	V^i	$\text{Rank}_B(V^i)$
1	$\langle \emptyset, \{s, s_b, s_r\}, \{s, s_b, s_r\}, \{s, s_b, s_r\}, \{s, s_b, s_r\} \rangle$	13/4
2	$\langle \emptyset, \{t, s_b, s_r\}, \{t, s_b, s_r\}, \{t, s_b, s_r\}, \{t, s_b, s_r\} \rangle$	15/8
3	$\langle \emptyset, \{c_0, c_1, c_2, s_b, s_r\}, \{c_0, c_1, c_2, s_b, s_r\}, \{c_0, c_1, c_2, s_b, s_r\},$ $\{c_0, c_1, c_2, s_b, s_r\} \rangle$	13/8
4	$\langle \emptyset, \{s_b, s_r\}, \{s_b, s_r\}, \emptyset, \{t\} \rangle$	11/8
5	$\langle \emptyset, \{c_0, c_1, c_2\}, \{s_b, s_r\}, \emptyset, \emptyset \rangle$	5/4
6	$\langle \emptyset, \{s_b, s_r\}, \emptyset, \emptyset, \{t\} \rangle$	7/8
7	$\langle \emptyset, \emptyset, \emptyset, \{s, s_b, s_r\}, \{s, s_b, s_r\} \rangle$	3/2
8	$\langle \emptyset, \{c_0, c_1, c_2, s_b, s_r\}, \emptyset, \emptyset, \emptyset \rangle$	9/8
9	$\langle \emptyset, \emptyset, \{s_b, s_r\}, \emptyset, \{t\} \rangle$	1

According to our criteria, we should choose the recovery V^6. Actually, we implemented a computer program computing formula-variable relevance.

5   Related Work and Conclusions

In the existing literature, relevance between a variable and a formula is concerned, but in a "binary" way, i.e., there are only two cases – "relevant" and "irrelevant", not concerning the measurement of the "degree" of relevance. For example, Lang, Liberatore and Marquis [2] defined various kinds of Literal-/Variable-independence, and the authors also showed the connections between independence and forgetting. In the paper where forgetting was firstly proposed in the field of artificial intelligence [1], Lin and Reiter defined a kind of irrelevance as follows: Let φ be a propositional formula, q a query, and p a ground atom, then we say that p in φ is *irrelevant* for answering q iff φ and $\text{ForgetVar}(\varphi, p)$ are equivalent w.r.t. q. This definition is based on relevance with respect to a certain query q.

Our work is inspired by the independence defined in the existing literature above, and adds quantitative measurement of formula-variable relevance.

In conclusion, our paper, based on the traditional idea of forgetting and relevance, proposed an approach to evaluate how relevant a variable and a formula are. Also, we have applied this approach to define preference relations in a forgetting-based approach of inconsistency resolving in the existing literature.

For the future work, note that Lang, Liberatore and Marquis [2] discussed literal forgetting, and also, Xu and Lin [6] discussed formula forgetting. We may employ the methods in these papers to define "formula-literal relevance" and the relevance between formulas. Another noticeable issue is that forgetting has been defined in other logics [7,8,9]. Quantitatively defining relevance by forgetting in these logics is a topic worth studying.

Acknowledgements. The authors wish to thank Chaosheng Fan, Kedian Mu, Geng Wang, Dai Xu, Xiaowang Zhang, and other people with whom we have had discussions on this topic. We would also like to thank the anonymous reviewers for the helpful comments. This work is supported by the program of the National Natural Science Foundation of China (NSFC) under grant number 60973003, the Research Fund for the Doctoral Program of Higher Education of China, and the Graduate School of Peking University.

References

1. Lin, F., Reiter, R.: Forget it! Working Notes of AAAI Fall Symposium on Relevance, pp. 154–159 (1994)
2. Lang, J., Liberatore, P., Marquis, P.: Propositional independence. Journal of Artificial Intelligence Research 18, 391–443 (2003)
3. Lang, J., Marquis, P.: Resolving inconsistencies by variable forgetting. In: International Conference on Principles of Knowledge Representation and Reasoning, pp. 239–250. Morgan Kaufmann Publishers (2002)
4. Lang, J., Marquis, P.: Reasoning under inconsistency: A forgetting-based approach. Artificial Intelligence 174(12-13), 799–823 (2010)
5. Bertossi, L., Hunter, A., Schaub, T. (eds.): Inconsistency Tolerance. LNCS, vol. 3300. Springer, Heidelberg (2005)
6. Xu, D., Lin, Z.: A prime implicates-based formulae forgetting. In: 2011 IEEE International Conference on Computer Science and Automation Engineering (CSAE), vol. 3, pp. 128–132. IEEE (2011)
7. Wang, Z., Wang, K., Topor, R., Pan, J.Z.: Forgetting concepts in DL-lite. In: Bechhofer, S., Hauswirth, M., Hoffmann, J., Koubarakis, M. (eds.) ESWC 2008. LNCS, vol. 5021, pp. 245–257. Springer, Heidelberg (2008)
8. Eiter, T., Wang, K.: Semantic forgetting in answer set programming. Artificial Intelligence 172(14), 1644–1672 (2008)
9. Zhang, Y., Zhou, Y.: Knowledge forgetting: Properties and applications. Artificial Intelligence 173(16), 1525–1537 (2009)

An Ensemble Method Based
on AdaBoost and Meta-Learning

Xuan Liu[1], Xiaoguang Wang[1], Nathalie Japkowicz[1,2], and Stan Matwin[3]

[1] School of Electrical Engineering & Computer Science, University of Ottawa, Ottawa, Canada
xliu107@uottawa.ca
bwang009@eecs.uottawa.ca
[2] Department of Computer Science, Northern Illinois University, USA
nat@eecs.uottawa.ca
[3] Faculty of Computer Science, Dalhousie University, Halifax, Canada
stan@cs.dal.ca

Abstract. We propose a new machine learning algorithm: meta-boosting. Using the boosting method a weak learner can be converted into a strong learner by changing the weight distribution of the training examples. It is often regarded as a method for decreasing both the bias and variance although it mainly reduces variance. Meta-learning has the advantage of coalescing the results of multiple learners to improve accuracy, which is a bias reduction method. By combing boosting algorithms with different weak learners using the meta-learning scheme, both of the bias and variance are reduced. Our experiments demonstrate that this meta-boosting algorithm not only displays superior performance than the best results of the base-learners but that it also surpasses other recent algorithms.

Keywords: AdaBoost, ensemble learning, meta-learning, bias, variance.

1   Introduction

Empirical studies have shown that a given algorithm may outperform all others for a specific subset of problems, but there is no single algorithm that achieves best accuracy for all situations [1]. Therefore, there is growing research interest in combining a set of learning algorithms into one system, which is the so-called ensemble method.

There are two different approaches to create ensemble systems: homogeneous classifiers—which use the same algorithm over diversified data sets; heterogeneous classifiers—which use different learning algorithms over the same data [2]. One example of homogeneous classifiers is boosting [3] and one of the examples for heterogeneous classifiers is Meta-learning [4].

Our method is inspired by the AdaBoost Dynamic algorithm that was introduced in [5]. As the conventional boosting algorithm only boosts one weak learner, AdaBoost Dynamic tries to improve the AdaBoost.M1 algorithm by calling different weak learners inside the boosting algorithm. In this paper, instead of combining different

O. Zaïane and S. Zilles (Eds.): Canadian AI 2013, LNAI 7884, pp. 278–285, 2013.
© Springer-Verlag Berlin Heidelberg 2013

weak learners using weighted majority voting inside the boosting algorithm, we combine the predictions of the boosting algorithm with different weak learners using a meta-learner.

This paper is organized as follows: section 2 provides the background information about AdaBoost.M1 and meta-learning; section 3 presents our proposed method and its advantages; section 4 discusses the experimental setup and results; section 5 gives the overall conclusions followed by references.

2   Background

2.1   AdaBoost

Here we discuss one of the widely used variants of boosting: AdaBoost. There are two implementations of AdaBoost: with reweighting and with resampling. AdaBoost.M1 was designed to extend AdaBoost from handling the original two classes case to the multiple classes case. In order to let the learning algorithm deal with weighted instances, an unweighted dataset can be generated from the weighted dataset by resampling. For boosting, instances are chosen with probability proportional to their weight.

The AdaBoost algorithm generates a set of hypotheses and they are combined through weighted majority voting of the classes predicted by the individual hypotheses. To generate the hypotheses by training a weak classifier, instances drawn from an iteratively updated distribution of the training data are used. This distribution is updated so that instances misclassified by the previous hypothesis are more likely to be included in the training data of the next classifier. Consequently, consecutive hypotheses' training data are organized toward increasingly hard-to-classify instances. It was proven in [3] that a weak learner—an algorithm which generates classifiers that can merely do better than random guessing— can be turned into a strong learner using Boosting.

2.2   Meta-Learning

Meta-learning can be loosely defined as learning from information generated by a learner(s). In the inductive learning case meta-learning means learning from the classifiers produced by the learners and the predictions of these classifiers on training data. One of the advantages of this approach is that individual classifiers can be treated as black boxes and in order to achieve a final system, little or no modifications are required on the base classifiers. Meta-learning can be used to coalesce the results of multiple learners to improve accuracy.

3   Proposed Method

3.1   Combining AdaBoost Using Meta-Learning

It was proposed that by calling a different weak learner in each iteration of AdaBoost.M1 this algorithm could be improved [5]. Instead of combining the results of different weak learners inside the AdaBoost.M1 algorithm the way that was done by Adaboost Dynamic, here, we propose to combine them outside this algorithm applying a combiner strategy. To make the final decision, we choose meta-learning.

Fig.1 describes the different stages for training the Meta-boosting algorithm: combination of AdaBoost algorithm and meta-learning. To implement this algorithm we split the dataset into training (81% of the dataset), validation (9% of the dataset) and test (10% of the dataset) datasets. First, different base classifiers (level-0 classifiers) are obtained by training the AdaBoost.M1 algorithm with different weak learners using the training datasets. Second, predictions are generated by the learned classifiers on the validation dataset. Third, the true labels of validation dataset and the predictions generated by the base classifiers are collected as the labels and features for the meta-level training set. For multi-class problems, we choose the prediction of the class which has the highest probability. Lastly, the final classifier (meta-classifier) is trained from the meta-level training set.

To test this meta-classifier, predictions are generated on the base classifiers using the test dataset. These predictions are set as the test data for this meta-classifier. To obtain the average accuracy, this algorithm is run a few times on different samplings of training, validation and test datasets.

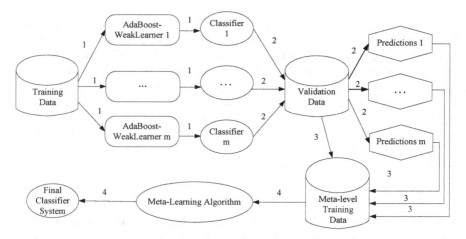

Fig. 1. The model for training Meta-boosting algorithm

3.2   Algorithm

Table 1 shows the pseudo code for the proposed Meta-boosting algorithm. The train ensemble model gets the predicted probabilities of each base learner and uses all the predictions and the true labels in the validation dataset to train the meta-learner. The test ensemble model gets the predictions of all the base learners on the test data and uses these predictions to test the already trained ensemble model. The output accuracy is obtained by comparing the test results with the true labels of the test data set.

Table 1. Pseudo code for Meta-boosting algorithm

Input:
- Training data S_T; Validation data S_V; Test data S_t
- n, number of base learners

Output:
- y, accuracy of ensemble learning

Calls:
- L, List of the base learners

Train ensemble model
1. For $i=1...n$
2.   training using Adaboost.M1 with $L[i]$ as the base learner on S_T
3.   test the trained classifier on S_V and get the output probabilities of the class which which has the highest predicted probability
4. End for
5. Create meta-level training data M_T with the set of output probabilities generated from from step 1-4 as features and labels of S_V as labels
6. Train meta-learning algorithm with T and output trained ensemble model E

Test ensemble model
- Repeat step1-5 in train ensemble model to generate the meta-level test dataset M_t using S_T as the training dataset and S_t as the test dataset
- Test E using M_t and get y

3.3   Advantages of Our Algorithm

The advantages of our algorithm are explained in terms of bias and variance. Ensemble errors are often decomposed into bias and variance terms to find out the reasons why an ensemble classifier outperforms any of its components. The definitions of bias and variance are given in [6]. Bias is the measure of the difference between the true distribution and the guessed distribution. Variance represents the variability of the classifier's guess regardless of the true distribution. Bias is related with under fitting the data while variance is associated with over fitting [7].

It was often found that very simple learners perform very well in experiments, and sometimes even better than more sophisticated ones [8-9]. The reason for this phenomenon is that there is a trade-off between bias and variance. To overcome this shortcoming, ensembles of models are designed which outperforms the results of a single model. Although intensive searching for a single model is susceptible to increased variance, the ensembles of multiple models can often reduce it.

There is no general theory about the effects of boosting on bias and variance. From the experimental studies we know that AdaBoost has much stronger reduction of variance with respect to the reduction of bias. Thus, variance reduction is the dominant effect for AdaBoost [10-12].

For meta-learning the goal of using the estimated probabilities of those derived from base classifiers as the input to a second learning algorithm is to produce an unbiased estimate. Although there is no detailed theoretical analysis of meta-learning's effect on bias and variance. It is often regarded as a bias reducing technique [13].

Therefore, the advantage of our algorithm is that it has both reduced variance and bias as it is an integration of AdaBoost and meta-learning.

4   Experimental Setup and Results

The experiments were performed on 17 datasets taken from the UCI repository [14]. A summary of the datasets used is presented in Table 2.

Table 2. Datasets used in the experiments

Index	Name	# of Instances	# of Attributes	# of Classes
1	Contact-lenses	24	5	3
2	Post-operative	90	9	3
3	Ionosphere	351	35	2
4	Diabetes	768	9	2
5	Blood	748	5	2
6	Crx	690	16	2
7	Au1	1000	21	2
8	Soybean	683	36	19
9	Iris	150	5	3
10	Segment	1500	20	7
11	Breast-tissue	106	10	6
12	CTG	2126	23	10
13	Car	1728	7	4
14	Cmc	1473	10	3
15	Glass	214	10	6
16	Zoo	101	18	7
17	Balance-scale	625	5	3

For this work, we chose three common base learners that use distinct learning strategies: Ada_NB—AdaBoost.M1 with Naïve Bayes [9] as the learner, Ada_KNN—AdaBoost.M1 with k-nearest neighbor [15] as the learner and Ada_tree—AdaBoost.M1 with J48 [16] as the learner. The meta-learner used is multi-response linear regression (MLR) which was proven to be the best meta-level learner [17].

4.1   Comparison between Meta-boosting and Base Learners

In Table 3, we present the percentage of correctly classified instances for each of the base learners and the Meta-boosting algorithm. The presented results were obtained with 10 fold cross validation. It is obvious from the results that our proposed ensemble method improves the accuracy of the base learners.

It can be seen from this table that meta-boosting is the best for 14 datasets, Ada_NB is the best for 2 datasets and Ada_KNN is the best for 1 dataset. As Friedman's Test [18] is a non-parametric statistical test for multiple classifiers and multiple domains, we performed this test on the results in table 3. The null hypothesis for this test is that all the classifiers perform equally.

Table 3. The accuracy of all the base learners and the meta-boosting algorithm (the method that has the best performance for a given dataset is marked in bold)

Dataset	Meta-boosting	Ada_NB	Ada_KNN	Ada_tree
Contact-lenses	**83.33±22.22**	70.83±25.09	79.17±33.38	70.83±35.18
Post-operative	**71.11±5.74**	63.33±5.37	60.00±9.37	58.88±9.15
Soybean	**94.58±2.79**	92.82±3.91	91.06±2.72	92.82±3.17
Ionosphere	**94.02±2.77**	92.02±4.37	86.32±4.59	93.16±3.57
Diabetes	**76.30±5.69**	76.17±4.69	70.18±4.69	72.39±4.86
Blood	76.07±2.43	**77.13±1.83**	69.25±4.16	76.07±3.70
Crx	**87.53±2.08**	81.59±2.47	81.15±4.83	84.20±3.09
Au1	**76.00±3.81**	72.80±1.40	66.70±4.72	70.80±3.43
Iris	**95.33±5.49**	93.33±7.03	**95.33±5.49**	93.33±7.03
Segment	**98.33±0.96**	81.06±2.33	96.20±1.63	97.46±1.43
Breast tissue	70.82±11.40	70.75±13.32	**71.69±13.94**	69.80±13.90
CTG	**90.78±2.44**	77.23±2.53	83.96±2.85	90.45±2.73
Car	**96.82±1.71**	90.16±2.30	93.52±1.32	96.12±1.85
Cmc	**51.93±4.06**	50.78±5.74	44.19±3.53	50.78±2.86
Glass	**75.56±15.85**	49.06±7.14	70.56±7.27	74.29±7.30
Zoo	96.18±6.54	**97.02±4.69**	96.04±6.54	95.04±8.15
Balance-scale	**93.28±1.80**	91.04±2.44	86.56±2.70	78.88±3.78
Average accuracy	**84.00**	78.06	78.93	80.31

The result for the Friedman's test are Friedman chi-squared=23.3963, $df=3$, p-value=3.339e-05. As the critical values for the chi-square distribution for $k=4$ and $n=17$ is 7.8 for a 0.05 level of significance for a single-tailed test, while 23.3963 is larger than 7.72, we can reject the hypothesis.

Following Friedman's test, in order to determine whether Meta-boosting's result is significantly better than its base learners we applied Nemenyi's post-hoc test [18] on the results in table 3. If we let 1, 2, 3 and 4 represent algorithms meta-boosting, Ada_NB, Ada_KNN and Ada_tree respectively, then we have $|q_{12}| = 58.71$, $|q_{13}| = 73.4$, $|q_{14}| = 62.1$. As $q_\alpha = 3.79$ for α=0.05, $df=48$, number of groups $k=4$, $k=4$, $q_\alpha/\sqrt{2} = 2.68$. Since the absolute value of q_{12}, q_{13}, q_{14} are all larger than 2.68, the null hypothesis that meta-boosting performs equally with Ada_NB, Ada_KNN and Ada_tree is rejected. Therefore, we conclude that meta-boosting performs significantly better than all the base learners.

The running time for datasets with indexes from 1 to 17 in the unit of second are 0.08, 0.36, 29.37, 26.01, 16.5, 38.44, 7.54, 40.84, 0.46, 43.97, 1.55, 193.84, 9.45, 69.78, 6.6, 0.24 and 2.39.The average running time is 28.67 sec.

4.2   Comparison between Meta-boosting and AdaBoost Dynamic

To compare our results with AdaBoost Dynamic [5], we used 13 of the 18 datasets in [5]. We omitted the datasets which are not the same as the original ones at UCI repository. Table 4 shows the comparison between our algorithm and the AdaBoost Dynamic in [5]. It can be seen that our algorithm wins 10 out of 13 datasets.

To test the significance of these results, we then applied Wilcoxon's signed rank test [18] to these two algorithms and the result is p=0.03271. As p<0.05, we conclude that at 0.05 significance level we reject the hypothesis that meta-boosting performs equally with AdaBoost Dynamic [5].

Table 4. Comparisons between Meta-boosting and AdaBoost Dynamic algorithm

Dataset	Meta-boosting	AdaBoost Dynamic
Contact-lenses	**83.33±22.22**	74.17±28.17
Post-operative	**71.11±5.74**	56.11±13.25
Soybean	**94.58±2.79**	93.45±2.82
Blood	76.07±2.43	**77.93±3.56**
Crx	**87.53±2.08**	81.42±4.49
Au1	**76.00±3.81**	72.25±3.79
Iris	**95.33±5.49**	95.13±4.63
Segment	**98.33±0.96**	95.81±2.06
Breast-tissue	**70.82±11.40**	64.28±11.46
Car	96.82±1.71	**99.14±0.92**
Cmc	51.93±4.06	**54.08±3.81**
Zoo	**96.18±6.54**	95.66±5.83
Balance-scale	**93.28±1.80**	89.71±4.72
Average accuracy	**84.00**	80.70

5   Conclusions

In this work we present a new method for constructing ensembles of classifiers by combining AdaBoost.M1 algorithms with the meta-learning framework. The experimental results demonstrate that our algorithm outperforms all its base learners. Friedman's test and Nemenyi's post-hoc test further confirmed that the superior performance of our algorithm over all the base learners is statistically significant. We also compared our results with AdaBoost Dynamic [5]. Both the experimental accuracies and statistical Wilcoxon's signed rank test demonstrated that our algorithm displayed better performance. For the future work, we intend to research the applications of our algorithm. We also plan to increase the efficiency of our algorithm by using parallel computation. Moreover, it would be interesting to see the application of our algorithm on class imbalance problem.

References

1. Wolpert, D., Macready, W.: No free lunch theorems for optimization. IEEE Trans. Evol. Comput. 1, 67–82 (19 97)
2. Gama, J.: Combining classification algorithms. PhD Thesis, University of Porto (1999)
3. Schapire, R.E.: The strength of weak learnability. Machine Learning 5, 197–227 (1990)
4. Philip, K.C., Salvatore, J.S.: Meta-learning for multistrategy and parallel learning. In: Proc. Second Intl. Work. Multistraegy Learning, pp. 150–165 (1993)
5. de Souza, É.N., Matwin, S.: Extending AdaBoost to Iteratively Vary Its Base Classifiers. In: Butz, C., Lingras, P. (eds.) Canadian AI 2011. LNCS, vol. 6657, pp. 384–389. Springer, Heidelberg (2011)
6. Kohavi, R., Wolpert, D.H.: Bias plus variance decomposition for zero-one loss functions. In: 13th International Conference on Machine Learning, pp. 275–283 (1996)
7. Dietterich, T.G.: Bias-variance analysis of ensemble learning. In: 7th Course of the International School on Neural Networks, Ensemble Methods for Learning Machines (2002)
8. Holte, R.C.: Very simple classification rules perform well on most commonly used datasets. Machine Learning 11, 63–91 (1993)
9. Domingos, P., Pazzani, M.: On the optimality of the simple Bayesian classifier under zero-one loss. Machine Learning 29, 103–130 (1997)
10. Bauer, E., Kohavi, R.: An empirical comparison of voting classification algorithms: Bagging, boosting, and variants. Machine Learning 36, 105–139 (1999)
11. Domingos, P.: A unified bias-variance decomposition. In: Proceedings of 17th International Conference on Machine Learning. Morgan Kaufmann, Stanford (2000)
12. Webb, I., Zheng, Z.: Multistrategy ensemble learning: Reducing error by combining ensemble learning technique. IEEE Tran. on Knowledge and Data Engineering 16, 980–991 (2004)
13. Philip, K.C., Salvatore, J.S.: Scaling learning by metalearning over disjoint and partially replicated data. In: Proc. Ninth Florida AI Research Symposium, pp. 151–155 (1996)
14. Frank, A., Asuncion, A.: UCI machine learning repository (2010)
15. Aha, D.W.: Lazy learning. Kluwer Academic Publishers (1997)
16. Quinlan, J.R.: C4.5: Programs for machine learning, vol. 1. Morgan Kaufmann (1993)
17. Kotsiantis, S.B., Zaharakis, I.D., Pintelas, P.E.: Machine learning: A review of classification and combining techniques. Artificial Intelligence Review 26, 159–190 (2006)
18. Japkowicz, N., Mohak, S.: Evaluating Learning Algorithms: A Classification Perspective. Cambridge University Press (2011)

Preference Thresholds Optimization by Interactive Variation

Mohamed Mouine and Guy Lapalme

RALI-DIRO Université de Montréal
CP 6128, Succ Centre ville
Montréal (Québec) H3C 3J7
{mouinemo,lapalme}@iro.umontreal.ca

Abstract. In order to customize the display of meteorological data for different users, we use clustering to group similar users. We compute a rate of similarity between the current user and all others in the same cluster. We use this rate for weighting users' preferences and then compute an average to be compared with a threshold to decide to display this parameter or not. The optimization of this threshold is also discussed.

Keywords: information visualization, artificial intelligence, clustering, personalization, graph.

1 Introduction

We want to generate customized meteorological reports for users of the Environment Canada website and adapt the reports to the needs and preferences of users. Users can change the layout according to their taste, preferences and needs. We want improve the user experience with the site by learning to *guess* users' preferences. To make this presentation closer to each user preferences, we aggregate *user types* and give each user a visualization close to what s/he wants. The principle that we consider most important is the customization of the display for each user taking into account his profile to be automatically detected without interference with his privacy and personal information. Law in Canada regarding the use of cookies in the government website is strict. We quote here an extract from the Office of the Information and Privacy Commissioner of Ontario[1]: "*The collection of personal information by government organizations must be in accordance with section 38(2) of the Freedom of Information and Protection of Privacy Act (FIPPA). This section of the Act requires government organizations to collect only the personal information that is necessary for the administration of a government program*". For this reason, we will not use cookies.

The information we have about a user (Location, preferred language, date, local time, season) does not allow us to predict user preferences. The first step of the method [9] is to archive (anonymously) interactions of the users with the

---

[1] http://www.ipc.on.ca/English/Privacy-Policy/

O. Zaïane and S. Zilles (Eds.): Canadian AI 2013, LNAI 7884, pp. 286–292, 2013.
© Springer-Verlag Berlin Heidelberg 2013

visualization and especially the final settings chosen. We take for granted that the final display as chosen by the user reflects her preferences. Users are then grouped according to their information similarity. We then determine the cluster corresponding to the user. Visualizations corresponding to users belonging to the same group as the user will be used to identify similar users' preferences. We weight their similarity rates based on the distance between the feature vectors. The preferences of users are weighted according to the rate of similarity. To decide what preferences be taken into account, we compare the mean of each parameter with a threshold to determine if a parameter is taken into consideration or not. In the next section we describe, for the problem of customizing visualization, the need to know the user profiles. In Section 3, we detail the problem. The solution to this problem is described in Section 4. And finally, a summary is presented in the last section.

2   Need for User Profiles

Graphics are useful to summarize and communicate numerical information found in weather reports. The task of generating graphics cannot be reduced to the encoding of a mass of information, it must take into account the decoding to be done by the user. If the user cannot decode the information, graphics generation has failed. It should take human perception into account. Robbins [11] explains how to create better visualizations by taking into account all the parameters (choice of chart type, the amount of information, choice of style attributes ...) and perception towards graphs.

The problem of information visualization is not necessarily raised by technical obstacles. Chen [2] studied 10 unsolved problems in this area, the first three ones being problems from a user-centric perspective:

usability the information contained in the visualization should meet the needs of users; this explains the increasing number of usability studies and evaluations of visualization [3,10,4,16];

perception the principles of perception were incorporated into rendering algorithms in order to optimize rendering computation and produce an *ideal* visualization human point of view and not from a machine standpoint [7];

prior knowledge of the user can be considered a parameter of the second problem. In general, users need two types of knowledge before understanding the message conveyed in the information displayed: how to use a system for displaying information and how to interpret the content.

There are two types of user profiles: determined by a user who sets preferences and needs or built automatically using several techniques such as history, behavior, rules of associations, classification techniques and algorithms of clustering.

We want the system to learn the preferences of the users. Reports generated for each one are interactive. It is very important that users can modify visualization according to their tastes and needs[14]. A study was conducted to describe why users should interact[15]. They are based on different intentions of users

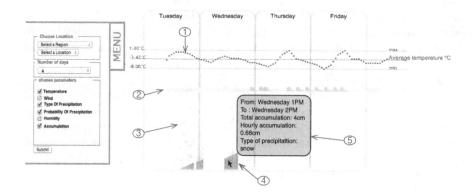

Fig. 1. Generated visualization showing some types of information that can be selected by the user. Circled numbers are added here for reference purposes. ① Temperature: how temperature is displayed gives an idea of the trend. Maximum and minimum line show limits of the temperature for the chosen period. ② Cloud cover. ③ TOP (Type Of Precipitation): rain or snow. Quantity displayed is proportional to the quantity provided. ④ Accumulation: shows the accumulation zone and total rainfall (mm) or snow (cm). ⑤ The user may have more details about aspect by putting the cursor over it (tooltip). Using the preference menu (in left), the user can select a region (Province) and location (city), modify number of the days to display and which parameter to display.

and introduced a list of categories (Select, Explore, Reconfigure, Encode, Abstract/Elaborate, Filter and Connect). The majority of these principles are used in our visualization.

3   Problem Statement

Environment Canada (EC) produces an enormous amount of weather information on a continuous (26 Mb twice a day). This information is used to provide Canadians with up to date information on weather conditions. Problem is that this amount of information must be summarized to be displayed. Weather reports prepared in advance may not contain all the information that all users hope to find. Already, more than 1,000 weather reports presenting the weather in Canada are issued twice per day. First solution is to prepare more weather reports in advance. But most of these reports will not likely be used. We thus propose to build a weather reports generator to answer users on demand in either English or French. Each report must meet the specific needs of the user for which it was generated. In order to summarize and analyze large amounts of information, we have presented a method that automatically generates a visual report (graph, image, text...).

To allow users to select the information to display, our system should be interactive. We have already proposed [9] generating reports based on the similarity of user profiles. Clustering was used to group similar users to produce a visualization corresponding to the needs of most of them. But visualization generation was based on an arbitrary threshold to decide the parameters to be taken into account. In this paper, we study the robustness of this threshold and present a sensitivity analysis.

4   Approach

To generate the visualization (see Fig. 1) based on the user's profile to learn more about their preferences, our system saves each final configuration chosen by a user. The system uses this data, for future customization of the weather reports according to user profiles. When a new report is generated, it is based on several parameters including: the user's profile and choice of former users similar to the current user. Clustering is used for determining the settings of similar users. The distance between the current user and others in a cluster will be used to compute the similarity. A user can modify the visualization settings using a menu (left in Fig. 1).

4.1   Clustering

K-means [5] is a simple algorithm for clustering. It classifies a set of data among a fixed number of clusters. The main idea is to define K centroids, one for each cluster. K being between 10 and 20 corresponding roughly to the number of provinces in Canada. The next stage is to take each user in our database for which a visualization has already been generated and to associate her with the nearest centroid. Variables of user profiles belonging to the same category represents the features for the clustering process.

4.2   Similarity Computing

Profile data and preferences for each user are saved (anonymously). We assign a weight, depending on the significance, to each criterion for users' profiles (see Table 1). The rate of similarity is given by:

$$R_j = \sum_{i=1}^{n} C_i * W_i \text{ with } j \in [1..m]$$

where: R is the rate of similarity, m number of user, n number of criterion, C is the criterion taken into account by the user, and W is its weight.

An example of the calculated similarity is shown in table 2. similarity rate is used to weight the preferences.

Table 1. Criterion weight

criterion	weight
Lang	0.2
Province	0.3
City	0.3
Season	0.1
Period	0.1

Table 2. Similarity rates

	Lang	Province	city	season	period	% of similarity
current user	fr	Qc	Montreal	winter	morning	-
user1	fr	Qc	Quebec	autumn	Afternoon	50%
user2	fr	Qc	Longueuil	winter	morning	70%
user3	en	Qc	Gaspé	autumn	morning	40%
user4	fr	Qc	Montreal	spring	morning	90%
user5	fr	Qc	Laval	autumn	evening	50%
user6	fr	Qc	Sutton	summer	evening	50%
user7	en	Qc	Delson	summer	morning	40%
user8	en	Qc	Montreal	spring	Afternoon	60%
user9	fr	Qc	Gatineau	winter	morning	70%
user10	fr	Qc	Montreal	winter	morning	100%
average						62%

4.3 Computing the Preferences

To predict the preferences of the current user, we rely on preferences data recorded in our database of similar user. In the database, we register number of day fixed in visualisation by user and for each parameter 1 if used, 0 if not. Preferences of each user is weighted by the level of similarity with the current user.

$$P_{wj} = P_j * R_j$$

4.4 Threshold Analysis

Threshold analysis is often used in multicriteria decision methods [1,6,12,13] in which a decision maker sets thresholds arbitrarily and then evaluates their robustness using sensitivity analysis [6,8] to modify their values. If a small modification results in a large change in the results, the threshold is considered sensible and the decision is not robust.

Our approach is inspired by this type of sensitivity analysis. A threshold determines whether to use the parameter or not. We analyzed its robustness to the user feedback with the visualization. We consider that:

- if the user changes the display and adds a new parameter that has not been used, that means that the threshold was very strict and led to the exclusion of the parameter in question; its value should be decreased;

Table 3. Result of visualization interaction

	temperature	wind	TOP	POP	humidity	accumulation
added	-	-	-	-	20%	40%
removed	0%	20%	10%	10%	-	-
kept	100%	80%	90%	90%	80%	60%
average	0.57	0.41	0.38	0.34	0.32	0.23
old threshold				0.33		
new threshold	0.33	0.35	0.34	0.34	0.31	0.29
retained (new)	yes	yes	yes	yes	yes	no

- if the user modifies the visualization and regenerates it by removing one or more parameters, it means that the value of the threshold for a removed parameter is very low and it value should be increased;
- if the user does not interact, the threshold is robust and corresponds to the wish of the user.

To assess the robustness of visualization, we analyze user interactions and compute the new thresholds according to the following formula:

1. Vary the threshold ± 0.01 for each 10 % of dissatisfied users.
2. Redo the experiment (Table 3).
3. Repeat step 1 and 2 until we haven't more dissatisfaction superior to 10 %.

Table 4 show the final result of our example. We stop the threshold variation after 3 iterations because we did not any dissatisfaction superior to 10 %.

Table 4. Final result of visualization interaction and threshold variation (after 3 itereation)

	temperature	wind	TOP	POP	humidity	accumulation
average	0.57	0.41	0.38	0.34	0.32	0.23
Final threshold	0.33	0.36	0.34	0.35	0.30	0.30
retained (new)	yes	yes	yes	yes	yes	no

It can be seen that the thresholds have been revised downward or upward according users disatisfaction. Although the threshold of the parameter *temperature* that satisfies users has not been changed. The thresholds for *wind, TOP* and *POP)* were increased because users were not satisfied with the visualization generated for them. The threshold parameters *humidity* and *accumulation* were lowered because multiple users wanted to have it in their visualizations that the former threshold does not allow.

5   Conclusion

Our job is to customize a visualization according to the profile of the user. We have little information about the users using our system. We rely on the history of user preferences similar to our current user. We set a minimum threshold for deciding which parameters will be used in the visualization. We use the same threshold to evaluate the level of satisfaction of the user. The interaction of the user with the visualization used for evaluation improves our system to better match preferences of users according to their profile.

References

1. Brans, J.P., Vincke, P., Mareschal, B.: How to select and how to rank projects: The promethee method. European Journal of Operational Research 24(2), 228–238 (1986)
2. Chen, C.: Top 10 unsolved information visualization problems. IEEE Computer Graphics and Applications 25(4), 12–16 (2005)
3. Elmqvist, N., Yi, J.S.: Patterns for visualization evaluation (2012)
4. Goldberg, J., Helfman, J.: Eye tracking for visualization evaluation: Reading values on linear versus radial graphs. Information Visualization 10(3), 182–195 (2011)
5. Hartigan, J.A., Wong, M.A.: Algorithm as 136: A k-means clustering algorithm. Applied Statistics, 100–108 (1979)
6. Maystre, L.Y., Pictet, J., Simos, J.: Méthodes multicritères ELECTRE: Description, conseils pratiques et cas d'application à la gestion environnementale, vol. 8. PPUR (1994)
7. McNamara, A., Mania, K., Gutierrez, D.: Perception in graphics, visualization, virtual environments and animation. In: SIGGRAPH Asia 2011 Courses, p. 17. ACM (2011)
8. Mena, S.B.: Une solution informatisée à l'analyse de sensibilité d'electre iii. Biotechnol. Agron. Soc. Environ. 5(1), 31–35 (2001)
9. Mouine, M., Lapalme, G.: Using clustering to personalize visualization. In: 2012 16th International Conference on Information Visualisation (IV), pp. 258–263. IEEE (2012)
10. Plaisant, C.: The challenge of information visualization evaluation. In: Proceedings of the Working Conference on Advanced Visual Interfaces, pp. 109–116. ACM (2004)
11. Robbins, N.B.: Creating more effective graphs. Wiley-Interscience (2012)
12. Roy, B.: Classement et choix en présence de points de vue multiples (la méthode electre). RIRO 2(8), 57–75 (1968)
13. Roy, B.: Electre iii: Un algorithme de classement fondé sur une représentation floue des préférences en présence de critères multiples. Cahiers du CERO 20(1), 3–24 (1978)
14. Ward, M., Grinstein, G., Keim, D.: Interactive data visualization: Foundations, techniques, and applications. AK Peters, Ltd. (2010)
15. Ware, C.: Information visualization: Perception for design. Morgan Kaufmann (2012)
16. Weaver, C.: Look before you link: Eye tracking in multiple coordinated view visualization. In: BELIV 2010: BEyond Time and Errors: Novel Evaluation Methods for Information Visualization, p. 2 (2010)

General Topic Annotation in Social Networks: A Latent Dirichlet Allocation Approach

Amir H. Razavi[1], Diana Inkpen[1],
Dmitry Brusilovsky[2], and Lana Bogouslavski[2]

[1] School of Electrical Engineering and Computer Science
University of Ottawa
{araza082,diana}@eecs.uottawa.ca
[2] Business Intelligence Solutions
dmitry@bisolutions.us

Abstract. In this article, we present a novel document annotation method that can be applied on corpora containing short documents such as social media texts. The method applies Latent Dirichlet Allocation (LDA) on a corpus to initially infer some topical word clusters. Each document is assigned one or more topic clusters automatically. Further document annotation is done through a projection of the topics extracted and assigned by LDA into a set of generic categories. The translation from the topical clusters to the small set of generic categories is done manually. Then the categories are used to automatically annotate the general topics of the documents. It is remarkable that the number of the topical clusters that need to be manually mapped to the general topics is far smaller than the number of postings of a corpus that normally need to be annotated to build training and testing sets manually. We show that the accuracy of the annotation done through this method is about 80% which is comparable with inter-human agreement in similar tasks. Additionally, using the LDA method, the corpus entries are represented by low-dimensional vectors which lead to good classification results. The lower-dimensional representation can be fed into many machine learning algorithms that cannot be applied on the conventional high-dimensional text representation methods.

Keywords: Latent Dirichlet Allocation (LDA), Automatic document annotation, Text representation, Topic extraction.

1 Introduction

Today, modern social networks and services have become an increasingly important part of how users spend their time in the online world. Social networking sites are now increasingly becoming social networking services, and they bring more and more information to the users through their available communication tools. In the meanwhile, in order to present the best set of features of a social network or service and also to have a proper control on such a vast interface, automatic social network analysis has an important role. In the same time, the concepts of social media are being actively adopted by the enterprises; many of them are implementing their own

O. Zaïane and S. Zilles (Eds.): Canadian AI 2013, LNAI 7884, pp. 293–300, 2013.
© Springer-Verlag Berlin Heidelberg 2013

enterprise social media platforms. The market of enterprise social media collaboration software is fast growing.

In this paper, we focused our efforts on the unstructured part of the social networks which is the textual postings and related comments. We present a semi-supervised method to extract general topics from a social network corpus and then annotate the postings using the general topics. The general topic annotation can be used for further conceptual analysis of the textual content of the social network. In the proposed method, we annotate a subset of the social network threads (posts and their comments) automatically and then we evaluate the annotation quality (by comparing it to the labels assigned manually) to show that is reliable enough to be used in the social network text analysis task. Our method applies Latent Dirichlet Allocation (LDA) on a corpus to initially infer some topical word clusters which will be then used for document annotation and representation at the next stage. Each extracted topical cluster is interpreted by a human judge and will be projected into a generic categorical topic which will be then the label of the document (a thread in our case).

2   Background

In 2003, Blei, Ng and Jordan presented the Latent Dirichlet Allocation (LDA) model and a Variational Expectation-Maximization algorithm for training their model. Those topic models are a kind of hierarchical Bayesian models of the applied corpus [1]. The model can unveil the main themes of the applied corpus, which can potentially use to organize, search, and explore the documents of the corpus. In LDA topic modeling, a "topic" is a distribution over a fixed vocabulary of the corpus and each document can be represented by several topics with different weights. The number of topics and the proportion of vocabulary that create each topic are considered as two hidden variables of the model. The conditional distribution of these variables given an observed set of documents is regarded as the main challenge of the model.

Collapsed Variational Bayes (CVB) inference [2] also analytically marginalizes the topic proportions and is regarded as an alternative deterministic inference for LDA. The proposed inference algorithm can improve the accuracy and efficiency of the standard Bayesian inference for LDA.

Griffiths & Steyvers [3] applied a derivation of the Gibbs sampling algorithm for learning LDA models. They showed that the extracted topics capture a meaningful structure of the data. The captured structure is consistent with the class labels assigned by the authors of the articles. The paper presents further applications of this analysis, such as identifying "hot topics" by examining temporal dynamics and tagging some abstracts to help exploring the semantic content.

Since then, the Gibbs sampling algorithm was shown as more efficient than other LDA training methods, e.g., variational EM and Expectation-Propagation [4]. This efficiency is attributed to a famous attribute of LDA namely, "the conjugacy between the Dirichlet distribution and the multinomial likelihood". This means that the conjugate prior is useful since the posterior distribution is the same as the prior, and it makes inference feasible and causes that when we are doing sampling, the posterior sampling become easier. Because of this, the Gibbs sampling algorithms was applied for inference in a variety of models which extend LDA [5] [6] [7].

Hoffman et al [8] introduced a new derivation named "Online LDA" which is a stochastic gradient optimization algorithm for topic modeling. The algorithm iteratively subsamples a small number of documents from the entire corpus and then updates the topics by the new inferences. Since in this method we do not need to store topic proportions for the entire corpus, it is much more memory conservative than the standard approach. Furthermore, the authors show that since the algorithm updates topics more frequently, it converges faster than the other methods. However, when it runs over a large corpus, it does not scale up to appropriate large numbers of topics. Adaptive scheduling algorithm [9] can also be regarded as applicable extensions of this model.

3   Data Set

In our research, we were looking for sources of social network data that include textual postings and related comments, in which the main posting could be connected to the corresponding comments in order to form a thread; the connection is done via parent/posting identifier (id) information items. It was challenging and rather time consuming to find such datasets. We selected a set of post/comment textual data that we extracted from the well-known "Friendfeed" social network media.

Initially, through a multi-level filtering task, a large amount of data (~ 23 GB in compressed format) was collected from "Friendfeed.com"; from which we extracted the information items useful to our research, including main postings and their related comments which are linked based on Post_id. Then we integrated the main postings (12,450,658) to their corresponding comments (3,749,890) in order to create same topic threads. At the next stage, we filtered out all the threads with no comments (with Null comments).

The source data was in more than 11 different languages; therefore we run a language identification tool in order to select a subset including only postings and related comments in English. There were many postings/comments mixed in English and another language; this represented another challenge at this stage. Hence, we decided to remove the threads that were partially commented in other languages and kept only threads that were entirely in English at the final stage. We also filtered out all threads smaller than 120 characters or with less than three comments.

The built data-set included more than 24000 usable threads as input for our topic detection task. A randomly selected subset of 500 threads was chosen to be manually annotated in order to be used for training/ testing a variety of classifiers as a proof for the applicability of our general topic annotation method of the threads. The class labels (general topics) were selected and generalized manually *based on the topics extracted automatically by the LDA method* [1]. The final set of general topics contained the following 10 categories:

```
consumers, education, entertainment, life_stories, lifestyle, politics, rela-
tionships, religion, science, social_life, technology.
```

---

[1] Will be explained more in the "LDA Topical Modeling" section.

Additionally, a random subset of ~4000 threads including the initial 500 threads plus 3500 unlabeled threads (that we call *background resource*) has also been selected for estimating the LDA models that are needed for the topic detection. (The method will be explained later in Section 4.2.)

4   Methodology

4.1   Preprocessing

In the preprocessing stage, initially all the different headers, internet addresses, email addresses and tags were filtered out. Then all the delimiters such as spaces, tabs or newline characters, have been removed from postings, whereas the expressive characters like: " - . ' ' ! ? " were kept. Punctuations (such as quotes, " ") could be useful for determining the scope of speaker's messages. This step considerably reduces the size of feature space and prevents the system from dealing with a large number of unrealistic tokens as features for our classifiers and LDA estimation/inferences.

Two types of stop-words removal were performed: static stop words removal and corpus based dynamically stop words removal. For the first one, we tokenized the posts/comments individually to be passed to the static stop-word removal step that is based on an extensive list of stop-words which has been already collected specifically for the applied dataset (i.e., social network).

In the second one, additional stop words were determined based on their frequency, distribution and the tokenization strategy over the corpus (i.e., unigrams, bigrams, 3 or 4 grams). We removed tokens with very high frequency relative to the corpus size where those appear in every topical class (i.e., those are almost useless for the topic identification task). The output of this stage passed to the stemming process through the Snowball[2] stemming algorithm.

4.2   LDA Topical Modeling

For our goal of general topic extraction from social network threads, we developed a method based on the original version of LDA [1]. LDA is a generative probabilistic model of a corpus. The basic idea is that the documents are represented as a weighted relevancy vector over latent topics, where a topic is characterized by a distribution over words. We applied and modified the code originally written by Gregor Heinrich [10] based on the theoretical description of Gibbs Sampling. A remarkable attribute of the chosen method is that lets a word to participate in more than one topical subset based on its different senses/usages in its context.

The subset that we used for running the LDA algorithm consisted of 4000 threads (500 labeled and 3500 background source) which already passed the preparation and filtration processes (the pre-processing). In this way, each thread is represented by a number of topics in which each topic contains a small number of words inside (i.e., each topic consists in a cluster of words); and each word can be assigned to more than one topic across the entire input data (e.g., polysemous words can be in more than one topic). Therefore, the number of topics and the number of words inside each topic are

---

[2] http://snowball.tartarus.org/

two parameters of the method that can be adjusted according to the input data. In this research, the values of the parameters have been empirically set to 50 topic clusters, and maximum 15 words in each cluster. Then, the LDA method assigns some groups of words (the 50 groups of 15 words inside each group) as topics, with different weights, for each text (in our case each thread).The topical cluster of words are interpreted and assigned to a real topical phrase/word, manually.

For example, the following topical cluster: {"Google", "email", "search", "work", "site", "services", "image", "click", "page", "create", "contact", "connect", "buzz", "Gmail", "mail"} which is a real example extracted by the LDA model estimation process from the explained corpus, initially has been interpreted (manually) as "Internet" topic and at the next level of the *topic generalization* was placed under the "technology" and "social_life" categories.

Similarly to the above example, all the 50 topical clusters extracted by the LDA method were manually mapped to the previously listed 10 generic and human-comprehensible topics. We observed that the 10 class labels (general topics) are distributed unevenly over the dataset of 500 threads, in which we had 21 threads for "consumers", 10 threads for "education", 92 threads for "entertainment", 28 threads for "incidents", 90 threads for "lifestyle", 27 threads for "politics", 58 threads for "relationships", 31 threads for "science", 49 threads for "social_activities", and 94 threads for "technology". Thus, the baseline of any classification experiment over this dataset may be considered as 18.8%, for a trivial classifier that puts everything in the most frequent class, "technology". However, after balancing the above distribution through over/under sampling techniques, the classification baseline lowered to 10%. The last step was performed via the Synthetic Minority Oversampling Technique (SMOTE) [11] over the class labels with frequencies lower than average, and random under-sampling method over those which have frequencies higher than average. We sustained those extra steps in order to obtain an evenly distributed dataset and do not deal with an unbalanced data classification task and its side effects.

Since the LDA modeling does not assign a single general topic (e.g., "entertainment") to each tread, the assignment of the general topics (i.e., one of the 10 class labels) is a further task that will be done through a separate classification process.

4.3   Topic Classification

As mentioned before, the training/testing dataset for the supervised classification task consisted of 500 manually annotated threads annotated with the 10 general categories enumerated in section 3. For this dataset, we initially applied a variety of Bag of Word (BOW) representations (i.e., binary, frequency and TF-IDF[3] based methods) in order to create the best discriminative representation over the entire 500 threads dataset. After removing stop-words and stemming as explained in section 4.1., we obtained 6573 words as the feature set for the general topic classification task.

---

[3] The TF-IDF (term frequency versus inverse document frequency) method was selected which is a classic method that gives higher weights to terms that are frequent in a document but rare in the whole corpus.

As the second and axillary representation of the same data, we used the topical cluster relevancy vector of the each thread[4] (calculated using the LDA technique) to obtain a low-dimensional representation of the threads. We evaluated that representation of the data and reserved for the complementary comparison between the two representations. Then we integrated the two representations mentioned above into one representation, which consisted of 6623 features (words and 50 topics) to test the classification (automatic annotation) performance over the integrated representation. As part of the supervised learning core of the system, we trained a variety of classifiers, in order to evaluate the general topic annotation performance of the method.

5   Results and Discussion

We run our comparing classification experiments on the 500 filtered Friendfeed threads. We conducted the classification evaluations using stratified 10-fold cross-validations (this means that the classifier is trained on nine parts of the data and tested on the remaining part, then this is repeated 10 times for different splits, and the results are averaged over the 10 folds). We performed several experiments on a range of classifiers and parameters for each representation to check the stability of a classifier's performance. We changed the "Seed", random parameter of the 10-fold cross-validation in order to avoid the accidental "over-fitting". In order to resolve any conjecture of over-fitting, the final evaluation of the method has been performed on a set of four pre-set classifiers included: Complementary Naïve Bayes (NB), Multinomial Naïve Bayes, Support Vector Machine (SVM) (SMO in Weka) and Decision Trees (DT) (J48 in Weka). They were chosen because Naïve Bayes is known to work well with text, because SVM is a very good performer in general, and because DT's output in readable for humans.

Table 1. Comparison of the classification evaluation measures for different representation methods

Evaluation measure → / Representation/ Classifier used ↓	TP Rate Wtd. Avg.[8]	FP Rate Wtd. Avg.	Precision Wtd. Avg.	Recall Wtd. Avg.	F-Measure Wtd. Avg.	Accuracy %
BOW(TF-IDF)/ CompNB	0.772	0.025	0.744	0.772	0.743	77.22
LDA Topics/ Adaboost (j48)	0.693	0.034	0.679	0.693	0.684	69.33
BOW(TF-IDF) +LDA/ SVM(SMO)	0.8	0.022	0.786	0.8	0.79	80.00

---

[4] Each vector contains only 50 features corresponding to the 50 LDA clusters.

The evaluation measures calculated by the most stable classifier over the three re-presentations are shown in table 1. This performance is acceptable, considering that manual general topic annotation is an uncertain task (even for the human beings). The uncertainty has roots in the following three aspects: 1) the topics in our list of 10 cat-egories are sometimes too general; 2) the nature of the social network scattered post-ings (informal text using abbreviations that are not clear for everybody, etc.); 3) the subjectivity of the manual annotations; the reasons for some discrepancies between human annotations (with the same problem definition) could be tracked in their dif-ferent personality, mood, background and some other subjective conditions. Human judgment is subjective and is not necessarily the same, among different people upon the same case. According to the related literature, when documents are annotated by more than one human annotator the expected agreement between judges is normally around 60-85% on different datasets [12], [13], [14]. Therefore, it is helpful to have a standard annotation system that always annotates based on some constant definitions, patterns and rules, as our automatic system does.

Our "general topic detection" method can be applied for trend detection purposes in any collaborative writing web sites in which people add or modify contents, in the style of posts/comments. It could also be handy for some web-logs or some specialist forums. It could also be adapted for some kinds of message categorization or even spam detection for any type of text messaging services on the internet or even on cellular phones.

6   Conclusion and Future Work

We designed and implemented an efficient "general topic detection" method over the "Friendfeed" social network textual dataset. The system applies LDA topical model-ing estimation/inference for the topic detection purpose. The method also gets benefit from some classification algorithms for the purpose of general topic detection. The system is useful as standard general topic annotation applications, mostly in messag-ing services and collaborative writing web sites. Moreover, the performance of the system is similar to a range of comparable tasks.

There are many advantages of our method, including:

1) The LDA method automatically assigns topics to the posts/comments (via a small group of words clustered together). Then we manually interpret and generalize the clusters into small number (e.g., 10) of high-level classes (showed in section 4.2.). The remarkable advantage of this method is that the number of topical groups that need to be manually mapped to the general topics are far smaller than the number of postings of a social network corpus (or any corpus in general) that would need to be annotated to build training and testing sets manually.

2) In the LDA representation each document (thread) is represented by the LDA weighted membership distribution of the topical word clusters; hence any other high dimensional vector representation of any collection of documents can be also replaced by its LDA weighted membership distribution in order to reduce the dimensionality and consequently dealing with the curse of dimensionality. The lower dimensional representation can be used for any supervised/unsupervised machine learning algo-rithm which cannot be applied on high-dimensional data.

3) We observed that the quality of the topical clusters of the LDA algorithm improves simply by adding the 3500 background source data (threads extracted from the same corpus) to the original 500 threads selected for the supervised learning. This means that consequently the performance of our automatic general topic detection method is improved using *unlabeled* background source data.

One limitation of the current design is that it is *case insensitive*; it could be developed based on *case sensitive* texts in order to extract more specific topical keywords/phrases of the contents.

In future work, we are planning to replace the manual interpretation of the LDA topical word clusters with an automatic topic assignment. This idea could be realized by getting benefits from resources such as "Wordnet Domains".

References

1. Blei, D., Ng, A., Jordan, M.: Latent Dirichlet allocation. Journal of Machine Learning Research 3, 993–1022 (2003)
2. Teh, Y.-W., Newman, D., Welling, M.: A collapsed variational Bayesian inference algorithm for latent Dirichlet allocation. In: Procs. NIPS (2006)
3. Griffiths, T.L., Steyvers, M.: Finding scientific topics. Proceedings of the National Academy of Sciences 101, 5228–5235 (2004)
4. Minka, T., Lafferty, J.: Expectation propagation for the generative aspect model. In: Proceedings of UAI (2002)
5. Wang, X., McCallum, A.: Topics over time: A non-markov continuous-time model of topical trends. In: Proceedings of KDD (2006)
6. Blei, D.M., McAulie, J.: Supervised topic models. In: Procs. of NIPS (2007)
7. Li, W., McCallum, A.: Pachinko allocation: Dag-structured mixture models of topic correlations. In: ICML (2006)
8. Hoffman, M., Blei, D., Bach, F.: Online learning for latent Dirichlet allocation. In: Proceedings of NIPS (2010)
9. Wahabzada, M., Kersting, K.: Larger residuals, less work: Active document scheduling for latent dirichlet allocation. In: Gunopulos, D., Hofmann, T., Malerba, D., Vazirgiannis, M. (eds.) ECML PKDD 2011, Part III. LNCS, vol. 6913, pp. 475–490. Springer, Heidelberg (2011)
10. Heinrich, G.: Parameter estimation for text analysis, Technical Report (For further information please refer to JGibbLDA at: http://jgibblda.sourceforge.net/)
11. Chawla, N.V., Bowyer, K.W., Hall, L.O., Kegelmeyer, W.P.: Synthetic Minority Oversampling Technique. Journal of Artificial Intelligence Research 16, 321–357 (2002)
12. Wang, A., Hoang, C.D.V., Kan, M.-Y.: Perspectives on crowdsourcing annotations for natural language processing. Language Resources and Evaluation, 1–23 (2012)
13. Ferschke, O., Daxenberger, J., Gurevych, I.: A Survey of NLP Methods and Resources for Analyzing the Collaborative Writing Process in Wikipedia (2012)
14. Fleischmann, K.R., Templeton, C., Boyd-Graber, J., Cheng, A.-S., Oard, D.W., Ishita, E., Koepfler, J.A., Wallace, W.A.: Explaining Sentiment Polarity: Automatic Detection of Human Values in Texts (2012) (to appear)

Classifying Organizational Roles
Using Email Social Networks

Abtin Zohrabi Aliabadi*, Fatemeh Razzaghi*,
Seyed Pooria Madani Kochak, and Ali Akbar Ghorbani

University of New Brunswick, Faculty of Computer Science,
550 Windsor Street, Fredericton, NB E3B 5A3
{Abtin.Zohrabi,F.Razzaghi,Pooria.Madani,Ali.Ghorbani}@unb.ca
http://ias.cs.unb.ca/

Abstract. This paper addresses the problem of role classification, which is related to classifying and grouping email users into a collection of organizational roles. This classification can be used in designing modern email clients by adding an Inbox prioritizing feature that can predict the role of a sender to the recipient of an email. A comprehensive study has been done on the social network of the Enron dataset. For classifying organizational roles, a feature vector containing a set of social network metrics and interaction-based features reflecting users' *engagingness* and *responsiveness* in their community is created. After representing each role in this feature space, Expectation Maximization (EM) algorithm has been applied to evaluate the extracted feature set. In turn, a Neural Network classifier has been built based on the extracted features for classifying organizational roles that resulted in 63.57% of accuracy.

Keywords: Email mining, Enron dataset, Role Classification, Social network analysis, Community discovery.

1   Introduction

This paper introduces a problem in the area of email mining known as role classification. Role classification is defined as finding the role of a user in an organizational setting from a communication network (i.e email). Email clients can take advantage of this technique in order to prioritize the incoming emails based on the role of the email's sender within the specific organizational setting. From analytic or investigative perspective, this technique can serve intelligence and security services by extracting underlying roles within the communities that are formed in an organization.

A series of questions arise: Can we extract some numerical measures out of an organization's email interactions that can contribute to classifying roles of the users? Are these features meaningful enough for accurate discovery and prediction of roles using only small amounts of labeled training data and a limited

---

* The first two authors contributed equally to this work.

O. Zaïane and S. Zilles (Eds.): Canadian AI 2013, LNAI 7884, pp. 301–307, 2013.

number of interaction traces?. This paper has tackled these fundamental problems by an extensive study on the Enron email dataset. A feature set has been defined and evaluated through a series of experiments. Our contributions in this paper include:

- Constructing the email's social network; then, using the Newman clustering algorithm [11], the underlying community structures are extracted and evaluated for their validness among other communities using topic discovery. Roles are evaluated in terms of social network metrics and best features have been chosen. Moreover, the EM algorithm is used to evaluate the whole feature set.
- Selecting a feature set from the social network metrics combined with interaction based features. Most of those metrics have been used for personalized email prioritization [15], or spam filtering although none of them has been used in depth for role prediction in the email domain.
- Building a *Neural Network classifier* for classifying email's sender into organizational role. This is known as the *Role Classification* procedure.

2   Related Work

There has been a stream of research in support of email-based prediction tasks for recipient reminding [2], action-item identification [3], spam identification [9], folder recommendation [7], and social group analysis [8] based on statistical learning techniques including supervised, unsupervised, and semi-supervised methods. Most previous research that are conducted on the Enron email corpus have focused on Natural Language Processing of the data for classification of the emails [7], dataset mapping of Enron's users [4], and quantitative analysis inside the Enron dataset [13].

From another perspective, the research community is exploring the Enron dataset from the network analytic view. In [6], Diesner et al. applied various network analysis techniques in order to find key players based on time. Recently in [14], Wang et al. have developed a classification technique to classify enterprize usage of emails versus personal. They emphasized social features more than the content-based features to solve email classification problems. Neustaedter et al. [10], focused on the problem of manhandled email and discussed using social meta-data to enhance the email triage experience. They also have used activity-based features based on user logs without taking into account the capability of social networks measures. As the novel contribution of this paper we are interpreting the interaction-based features effects simultaneously with social network analysis potential metrics.

3   Problem Definition and Proposed Solutions

Given a user, described by a set of features, the goal is to build a predictive model which can classify a user's role in an organization. Through a series of experiments, first the social network of the Enron users is analyzed for selecting a set of feature measures. Then, these social network based features along with interaction-based features are evaluated by a cluster analysis process in order to confirm their discriminative power. Finally, using these selected features a model is built for organizational role classification.

3.1 Features

Two sets of features have been used for the purpose of Organizational Role Classification illustrated in Table 1:

- **Social network-based features** define the social centrality of each node commonly used in social network analysis.
- **Interaction-based features** indicates personal contact network and illustrate the level of responsiveness and engagingness of each contact [12].

Table 1. Selected Features

(1) In-degree	$\frac{1}{|C|}\sum_{j=1}^{i} R_{ji}$,where $R_{ji} \in \{0,1\}$, and $|C|$ is the total number of contacts.
(2) Out-degree	$\frac{1}{|C|}\sum_{j=1}^{i} R_{ij}$
(3) Total-degree	$\frac{1}{|C|}\sum_{j=1}^{|C|} \lceil \frac{R_{ij}+R_{ji}}{2} \rceil$
(4) Clustering Coefficient	$\frac{1}{v}\sum_{i\in Nbr(v)}\sum_{j\in Nbr(n),j\neq i} R_{ij}$
(5) Betweenness	$\frac{1}{(v-1)(v-2)}\sum_{j=1,j\neq i}^{|n|}\sum_{k=1,k\neq j,k\neq i}^{|n|} \frac{\sigma_{jk}(i)}{\sigma_{jk}}$, where σ_{jk} is the number of shortest path between j and k that goes through i[15].
(6) Email Reply-Sent count	$\frac{|RepT(u_i)|}{|Sent(u_i)|}$, where $RepT(u_i)$ are the emails replied to u_i earlier emails and $Sent(u_i)$ is the total number of emails sent by this user.
(7) Email Reply-Received count	$\frac{|RepB(u_i)|}{|Recieved(u_i)|}$, where $RepB(u_i)$ are the emails replied by u_i and $Received(u_i)$ is the total number of emails received by this user.
(8) HITS Authority	Measures the global ratings of a contact inside the whole network

3.2 Dataset

The only publicly available dataset which has been used in email research is the Enron corpus [1]. To our knowledge there exists only one list[1] that indicates the roles of Enron users: 103 out of 149 users' email addresses and the organizational roles are specified. Some statistics about the selected dataset illustrated in Table 2. For the preprocessing, the names and email addresses were examined and found that 60 of selected users had second email addresses. We extracted all of messages related to these addresses and unified the whole dataset. Note that no additional information (e.g. user signatures) were available in the body of emails.

---

[1] http://cis.jhu.edu/~parky/Enron/employees

Table 2. Number of users in each role category

Role	CEO	Director	Employee	IHL	Manager	MD	President	Trader	VP
Total #of Sent and Received	3359	3189	28298	2102	2476	1213	2802	16997	13485
#of Users	4	13	34	3	10	3	4	12	19

4   Social Network Analysis on Enron Dataset

A comprehensive analysis is done in this section to select the most informative social network features. First, social communities are discovered and role communications are studied in this context. To evaluate if these communities are meaningful and distinguishable in the Enron organizational setting, a topic discovery is made to evaluate these communities.

To discover communities in this social network the Newman clustering algorithm has been applied [11]. Applying, it divided the whole network into five different communities. Each community contains a variety of different roles, meaning that each cluster of users gathers around a centric organizational concept, i.e same project or topic. Even the smallest cluster(*Cluster 5*) consists of employees, a trader, managers, directors, and vice presidents. These communities may have different interpretations in the real world; they may be different parts of the Enron company working on different projects with different topics or issues. To evaluate the quality of the communities, emails in each community have been analyzed in isolation in order to find the most frequently discussed terms in that community. In this process NLP techniques have been used for extracting terms from the emails and building a dictionary of most frequent terms for each community (a.k.a cluster). As illustrated in Table 3, each term is unique to its related cluster, and it is argued that these terms can describe the clusters uniquely in real world. As is evident from the given term sets bellow, the topic of discussion for each cluster is totally different and each cluster can be differentiated using their term sets.

Table 3. Top most frequent terms and topics for each community

Cluster1		Cluster2		Cluster3		Cluster4		Cluster5	
term	frequency	term	frequency	term	frequency	term	frequency	term	frequency
november	187	power	1710	blackberry	224	publish	133	schedule	2494
west	171	california	1627	capacity	200	germany	128	portland	2251
desk	137	energy	1565	rate	195	PVI	110	scheduling iso	2251
start	133	agreement	1303	pipeline	163	offer	94	datum	645
#of emails	1614		7913		1254		1356		2758

To study and evaluate whether this social network representation has some logical interpretation in the real world four different roles from the Enron social network have been selected. Based on the degree centrality traders and managers gain high scores while CEO and employee are ranked low. This is also true in real organizational settings: traders and managers are very active in communications since they tend to contact many individuals to manage a team or contact other traders and teams for marketing decisions. But, roles such as CEO or an employee are very isolated in their own community as they are focused on a specific project or environment. These behaviours are clearly reflected in their centrality and betweenness scores: 74% of managers, directors, and presidents and 81% of traders are at the top of degree centrality and betweenness list while 67% of CEOs and 62% of employees are at the bottom.

5   Feature Set Evaluation

In order to find out whether the feature set given in Section 3.1, can discriminate the dataset in terms of users' roles, EM algorithm is applied [5]. The algorithm is tuned to automatically find the optimized number of distributions (a.k.a. clusters) for the data. Using 10-fold, the EM algorithm found 5 clusters, as illustrated in Table 4, which is a close estimation of the number of predefined roles. Due to the small sample size of some role classes (i.e. CEO) distributions of

Table 4. EM clustering algorithm results over Enron dataset

	Employee	Director	VP	IHL	Manager	CEO	President	Trader	MD
Cluster1	0	2	1	2	3	0	2	10	0
Cluster2	5	6	2	0	4	4	0	2	2
Cluster3	23	0	0	1	1	0	0	0	0
Cluster4	4	5	3	0	2	0	0	1	0
Cluster5	2	0	13	0	0	0	2	0	1

these categories are not rich in the feature space. Hence, some of the role classes are assigned to other clusters. The results of this experiment shows that the Enron users' roles are separable in the proposed feature space although some of the classes with small sample size appear to be noisy since they don't fit completely in a single cluster.

6   Role Classification

As discussed in the Section 3.2, there is only a limited number of employees available in the Enron dataset in which their positions in the company are known. Consequently, this will reduce the sample size to only 102 instances. The other issue is that role categories are skewed. As illustrate in Fig.1 we merged categories based on their closeness in any organizational setting, in a hope this new categorization offers richer and more robust sample size.

Based on these newly formed classes, a Neural Network classifier is trained. Ten-fold cross validation is used to measure the average metrics of the classifiers. As illustrated in Table 5, the neural network model has produced the highest True Positive rate of 0.745 for Managing class while Top-level class is highest

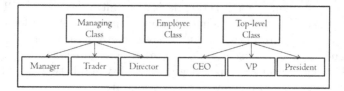

Fig. 1. Hierarchy of roles

in Precision and Receiver Operating Characteristic (ROC) area. The overall accuracy among all the classes is 63.57%. Considering the very small sample size in this application, this is reasonably good given that random classification gives an accuracy of 50%.

Table 5. Classification results by Neural Networks

	TP	FP	Precision	Recall	F-Measure	ROC
Managing Class	0.745	0.221	0.618	0.732	0.675	0.864
Employee Class	0.584	0.124	0.67	0.574	0.608	0.791
Top-level Class	0.568	0.095	0.642	0.55	0.596	0.895
Weighted Avg.	0.636	0.159	0.645	0.629	0.635	0.853

Correctly Classified 63.57%

7   Conclusion and Future Work

In this paper, first we analyzed the social network of the Enron users through SNA popular metrics such as degree centrality and Betweenness. Through these experiments, it is concluded that some social network's measures can be used for extracting informative behaviors of users for further analysis. After representing each Role in this feature space we applied EM algorithm to show that these clusters are meaningful since they can discriminate most classes in a satisfactory manner. Finally, we built a Neural Network classifier which was capable of classifying organizational roles from the new email instances with the reasonable accuracy of 63.57%.

There were some issues that have been addressed in this paper: small sample size for some categories made them hard to classify. The classification accuracy was not that high because these classes affected the whole model. However, we combined a few classes through a real organizational hierarchy which increased the overall accuracy.

For future work it will be interesting to analyze these communities in time spaces: one can study users and their corresponding feature values in time slices, hoping that these features are more informative in short period of time such that roles may express more robust behaviors in different time windows. For overcoming the problem of "small sample size" one can try semi-supervised classifiers like co-training to make use of the other 50 employees of the Enron dataset whose role positions were not available.

References

1. AbdelRahman, S., Hassan, B., Bahgat, R.: A new email retrieval ranking approach. CoRR, abs/1011.0502 (2010)
2. Balasubramanyan, R., Carvalho, V.R., Cohen, W.: Cutonce- recipient recommendation and leak detection in action (2008)
3. Bennett, P.N., Carbonell, J.G.: Combining probability-based rankers for action-item detection. In: HLT-NAACL 2007, pp. 324–331 (2007)
4. Corrada-Emmanuel, A. (n.d.).: Enron Email Dataset Research (2004)
5. Dempster, A.P., Laird, N.M., Rubin, D.B.: Maximum likelihood from incomplete data via the em algorithm. Journal of the Royal Statistical Society. Series B (Methodological) 39(1), 1–38 (1977)
6. Diesner, J., Carley, K.M.: Exploration of Communication Networks from the Enron Email Corpus. In: Proceedings of Workshop on Link Analysis, Counterterrorism and Security, SIAM International Conference on Data Mining 2005, pp. 3–14 (2005)
7. Klimt, B., Yang, Y.: The Enron Corpus: A New Dataset for Email Classification Research. In: Boulicaut, J.-F., Esposito, F., Giannotti, F., Pedreschi, D. (eds.) ECML 2004. LNCS (LNAI), vol. 3201, pp. 217–226. Springer, Heidelberg (2004)
8. McCallum, A., Wang, X., Corrada-Emmanuel, A.: Topic and role discovery in social networks with experiments on enron and academic email. Journal of Artificial Intelligence Research 30(1), 249–272 (2007)
9. Mobasher, B.: Web usage mining and personalization. CRC Press (2005)
10. Neustaedter, C., Bernheim Brush, A.J., Smith, M.A., Fisher, D.: The social network and relationship finder: Social sorting for email triage. In: CEAS 2005 - Second Conference on Email and Anti-Spam, Stanford University, California, USA, July 21-22 (2005)
11. Newman, M.E.J.: Modularity and community structure in networks. Proceedings of the National Academy of Sciences 103(23), 8577–8582 (2006)
12. On, B.-W., Lim, E.-P., Jiang, J., Purandare, A., Teow, L.-N.: Mining interaction behaviors for email reply order prediction. In: 2010 International Conference on Advances in Social Networks Analysis and Mining (ASONAM), pp. 306–310 (August 2010)
13. Shetty, J., Adibi, J.: The Enron email dataset database schema and brief statistical report. Information Sciences Institute Technical Report, University of Southern California (2004)
14. Wang, M.-F., Jheng, S.-L., Tsai, M.-F., Tang, C.-H.: Enterprise email classification based on social network features. In: Proceedings of the 2011 International Conference on Advances in Social Networks Analysis and Mining, ASONAM 2011, pp. 532–536. IEEE Computer Society, Washington, DC (2011)
15. Yoo, S., Yang, Y., Lin, F., Moon, I.-C.: Mining social networks for personalized email prioritization. In: Proceedings of the 15th ACM SIGKDD International Conference on Knowledge Discovery and Data Mining, KDD 2009, pp. 967–976. ACM, New York (2009)

Improved Arabic-French Machine Translation through Preprocessing Schemes and Language Analysis

Fatiha Sadat and Emad Mohamed

201 Avenue President Kennedy, Montreal
QC, Canada H2X 3Y7
Sadat.fatiha@uqam.ca, emohamed@umail.iu.edu

Abstract. Arabic is a morphologically rich and complex language, which presents significant challenges for natural language processing and machine translation. In this paper, we describe an ongoing effort to build a competitive Arabic–French statistical machine translation system using the Moses decoder and other tools. The results show a significant increase in terms of Bleu score by introducing some preprocessing schemes for Arabic in addition to other language analysis rules.

Keywords: Arabic morphology, statistical machine translation, comparable corpora, parallel corpora, pre-processing.

1 Introduction

Arabic is a morphologically rich and complex language, in which a word carries not only inflections but also clitics, such as pronouns, conjunctions, and prepositions. This morphological complexity also has consequences on NLP applications, such as machine translation and information retrieval. Developing an Arabic-French machine translation is not an easy task, although there is a vast amount of training data nowadays. In another side, dealing with the complexity and ambiguity of the source language plays a major role in boosting the efficiency of the translation system.

In previous research, it was shown that morphological pre-processing of a morphologically rich language, such as Arabic does provide a benefit, especially in the case of limited volume of training data [6, 7, 11,12]. In Statistical Machine Translation (SMT) context, Habash et Sadat [6], pre-processed Arabic texts using different segmentation schemes for translation into English and showed that the quality of translation is generally better than the baseline. In relation to Arabic-French SMT, few research and evaluations were reported, compared to Arabic-English SMT among other pairs of languages. One of the first statistically-driven machine translation systems for Arabic-French was reported by Hasan et al [11] during the second Cesta evaluation campaign[1]. The proposed SMT system used a simple stemming algorithm based on finite-state-automaton to split Arabic words into prefixes, stem and suffixes.

---

[1] http://www.technolangue.net/article.php3?id_article=199

O. Zaïane and S. Zilles (Eds.): Canadian AI 2013, LNAI 7884, pp. 308–314, 2013.

Nevertheless, this simple segmentation method showed a reduced OOV rate from 8.2% to 2.6% for the test data and thus a better quality of translation in terms of BLEU score [8].

In relation to improving an SMT system using some language analysis rules, such as re-ordering and Arabic as a source language, there was no reported research on Arabic-French SMT. However, Carpuat et al. [4] showed that post-verbal subject (VS) constructions are hard to translate because they have highly ambiguous reordering patterns when translated to English. They proposed to reorder VS construction into SV order for SMT word alignment only. This strategy significantly improves BLEU and TER scores of an Arabic-English SMT.

This paper is organized as follows. In section 2, we discuss the problems of Arabic in the scope of SMT, the data that we used through TRAD 2012 evaluation campaign[2] and the proposed solutions of pre-processing through segmentation and part-of-speech tagging. In section 3, we present the experiments on Arabic-French SMT, different evaluations and results. Section 4 concludes the present paper.

2   The Problem, Data and Methods

With Arabic being morphologically complex and rich, lexical scarcity comes as a natural result. In such cases it helps to reduce this morphological complexity in order to obtain better alignments and decoding for Statistical Machine Translation [6].

We participated in the 2012 TRAD[2] evaluation campaign, that was coordinated by the *Laboratoire National de métrologie et d'Essais (LNE)* and CASSIDIAN (*the defence and security subsidiary of the EADS group*), and was funded by the French General Directorate for Armament (DGA). The system is trained on 3.5 million words of French and their parallel text in Arabic in addition to 9700 parallel sentences that were extracted from the essentially comparable UN corpus of 2009. The training parallel Arabic-French corpus is a mixture of two parallel corpora, Trames and the news commentary. The development corpus contains 20,000 words, namely 40,000 words with the reference. The evaluation corpus contains 15,000 words with 4 references.

The common practice of extracting bilingual phrases from the parallel data usually consists of three steps: first, words in bilingual sentence pairs are aligned using state-of-the-art automatic word alignment tools, such as GIZA++ [13], in both directions; second, word alignment links are refined using heuristics, such as Grow-Diagonal-Final (GDF) method; third, bilingual phrases are extracted from the parallel data based on the refined word alignments with predefined constraints [13].

The trigram language models are implemented using the SRILM toolkit [1]. Moses[3] [3], an open source toolkit for phrase-based SMT system, was used as a decoder. These steps of building a translation system are considered as a common practice in the state-of-the-art of phrase-based SMT systems. Our research for improving the Arabic-French SMT system was emphasized more on the pre- and post-processing parts.

---

[2] http://www.trad-campaign.org/
[3] Available on http://www.statmt.org/moses/

2.1   Arabic Pre-processing

In order to perform Arabic pre-processing, we used a machine learning approach that performs word segmentation and POS tagging at the segment level. We then use rules to derive the different pre-processing schemes required for the machine translation experiments.

We use memory-based learning for both word segmentation and Part of Speech tagging. For segmentation, we use TiMBL [9] for POS tagging MBT, a memory-based tagger [9]. Memory-based learning is a lazy learning paradigm that does not abstract over the training data. During classification, the k-nearest neighbors to a new example are retrieved from the training data, and the class that was assigned to the majority of the neighbours is assigned to the new example. A Memory-Based Tagger (MBT) uses TiMBL as classifier; it offers the possibility to use words from both sides of the focus word as well as previous tagging decisions and ambitags as features. An ambitag is a combination of all POS tags of the ambiguity class of the word   [10].

2.2   Word Segmentation

Word segmentation is defined as a per-letter classification task: If a character in the word constitutes the end of a segment, its class is '+', otherwise '-' [10]. We use a sliding window approach with 5 characters before and 5 characters after the focus character, the previous decisions of the classifier, and the POS tag of the focus word assigned by the whole word tagger (cf. below) as features. The best results were obtained for all experiments with the IB1 algorithm with similarity computed as weighted overlap, relevance weights computed with gain ratio, and the number of k nearest neighbors equal to 1.

For example, the word "wllmhndsAt" (Eng. and for the female engineers) receives the segmentation w+l+l+mhnds+At. This form will then be passed, in context, to the POS tagger.

2.3   POS Tagging

For POS tagging, we use the full tagset, with information about every segment in the word, rather than the reduced tagset (RTS) used by Diab et al. [2] and Habash and Rambow [5, 6], since the RTS assumes a segmentation of words in which syntactically relevant affixes are split from the stem. The word w+y+bHv+wn+hA, for example, in RTS is split into 3 separate tokens, w, ybHvwn, hA. Then, each of these tokens is assigned one POS tag, Conjunction for w, Imperfective Verb for ybHvwn, and Pronoun for hA. The split into tokens makes a preprocessing step necessary, and it also affects evaluation since a word-based evaluation is based on one word [10].

3   Experiments on Statistical Machine Translation

The segmentation and POS tagging modules above give us a rich representation with enough information for almost any further required transformation. We run four different MT experiments based on variations on the output as described above.

3.1   Description of the MT Experiments

(1)   Basic. The Basic experiment is the baseline of all the work we are doing. In this experiment, the Arabic side undergoes minimal pre-processing in which we only separate the punctuation and remove the occasional diacritization (the short vowels). Short vowels do not normally occur in Arabic, but sometimes scattered ones are there mainly for disambiguation purposes; however since their use is not standardized and subjective, their removal usually leads to better agreement between the training and test sets.

(2)   Tokenized. In this context, tokenization means splitting the prefixes and suffixes that have a syntactic value and that usually stand as independent words in other languages. Examples of these include the possessive pronouns (-hm, -h, -y, -hA), conjunctions (w, f), and prepositions (l-, k-, t-). We have also chosen to split the Arabic definite article *Al* due to the perceived similarity in distribution between the Arabic and French definite articles. The process also normalized the definite article from *l* to *Al*, which is the more frequent form.

(3)   MorphReduced. In the morphologically reduced experiment, we reduce the morphology of Arabic to a level that makes it closer to that of the French language. An example of this is the dual form, which does not occur in French and has thus been transformed to the plural. The following table (Table 1) lists the most common examples of Arabic morphological reduction.

(4)   Swapped. The swapped experiment tries to introduce some structural matching between the source language (Arabic) and the target language (French). Two structural changes have been attempted, as follows:

(a) While Arabic possessive pronouns follow the nouns, we have made them precede the nouns in order to match the French. For example ktAb -y (book - my) has now become (-my book) to match "mon livre" (in French).

(b) Arabic object pronouns, which follow the verb, have been made to precede it. *>nA >ryd h* (I want it) is now *>nA h >ryd* with the purpose of matching the French structure "Je le veux".

3.2   The Effect of Tokenization on Data Sparseness

We have measured the effect of the pre-processing steps above on data sparseness, measured in the percentage of unknown unigrams (OOVs) on a development set (dev set). Table 2 summarizes the findings on the dev set. We give numbers in terms of tokens (the total number of words) and types (the number of unique words in the text, i.e. no-redundant words in the text).

It can be noticed that the tokenization has a major effect on combatting data sparseness and consequently improving the quality of translation as measured by the BLEU score. Morphological normalization, which is a layer on top of tokenization, improves things even further, and this is reflected in the difference between the baseline BLEU score and the MorphReduced BLUE score which is 8.6 absolute points. The swapped experiment leads the system output to deteriorate; which leads to a review of the introduced rules for the structural matching between the source Arabic and the target French languages, in the future.

Table 1. The most common rules for Arabic morphological reduction

Rules and Examples for the MorphReduced experiments
(1) Regular Plural Nominative → Regular Plural Accusative
Example: before applying the rule: mstwTn***wn*** After applying the rule: mstwTn***yn***
(2) dual Nominative → Regular Plural Accusative
Example: before applying the rule: lAEb***An*** After applying the rule: lAEb***yn***
(3)Jussive Mood → Indicative Mood
Example: before applying the rule: hm lm ylEb***wA*** hmA lm ylEbA After applying the rule: hn lm ylEb***wn*** hm lm ylEb***wn***

Table 2. Experiments on the development set

Experiment	% OOV (Types)	% OOV (Tokens)	BLEU score on Dev
Baseline	10.74	4.81	17.69
Tokenized	7.99	2.00	25.84
MorphReduced	7.87	1.98	26.33
Swapped	7.87	1.98	25.48

3.3   Results, and Discussion

Table 3 compares the results, in term of BLEU scores, of the 4 experimental settings in 3 evaluations schemes, as follows:

(a) ***Standard***, which includes performing re-casing and removing white space before punctuation,

(b) *Nopunct*, in which punctuation is stripped and evaluation is performed on the lexical text only, and

(c) *Nopunctcase* in which, in addition to removing punctuation, all words are lower-cased.

We can see from Table 3 that the Baseline experiment produces the lowest results, and that the tokenization scheme is a big leap with a 7.2 BLEU scores of improvement (25.9 vs. 33.1), which means that performing tokenization is a really a necessary step for translating from Arabic, an that the morphological complexity of Arabic could be a hindrance to quality automatic translation. While tokenization leads to considerable improvement, morphological reduction fares even better with a 7.4 BLEU score higher than the baseline. This could be due to the fact the morphological reduction reduces the number of unknown words even further than tokenization alone. Swapping elements to math the target language, which is built upon tokenization and morphological reduction, leads to the results to deteriorate a little as it cancels out the effect of the morphological reduction process.

It is still an open question whether the positive effect of pre-processing will still carry over with increasing the amount of training data and to what extent this will help.

Table 3. Results in term of BLEU score

	Base	Tokenized	MorphReduced	Swapped
Standard	25.9	33.1	33.3	33.1
Nopunct	23.8	31.5	31.7	31.4
Nopunctcase	25.8	34.1	34.1	34

4   Conclusion

We have presented an ongoing project on developing a competitive Arabic-French machine translation, using the methods and evaluations presented at TRAD 2012 evaluation campaign. We have introduced pre-processing schemes for the source language (Arabic) and some rules of language analysis related to the target language (French). Our method for POS tagging and segmentation of Arabic texts showed a significant improvement in terms of BLEU score; however it does not assume the best results. The introduced morphological rule that reduces the morphology of Arabic to a level that makes it closer to that of the French language, showed the best results. We have introduces extra swapping rules, that tries to introduce some structural matching between the source language (Arabic) and the target language (French); however there was no improvement in terms of Bleu score. Our future work is concentrated on the revision of these swapping rules and the introduction of more rule for the recognition and transliteration of named entities; which makes our translation system a hybrid rule-based and statistical SMT system.

References

1. Stolcke, A.: Srilm-An Extensible Language Modeling Toolkit. In: Proc. of the International al Conference on Spoken Language Processing (2002)
2. Diab, M., Hacioglu, K., Jurafsky, D.: Automatic Tagging of Arabic Text: From Raw Text to Base Phrase Chunks. In: Proc. of the North American Chapter of the Association for Computational Linguistics (NAACL), Boston, MA (2004)
3. Koehn, P., Shen, W., Federico, M., Bertoldi, N., Callison-Burch, C., Cowan, B., Dyer, C., Hoang, H., Bojar, O., Zens, R., Constantin, A., Herbst, E., Moran, C., Birch, A.: Moses: Open source toolkit for statistical machine translation. In: Proceedings of the ACL 2007 Interactive Presentation Sessions, Prague (2007)
4. Carpuat, M., Marton, Y., Habash, N.: Improving Arabic-to-English Statistical Machine Translation by Reordering Post-verbal Subjects for Alignment. Machine Translation, Special Issue on Machine Translation for Arabic 26(1-2), 105–120 (2012)
5. Habash, N.: Introduction to Arabic Natural Language Processing. Morgan & Claypool (2010)
6. Habash, N., Sadat, F.: Arabic Preprocessing Schemes for Statistical Machine Translation. In: Proceedings of NAACL 2006, New York, USA, June 5-7 (2006)
7. Lee, Y.: Morphological Analysis for Statistical Machine Translation. In: Proc. of NAACL, Boston, MA (2004)
8. Papineni, K., Roukos, S., Ward, T., Zhu, W.: Bleu: A Method for Automatic Evaluation of Machine Translation. Technical Report RC22176(W0109-022), IBM Research Division, Yorktown Heights, NY (2001)
9. Daelemans, W., Zavrel, J., Berck, P., Gillis, S.: MBT: A memory part speech tagger generator. In: Proceedings of the Fourth Workshop on Very Large Corpora, ACL 1996, Copenhagen, Denmark, pp. 14–27 (August 4, 1996)
10. Mohamed, E., Kübler, S.: Arabic part of speech tagging. In: Proceedings of LREC, Valetta, Malta (2010)
11. Hasan, S., Isbihani, A.E., Ney, H.: Creating a Large-Scale Arabic to French Statistical Machine Translation System. In: International Conference on Language Resources and Evaluation (LREC), Genoa, Italy, pp. 855–858 (May 2006)
12. El Isbihani, A., Khadivi, S., Bender, O., Ney, H.: Morpho-syntactic Arabic Preprocessing for Arabic to English Statistical Machine Translation. In: Human Language Technology Conf./North American Chapter of the Assoc. for Computational Linguistics Annual Meeting (HLT-NAACL), Workshop on Statistical Machine Translation, New York City, pp. 15–22 (June 2006)
13. Och, F.J., Ney, H.: A Systematic Comparison of Various Statistical Alignment Models. Computational Linguistics 29(1), 19–51 (2003)

An Application of Answer Set Programming for Situational Analysis in a Maritime Traffic Domain

Zahra Vaseqi and James Delgrande

School of Computing Science
Simon Fraser University
Burnaby, B.C. Canada V5A 1S6
{zvaseqi,jim}@cs.sfu.ca

Abstract. In this work, we investigate the use of Answer Set Programming (ASP) as a component in a multi-layered *situation awareness* system for the marine traffic domain. The State Transition Data Fusion (STDF) model which has been adopted for the situation assessment task enables performing each of the tasks at an appropriate level of abstraction. In this model, we delegate the lower-level analysis to an imperative modelling language called *CoreASM*; while the higher-level analysis for the impact assessment is handled through a reactive ASP system. The reactive answer set solver enables using dynamic input data to generate answer sets in an incremental fashion. Furthermore, ASP has a rich potential in representing domain rules as it is declarative and provides a compact and intuitive encoding of the domain expert's knowledge within a non-monotonic framework.

Keywords: Reactive Answer Set Programming, Situation Awareness, State Transition Data Fusion Model.

1 Introduction

Maritime traffic control is an example of a task that looks for methods to enhance the capabilities of human operators. Maritime operators, who watch over the oceans 24/7, are responsible for ensuring vessels comply with maritime regulations. In order to do this, an operator must analyze large amounts of data. The operators must use their expertise to combine the relevant data and infer additional situational facts in order to determine if further action is required. However, with the vast amount of information along with the numerous rules for safety compliance, we can see how the problem quickly becomes intractable for a human operator without some automated assistance to evaluate the large number of situational facts.

A *situation awareness*[1] system can be utilized to augment the human operators' expertise and analyze the available information to detect anomalous actions and events. As an illustration, if a ship is moving faster than the allowed speed, the system can consider it as being anomalous. Unfortunately, unlike speed limit

O. Zaïane and S. Zilles (Eds.): Canadian AI 2013, LNAI 7884, pp. 315–322, 2013.

checking, not all of the anomalous activities are easily observable. Some situational facts require expertise and abstraction of data in order to be inferred. Therefore, the situation assessment model needs to perform the analysis in different levels of abstraction from low-level input data analysis to higher level information analysis. In order to implement a system able to act upon its knowledge and information from the domain, we have adopted the State Transition Data Fusion (STDF)[2] model for situation awareness. The customized STDF model enables performing each of the anomaly detection tasks in an appropriate level of abstraction. Section 3 explains the STDF model in further details.

In this work, we investigate use of Answer Set Programming (ASP) as a component in a multi-layered *situation awareness* system for marine traffic domain. ASP has a rich potential in representing rule-based domains because it is declarative and provides a compact and intuitive encoding of the domain expert's knowledge within a non-monotonic framework. Moreover, marine traffic domain, like most other real world rule-based domains, is prone to policy (rule) changes. ASP is flexible enough to extend and can adapt the system as the policies change. On the other hand, marine traffic domain's highly dynamic environment requires us to maintain a history so we can detect anomalies that require analysis over a period of time. Recent advancements to implementations of ASP enable incremental evaluation of logic programs; this allows for the program to have a notion of history which is essential for a *situational awareness* system. The *history* of the system includes the raw information received from the environment as well as the knowledge inferred in the system. The history in the system grows throughout time as new situational facts are inferred. In order to handle the increasing information accumulating over time, efficient means to store the history will be beneficial.

In a nutshell, our contributions with respect to situation analysis in a rule-based domain are to: (1) present an intuitive encoding of the domain rules; (2) use the efficient means presented by ASP to handle the history of the domain; (3) provide a modular architecture to perform situation analysis in different levels of abstraction.[1]

2   Answer Set Programming: *Preliminary Notions*

Answer Set Programming (ASP) is a declarative language introduced by Gelfond and Liftchitz[4]. In ASP a given problem is represented as a *logic program* and the solution(s) correspond to the resulting *answer sets*. An ASP program is a logic program which is a collection of statements analogous to *if-then* rules. These rules have the form:

$$A_0 : - A_1, \ldots, A_m, \text{ not } A_{m+1}, \ldots, \text{ not } A_n.^2 \tag{1}$$

where A_0 is the head of the rule and $A_1, \ldots, A_m, \text{not } A_{m+1}, \ldots, \text{not } A_n$ is the body of the rule. Logic programming rules function as inference rules where

---

[1] This work is based on the M.Sc. thesis [3] from Simon Fraser University.

[2] The ': −' symbol divides the *head* and the *body* parts in a rule. Intuitively speaking, the head atom in a rule is inferred if the body holds.

intuitively A_0 is inferred if $A_1, \ldots,$ A_m hold and there is no evidence supporting any of $A_{m+1}, \ldots,$ A_n. The inferred atom, A_0, will no longer hold once we acquire evidence supporting at least one of the naf-literals[3] in the body of the rule. There are three basic type of rules defined in ASP, namely *regular rules*, *facts*, and *integrity constraints*. A regular rule appears in the form (1) where it contains a head atom and at least one literal or naf-literal in its body. Facts are rules with an empty body of the form (2).

$$A_0. \tag{2}$$

Lastly, integrity constraints are rules with empty heads. They prohibit co-occurrence of literals in the body of the rule in one answer set. Adding integrity constraints to the program may omit some of the answer sets. Integrity constraints, in general, are in the following form:

$$: - \ A_1, \ldots, \ A_m, \ \texttt{not} \ A_{m+1}, \ldots, \ \texttt{not} \ A_n. \tag{3}$$

Additionally, the *cardinality* and *sum* aggregates are useful features for our application. A *cardinality* aggregate is a literal of the form (4) that appears in the body of the rule.

$$\texttt{lower}\{A_0, A_1, \ldots, A_n\}\texttt{upper} \tag{4}$$

lower and *upper* are two integer values that indicate the boundaries. This literal takes truth value only if the number of atoms A_i, $0 \leq i \leq n$ that hold truth value add up to a number between the *lower* and the *upper* bound. The *sum* aggregate is a generalization of the cardinality aggregate where there is a weight associated to each atom:

$$\texttt{lower}[A_0 = w_0, A_1 = w_1, \ldots, A_n = w_n]\texttt{upper} \tag{5}$$

lower and *upper* are again the boundaries indicating the lower and the upper bounds for the sum of the weights of the satisfied atoms.

Reactive Answer Set Programming

Reactive answer set programming bridges the gap between the declarative paradigm used by answer set programming and a wide variety of applications dealing with dynamic domains. Gebser et al. [5] introduce this approach by incorporating online data streams into the logic program. They take advantage of incremental logic programs[6] to bring in dynamicity through a time component.

The reactive ASP-solver that we have used in this work is called *oclingo*[7]. *oclingo*'s input language is an *incremental logic program* which is different from the regular ASP programs in that it takes a dynamic parameter and enables incremental generation of the answer sets. Incremental logic programs are composed of three logic programs, namely *base*, *cumulative*, and *volatile* programs. The **base** program includes static knowledge, while the **cumulative** program is meant to track the knowledge that should be accumulated over time. Lastly, the **volatile** program contains the rules that are time dependent but transient after a specific number of steps. The online input for each step must be presented in the following format:

---

[3] Literals that are negated with *negation as failure*(naf).

```
#step k. external_predicate. #endstep.
```

The incremental grounder grounds the time dependent parts of the program as the time parameter increases. This approach in dealing with a dynamic parameter inside the ASP solver provides the means for handling history of the domain inside the ASP solver itself.

3   Application Model

In this work, we have adopted the STDF model [2] for situation awareness (SA). The STDF model elaborates Endsley's definition[4] of Situation Awareness [1]. STDF has a systematic approach to manage the complexity of SA systems through modularization. It achieves situational awareness by assessing the input at three levels of abstraction: (*i*) *object* assessment where the raw input is transformed into objects in the world model; (*ii*) *situation* assessment in which relations between objects are represented; and (*iii*) *impact* assessment in which the effects of relations are identified.

We perceive the maritime domain as being composed of two main agents: the *environment*, and the *base station*. The environment is where all the objects operate and actions are performed. An Automatic Identification System (AIS) system passes information obtained from the environment to the base station. The base station analyzes the data and tries to maintain a situation-aware state given the observations. In order to enable the base station to achieve situation awareness, the SA model is located inside the base station. The STDF model performs the situation analysis by understanding the world in terms of: (*i*) *objects*; (*ii*) interaction between them (*situations*); and (*iii*) *scenarios* composed of events occurring in the domain (a set of interacting situations) which are addressed in *Object*, *Situation*, and *Impact* assessment layers respectively.

The analysis from the first two layers, *object* and *situation* assessment, deals with numeric features of the domain and therefore opts for an imperative language to perform computations. However, the *impact* assessment layer deals with higher level statements and contains no numeric information; therefore we have used a declarative language for the analysis in the latter layer. Figure 1 depicts the main components of the SA system and their interaction. As shown in the diagram, the *CoreASM* modelling language[8] performs the analysis from the object assessment and the situation assessment layers; while it delegates the impact assessment analysis to the ASP component where the basic situational facts are supplied from the previous two layers. The ASP component is composed of an *incremental logic program*, where the expert knowledge is encoded, a *controller* program, that passes the external situational facts to the engine, and an ASP-solver engine performing the inference task.

The object assessment layer identifies each vessel and its status in the domain. Compliance of the vessel with basic rules in the domain is one of the factors that

---

[4] "The *perception* of the elements in the environment within a volume of time and space, the *comprehension* of their meaning and the *projection* of their status in the near future."

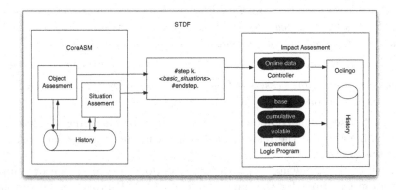

Fig. 1. Situation Awareness Model Using *CoreASM* and ASP

identify the status of the vessel in the domain. For example, one needs to check the vessel location to ensure that it is not in a prohibited area. Such situational facts are represented as first order propositions in the form *situation*(< *parameter_list* >). For example, in the situation where there is a vessel in a prohibited area we get the proposition *in_prohibited_area*(*vessel, area, time*).

After clarifying the status of each vessel, the interaction of the vessels with each other needs to be analyzed. The *situation assessment* layer takes the resulting object representations as well as the basic situational statements to discover the relations between the objects in the domain. This layer accesses the numerically based features of objects as well as propositions explaining status of objects in the domain and draws higher level situational facts identifying the relations between the objects. Resulting statements are first order propositions corresponding to events involving two or more objects in the domain. An example of such a situation is where two vessels are involved in a dangerous cargo coincidence; e.g. *dangerous_cargo_coincidence*(*vessel1, vessel2, location, time*).

Determining if there are any vessels carrying conflicting cargos in a close proximity is called a *dangerous cargo coincidence* and is a situation of interest in our application.

Impact Assessment Using Reactive Answer Set Programming

The *impact assessment* layer is built on the previous two layers to perform higher level analysis based on the inferred situational facts about the objects and their relations. The impact assessment analysis represents the world in terms of scenarios that may involve multiple situations. An impact assessment component should be capable of interpreting the situations, discovering the interacting situations, and predicting how a situation may evolve over time.

Running Example. As mentioned earlier, we define a scenario as a set of interacting situations which results in some effects of interest (impact) in the domain. As an illustration, let's consider a scenario in which a vessel would be considered as being suspicious if the following pattern is observed:

> ... *two vessels have been involved in more than one dangerous cargo co-incidence situations together in the past two months and at least one of them has not been moving outside of the area for more than a week.*

In this scenario the domain expert would take the scenario as being safe if:

> ... *at least one of the vessels is from a legitimate addressee or the vessels both have been inspected in the last two days.*

The impact assessment component is responsible for analyzing the history of situations, figuring out the relations between them, and the possible suspicious interactions of events.

Propositional statements, representing situations involving one or multiple vessels, form the input for the impact assessment ASP program. This input is supplied through the output from the object and situation assessment layers. The propositional statements from these two layers are encoded in terms of *oclingo*'s external input language. As described in Section 2, *oclingo* augments the *incremental logic program* with the *online data* at each incremental step and solves the resulting program. The incremental logic program represents the impact assessment scenario rules where the incoming situations, representing the current state of the world, need to be matched against the scenario rules in order to discover impacts of the courses of actions.

In order to encode a scenario, the impacts of the scenario are mapped to head predicates of the rules; while the basic situations are mapped to body predicates representing the conditions where an impact can be inferred. The rules encoding the scenarios need to be assessed at each step in order to infer possible impact; therefore, there is a notion of time in the rules encoding the scenarios. Thus they should be placed in one of the time-dependent program parts; i.e. *cumulative*, or *volatile*. The *cumulative* part allows persistent accumulation of the history; while the *volatile* part can be used to keep track of the history for a specific duration. The persistent rules need to be located under the tag "#cumulative t." indicating the cumulative program part; while the volatile rules are placed under "#volatile t : duration." with an arbitrary duration. The choice between these two program parts depends on how far we need to maintain the history of the inferred facts in the system. For example, one may assume that the history of the vessels that have been recognized as being suspicious should never be removed from the system and therefore be placed in the *cumulative* part. However, this may not be a good decision since the persistent atoms can occupy a large amount of memory in the long run, and result in regressive performance of the system.

The history for the basic situations which are being fed to the system through the *controller* program are handled likewise. For the input predicates, we have the choice of handling the history at the rule level and the input level. For example, in the suspicious coincidence of vessels scenario, the history of the basic situational fact indicating dangerous cargo coincidences needs to stay in the system for two months; however, the history of vessels to be inspected would only need to be kept valid for two days. Therefore, they can be defined to be volatile at the input level for a specific duration of time.

Program 1. Suspicious Coincidence Scenario Rule Encoding

```
01  #cumulative t.
02  suspicious_coincidence(V1,V2,t)  :-
03              dangerous_coincidence(V1,V2,T1),
04              dangerous_coincidence(V1,V2,t),
05              T1!=t,
06              not vessel_area_changed(V1;V2,T):time(T):T<=t,
07              [ legitimate_vessel_addressee(V1)=2,
08                legitimate_vessel_addressee(V2)=2,
09                inspected_vessel(V1,T):time(T):T<=t = 1,
10                inspected_vessel(V2,T):time(T):T<=t = 1]1.
```

#volatile : duration. ext_vessel_inspected(vessel1,t).

Program 1 illustrates the encoding of the suspicious coincidence scenario, described earlier. In this scenario a predicate "suspicious_coincidence" is mapped to an impact in which two vessels are considered as being suspicious if the available basic situations match to the scenario pattern. In this case we would like the history of the suspicious vessels be persistent; therefore we defined them under the cumulative label in the program. Lines 6-10 encodes the condition: "*at least one of the vessels is from a legitimate addressee or the vessels both have been inspected recently*" by using a sum aggregate and assigning a weight of 2 to each vessel being from a legitimate addressee and assigning a weight of 1 to their being inspected in the last two days. The sum aggregate returns the sum of the weights for the literals that hold. Setting the upper bound to one ensures that the aggregate statement would evaluate to false if the criteria for the scenario's being safe is met. In other words, if the aggregate statement in the body evaluates to false, the case would not be considered as being a suspicious coincidence, as it means that either at least one of the vessels is from a legitimate addressee or they have both been inspected in the last two days.

As we can see in Program 1, the criteria for the vessels being inspected in the "*last two days*" is not mentioned anywhere in the rules. The reason is that the history of the external situational facts for inspected vessels has been handled in the input level by setting a volatile duration of two days. This means that the atoms "inspected_vessel(V1, T)" and "inspected_vessel(V2, T)" would not be valid unless they occurred in the past two days. In this example, negation-as-failure is used to check the absence of information regarding the vessels' being outside of the area in the past week. Similar to the previous case, the history for atoms "vessel_area_changed(V1, T)" and "vessel_area_changed(V2, T)" has also been handled in the input level.

As demonstrated in the example, ASP offers an intuitive and natural encoding for the domain expert's knowledge in a descriptive fashion. Its available features allows us to represent the rules in a succinct way. Furthermore, the facilities provided by the recent reactive implementation of ASP enables handling the history of the domain in both input and the rule level. At the input level, we can specify an expiry time for the raw input information provided; while at the rule level one can manage the history of the inferred predicates.

4   Conclusions and Future Directions

In this work, we highlighted the use of ASP for high-level analysis in the rule-based dynamic domains. We investigated the use of ASP as a component in a situation awareness system where we need to perform a large number of inference tasks in order to achieve a state of situation awareness. We have located our ASP component in a multi-layered situation awareness model called the STDF model. The STDF model offers a modular model where the situational analysis tasks can be delegated to appropriate components.

We demonstrated how ASP offers a powerful and intuitive way of encoding the expert knowledge in terms of rules. The reactive answer set solver provides a seamless way of handling the history inside the ASP system. It enables a simple means to discard the information which is no longer useful. The means to manage the history from the previous time-steps makes it useful for dynamic domains.

One of the tracks for the future work is an extensive analysis of how the system compares to an alternative implementation where the history is handled using a database management system. This comparison can be made along quantitative and qualitative aspects. Quantitatively, the response time can be measured to examine how does the system scale as the input data grows. In the qualitative aspect we can investigate the flexibility of the two systems with regard to policy changes, expressiveness, and the extensibility of the systems.

References

1. Endsley, M.: Toward a Theory of Situation Awareness in Dynamic Systems. Human Factors: The Journal of the Human Factors and Ergonomics Society 37(1) (1995)
2. Lambert, D.A.: STDF Model based Maritime Situation Assessments. In: 2007 10th International Conference on Information Fusion. IEEE (2007)
3. Vaseqi, Z.: A Prototype Implementation for Situation Analysis using ASP and Core-ASM. Master's thesis, Simon Fraser University (2012)
4. Gelfond, M., Lifschitz, V.: The Stable Model Semantics for Logic Programming. In: Proceedings of the 5th International Conference on Logic programming, vol. 161 (1988)
5. Gebser, M., Grote, T., Kaminski, R., Schaub, T.: Reactive Answer Set Programming. In: Delgrande, J.P., Faber, W. (eds.) LPNMR 2011. LNCS, vol. 6645, pp. 54–66. Springer, Heidelberg (2011)
6. Gebser, M., Kaminski, R., Kaufmann, B., Ostrowski, M., Schaub, T., Thiele, S.: Engineering an incremental asp solver. In: Logic Programming (2008)
7. Gebser, M., Kaminski, R., Kaufmann, B., Ostrowski, M., Schaub, T., Thiele, S.: A Users Guide to gringo, clasp, clingo, and iclingo. University of Potsdam. Tech. Rep. (2008)
8. Farahbod, R.: CoreASM: An Extensible Modeling Framework & Tool Environment for High-level Design and Analysis of Distributed Systems. PhD thesis, Simon Fraser University (2009)

Preference Constrained Optimization under Change

Eisa Alanazi[*]

Department of Computer Science
University of Regina
Regina, Canada
alanazie@cs.uregina.ca

Abstract. The problem of finding the set of Pareto optimal solutions for constraints and qualitative preferences together is of great interest to many real world applications. It can be viewed as a preference constrained optimization problem where the goal is to find one or more feasible solutions that are not dominated by other feasible outcomes. Our work aims to enhance the current literature of the problem by providing solving methods targeting the problem in static and dynamic environments. We target the problem with an eye on adopting and benefiting from the current constraint solving techniques.

Keywords: Decision Making, Qualitative Preferences, Constraint Satisfaction, Optimization.

1 Introduction

Preference reasoning is a topic of great interest to many domains including Artificial Intelligence (AI), economics and social science [8]. Mostly, this is due to the fact that preferences provide an intuitive mean to reason about user desires and wishes in the problem. This makes it a fundamental part in the decision making process. Most of the work done in the literature adopts the quantitative (numeric) measurement of the preference. Examples of this line of work include utility functions, Multi Attribute Utility Theory (MAUT) and soft constraints. However, the last decade shows a great interest in adopting qualitative preferences instead of the numeric ones [7]. This was derived by the observation that users usually face difficulties in specifying their preferences quantitatively. Therefore, different preference representations have been proposed to remove this burden from the users and handle qualitative preferences adequately. One notable representation for handling qualitative preferences is the Conditional Preference Networks (CP-Nets) [3]. A CP-Net is a graphical model exploiting conditional qualitative preferences independencies in a way similar to the Bayesian Network (BN) [5] representation for the conditional probabilistic independencies. Constraint processing, on the other hand, is a well established research topic in AI

---

[*] The author is supported by Ministry of Higher Education, Saudi Arabia.

O. Zaïane and S. Zilles (Eds.): Canadian AI 2013, LNAI 7884, pp. 323–327, 2013.
© Springer-Verlag Berlin Heidelberg 2013

community. A Constraint Satisfaction Problem (CSP) is an intuitive framework to represent and reason about constrained problems [9].

Preferences and constraints co-exist naturally in different applications [4,12]. For example, in product configurations and recommender systems. Thus, handling both is of great interest to many applications. Preference constrained optimization [4] concerns studying such problems and efficiently finding Pareto solutions (or outcomes) that are satisfied by the set of constraints and optimal according to the given preferences. A feasible solution is Pareto optimal if it is not dominated by any other feasible solution. Finding the set of Pareto optimal for such problems is known to be a hard problem in general [11].

The next section discusses my research problem and some research questions related to it. In section three, we briefly mention the current state of the art for the preference constrained optimization problem. Section four addresses the progress of the work done so far. The future work and remaining challenges are reported in section five. Finally, concluding remarks of the research are presented in section six.

2   Preference Constrained Optimization

The problem of finding assignments that satisfy a set of constraints and maximize the corresponding set of qualitative preferences is what we are tackling in this work. Initially, we assume a static environment where constraints and preferences are represented through CSPs and CP-nets respectively. Then, we study the problem in a dynamic setting where variables are expected to be included or excluded from the problem. Specifically, in our research, we are trying to answer the following questions:

- How could we benefit from the existing constraint solving techniques in simplifying and efficiently solving the constrained CP-Net problem[1]?
- How could we handle the problem in a dynamic setting?
- Are metaheuristics (evolutionary techniques, SLS...etc) applicable in practice for these types of problems? if yes, under what settings?

In the first question, our goal is to benefit from the existing techniques available in constraint processing literature and verify their usefulness in the context of constrained CP-Net. For instance, it has been shown that using propagation techniques over the problem can, in some cases, drastically reduce the search space [1]. Also, we aim to study different heuristic functions to prune unpromising branches in the search space and guide the search effectively towards the set of Pareto optimal solutions.

In the second question, we are interested in extending the current semantics of CP-Net to handle changes over the network. In order to do so, we first assume the dynamic aspect is simply mapped to variable inclusion and exclusion and the

---

[1] We use both terms *Preference Constrained Optimization* and *constrained CP-Net* interchangeably.

set of changes are known in advance. This naturally arises in the configuration problems where different possible combination requirements are known before the process starts. Then we will study the problem of temporal reasoning over the CP-Nets. This requires finding a set of conditions under which the consistency of the preference information is preserved. The goal from studying and trying to answer the last question stems from the fact that non systematic searching methods have proved, in practice, their usefulness in many domains.

3   Related Work

Several methods have been proposed to handle the constrained CP-net problem [4,13,11,6]. Some of the methods attempt to transform the CP-Net into a CSP where the solutions of the CSP are the optimal of the CP-Net [11]. Other attempts include approximating the CP-Net into soft constraints framework [6]. In [4], a recursive optimization algorithm to handle the problem has been proposed. However, the current literature lacks a comprehensive overview over the proposed techniques tackling this problem. Also, utilizing the underlying structure of the constraints have been neglected in most of the methods. Moreover, all the proposed methods assume a static environment exist over the preference information.

4   Progress

Our work so far considers two aspect of the problem. First, we studied the problem of propagating consistent values over the CP-Net structure. This results in simplifying the problem and reducing the search space needed when looking for the optimal outcome. Therefore, in [1], we proposed a method to remove inconsistent values from the CP-Net based on the Arc Consistency (AC) technique [10]. The result of the method is a new CP-net where some domain values have been removed from the network. Experimental tests over randomly generated problems with and without applying the AC technique shows a large savings in finding the optimal outcome.

Second, we consider extending the CP-Net semantics to handle dynamic settings. A CP-Net is a fixed representation for reasoning about qualitative conditional preferences. Given a decision problem P involving n attributes, the CP-Net N over P is always the same (i.e. the set of variables $v \subseteq n$ participating in N are fixed beforehand). In other words, the solutions for the CP-Net N are always defined over the same domain space. While this is acceptable on some static problems, it is not the case in interactive and configuration problems. In the latter, users usually interested in different subsets of n satisfying certain requirements. Moreover, the user interests in one attribute might be conditional upon the existence of other attributes. For example, consider a computer configuration problem where the user explicitly stating her preferences qualitatively. Assume the user is interested in the type of screen only if high performance graphic card is chosen as part of the configuration. In this case, it is clear that there is no need to include the screen type preference for all configurations. Therefore, in [2], we proposed a framework (Preference Conditional Constraints Satisfaction Problem (PCCSP)) which extends the CP-Net to handle activity constraints defined through a conditional CSP instance. A direct application to the PCCSP framework is configuring the webpage content where qualitative preferences and constraints co-exist over different webpage components.

5   Conclusion and Future Work

This research concerns the problem of constraints and qualitative preferences co-existence over static and dynamic settings. The problem is an optimization problem guided by the set of qualitative preferences. Although the problem has been studied during the last decade, much work remains to be done. Examples include examining different heuristic methods to quickly find the Pareto optimal, utilizing the constraint structure to find a good variable ordering over the CP-Net structure and extending the semantics of CP-nets to handle dynamic settings. Our research goal is to contribute to the current literature through advanced techniques and algorithms to solve the constrained CP-Net problem effectively.

The initial results were promising and we aim to continue working on different ideas mentioned in this paper towards successfully finishing the thesis work. In the near future, we plan to empirically evaluate different existing methods for the constrained CP-Net problem. We investigate the problem trying to find out under which CP-Net and CSP structures does one method outperforms another. Moreover, the response time is an important factor for many constrained CP-Net applications. For example, the response time is very important in interactive applications under constraints and preferences. This motivates us to investigate the applicability of applying different evolutionary algorithms to the problem and examine its usefulness. Also, investigating the problem under uncertainty is one of our planned research directions. This might result in a new representation where some variables in the CP-Net are associated with probability distributions and potentially incorporate inference algorithms to reason about their values.

Acknowledgments. I would like to thank my supervisor Prof. Malek Mouhoub for his advice, suggestions and insightful criticism over this research work.

References

1. Alanazi, E., Mouhoub, M.: Arc consistency for cp-nets under constraints. In: FLAIRS Conference (2012)
2. Alanazi, E., Mouhoub, M.: A framework to manage conditional constraints and qualitative preferences. In: FLAIRS Conference (to appear, 2013)
3. Boutilier, C., Brafman, R.I., Domshlak, C., Hoos, H.H., Poole, D.: Cp-nets: A tool for representing and reasoning with conditional ceteris paribus preference statements. J. Artif. Intell. Res. (JAIR) 21, 135–191 (2004)

4. Boutilier, C., Brafman, R.I., Hoos, H.H., Poole, D.: Preference-based constrained optimization with cp-nets. Computational Intelligence 20, 137–157 (2001)
5. Darwiche, P.A.: Modeling and Reasoning with Bayesian Networks, 1st edn. Cambridge University Press, New York (2009)
6. Domshlak, C., Rossi, F., Venable, K.B., Walsh, T.: Reasoning about soft constraints and conditional preferences: complexity results and approximation techniques. CoRR abs/0905.3766 (2009)
7. Doyle, J., Thomason, R.H.: Background to qualitative decision theory. AI Magazine 20 (1999)
8. Goldsmith, J., Junker, U.: Preference handling for artificial intelligence. AI Magazine 29(4), 9–12 (2008)
9. Kumar, V.: Algorithms for constraint satisfaction problems: A survey. AI Magazine 13(1), 32–44 (1992)
10. Mackworth, A.K.: Consistency in networks of relations. Artificial Intelligence 8(1), 99–118 (1977)
11. Prestwich, S., Rossi, F., Venable, K.B., Walsh, T.: Constrained cpnets. In: Proceedings of CSCLP 2004 (2004)
12. Rossi, F., Venable, K.B., Walsh, T.: Preferences in constraint satisfaction and optimization. AI Magazine 29(4), 58–68 (2008)
13. Wilson, N.: Consistency and constrained optimisation for conditional preferences. In: ECAI, pp. 888–894 (2004)

Learning Disease Patterns from High-Throughput Genomic Profiles: Why Is It So Challenging?

Mohsen Hajiloo

Alberta Innovates Center for Machine Learning, Department of Computing Science,
2-21 Athabasca Hall, University of Alberta, Edmonton, Alberta, T6G 2E8, Canada
hajiloo@ualberta.ca

Abstract. In the 20th century, genetic scientists anticipated that shortly after availability of the whole-genome profiling technologies, the patterns of complex diseases would be decoded easily. However, we recently found it extremely difficult to predict women's susceptibility to breast cancer based on their germline genomic profiles and achieved an accuracy of 59.55% over the baseline of 51.52% after applying a wide variety of biologically-naïve and biologically-informed feature selection and supervised learning methods. By contrast, in a separate study, we showed that we can utilize these genomic profiles to accurately predict ancestral origins of individuals. While there are biomedical explanations of accurate predictability of an individual's ancestral roots and poor predictability of her susceptibility to breast cancer, my research attempts to utilize the computational learning theory framework to explain what concepts are learnable, based on the three common characteristics of biomedical datasets: the high dimensionality, the label heterogeneity, and the noise.

Keywords: genomics, disease, breast cancer, ancestral origin, computational learning theory, high dimensionality, label heterogeneity, noise.

1 Introduction: Genomics as a New Lens to Monitor Diseases

From the earliest time that human beings started living on the earth, diseases that cause pain, dysfunction, distress, social problems, and death were also present. Although progress in the biomedical sciences in the recent centuries has shed light on some of these diseases and has provided high level definitions for them using an unaided eye, the ability to explore areas of micrometer (μm), nanometer (nm), and picometer (pm) size has enabled scientists to identify new cellular and molecular players in the disease environment. Upon the completion of the Human Genome Project in 2003 [1], the scientific era of omics technologies (such as genomics, transcriptomics, epigenomics, proteomics, and metabolomics) emerged with the promise of revolutionizing our understanding of life-threatening diseases such as cancer, diabetes, cardiovascular disease, stroke, and Alzheimer's disease [2].

Among many rising omics fields, genomics, which studies the genome (DNA) of organisms, has obtained the highest attention because of the static nature of the genome compared to the dynamic nature of transcriptome, proteome, and metabolome

O. Zaïane and S. Zilles (Eds.): Canadian AI 2013, LNAI 7884, pp. 328–333, 2013.

and availability of the relevant high-throughput measurement technologies such as microarrays and next generation sequencers for the genomics measurements. The human genome consists of approximately 3 billion units called nucleotides. Each DNA segment that carries genetic information is called a gene. The total number of human genes is estimated to be around 30,000 and less than 2% of the genome codes for genes. The other segments of the genome have structural purposes or are involved in regulating the use of genes. Single nucleotide polymorphisms (SNPs) are the substitutions of single nucleotides at a specific position on the genome, observed at frequencies above 1% in a human population. While SNP microarrays provide profiles of about 1-2 million SNPs simultaneously, next generation sequencers provide profiles of the all 3 billion nucleotides.

2   Methods: Predictive Study as a Key to Find Disease Patterns

Given a dataset of subjects, each represented by a set of features (here, a genomic profile), and a label that specifies a phenotype in each individual, we can conduct various types of studies including associative, risk modeling, and predictive. A predictive study aims to build a predictor to be used later to forecast the class label of unlabeled subjects. To conduct a predictive study, we first filter the dataset by removing the subjects that do not belong to the population under study and/or features that do not pass the quality control criteria. Then we apply a combination of feature selection and learning algorithms from the field of machine learning to learn a predictor from the dataset. Then we test the quality of the predictor using an evaluation strategy and performance metric [3]. We would like to answer a set of significant questions using the predictive study framework such as:

1. Is an individual susceptible to a disease? (prevention)
2. Does an individual have a disease? (diagnosis)
3. What is the best treatment for an individual diagnosed with a disease? (treatment)
4. Will an individual survive from a disease given a specific treatment? (prognosis)

3   Results: Two Case Studies

3.1   Case Study 1: Breast Cancer Prediction

Given the genotypes of 696 female subjects (348 breast cancer cases and 348 apparently healthy controls), predominantly of Caucasian origin from Alberta, Canada using Affymetrix Human SNP 6.0 arrays which measures 906,600 SNPs simultaneously, we filtered 73 subjects not belonging to the Caucasian population and any SNP that had any missing calls, whose genotype frequency was deviated from Hardy-Weinberg equilibrium, or whose minor allele frequency was less than 5%. Then, we applied a combination of MeanDiff feature selection method and KNN learning method to this filtered dataset to produce a breast cancer prediction model. Leave-one-out cross validation (LOOCV) accuracy of this classifier was 59.55%. Random permutation tests showed that this result was significantly better than the baseline accuracy of 51.52%. Sensitivity analysis showed that the classifier is fairly robust to the number of MeanDiff-selected SNPs. External validation on the CGEMS breast cancer dataset, the only other publicly available breast cancer dataset, showed that the combination of MeanDiff and KNN would lead to a LOOCV accuracy of 60.25%, which was significantly better than its baseline of 50.06%. Furthermore, we considered a dozen different combinations of feature selection and learning methods, but found that none of these combinations would produce a better predictive model than our model (see Table 1). We also considered various biological feature selection methods like selecting SNPs reported in recent genome wide association studies to be associated with breast cancer, selecting SNPs in genes associated with KEGG cancer pathways, or selecting SNPs associated with breast cancer in the F-SNP database to produce predictive models, but again found that none of these models had a better than baseline accuracy [4].

Table 1. 10-fold CV accuracy of various feature selection and learning algorithms [4]

| | | Feature Selection Methods | | | |
		Inf. Gain	MeanDiff	mRMR	PCA
Learning Methods	**Decision Tree**	50.88%	52.06%	51.20%	51.69%
	KNN	56.17%	58.71%	57.78%	51.36%
	SVM-RBF	55.37%	57.30%	56.18%	51.84%

3.2   Case Study 2: Ancestral Origin Prediction

We proposed a novel machine learning method, ETHNOPRED, which used the genotype and ancestry data from the HapMap project to learn ensembles of disjoint decision trees, capable of accurately predicting an individual's continental and subcontinental ancestry. To predict an individual's continental ancestry, ETHNOPRED produced an ensemble of 3 decision trees involving a total of 10 SNPs, with 10-fold cross validation accuracy of 100% using HapMap II dataset genotyped on Affymetrix Human SNP 6.0 arrays which measure about 906,600 SNPs simultaneously. We extended this model to involve 29 disjoint decision trees over 149 SNPs, and showed that this ensemble has an accuracy of $\geq 99.9\%$, even if some of those 149 SNP values were missing. On an independent dataset, predominantly of Caucasian origin, our continental classifier showed 96.8% accuracy. We next used the HapMap III dataset, genotyped on arrays that measure 1,458,387 SNPs, to learn classifiers to distinguish European subpopulations (North-Western vs. Southern), East Asian subpopulations (Chinese vs. Japanese), African subpopulations (Eastern vs. Western), North American subpopulations (European vs. Chinese vs. African vs. Mexican vs. Indian), and Kenyan subpopulations (Luhya vs. Maasai). In these cases, ETHNOPRED produced ensembles of 3, 39, 21, 11, and 25 disjoint decision trees, respectively involving 31, 502, 526, 242 and 271 SNPs, with 10-fold cross validation accuracy of 86.5%±2.4%, 95.6%±3.9%, 95.6%±2.1%, 98.3%±2.0%, and 95.9%±1.5% [5].

4   Discussions and Future Works: Excavating the Challenge

These studies reveal that: while SNPs can accurately determine an individual's ancestral origins, they can only weakly predict breast cancer susceptibility. From a biological point of view, this can be explained in part by two facts: (1) ancestral origins depends exclusively on genetic factors (including SNPs), but (2) breast cancer is also influenced by non-heritable environmental and lifestyle factors, which are not represented in germline SNPs as well as, other genomic changes like point mutations, copy number variations, and structural changes of the genome. Motivated by these results, by utilizing computational learning theory framework, my research attempts to understand what concepts are learnable, based on these three factors: the high dimensionality of the data (relative to the relatively small sample size), the heterogeneity of the label, and the noise (i.e., a bound on the best accuracy possible) [6,7]. These are standard characteristics of many biomedical datasets. The high dimensionality issue, discussed extensively in the literature, mentions that: predictive modeling of high dimensional datasets (having small sample size and large feature size) conventionally results in overfitting, in which a model performs well on training data and performs poor on test data. However, little attention is given to the label heterogeneity which highlights that there are several possible factors, any of which could lead to a disease. This means a disease is formulated as follows:

$$\text{Disease} = F_1(x_1,x_2,x_3) \lor F_2(x_4,x_5,x_6) \lor F_3(x_2,x_5,x_7,x_8,x_9) \lor F_4(x_{10},x_{11}) \lor \dots \quad (1)$$

Where the function $F_1(x_1,x_2,x_3)$ is sufficient for Disease to be true, as is $F_2(x_4,x_5,x_6)$, etc. This means there is no simple set of features that is sufficient for explaining the phenotype and as a result the learning is much more complicated. The noise issue also complicates learning as in the standard PAC learning framework, it is known that the sample complexity is only $O(1/\varepsilon \dots)$ (ε is the error), when there is no noise in the data, but becomes $O(1/\varepsilon^2 \dots)$ if there is a bound on the best achievable accuracy (noise). We will attempt to quantify how this noise affects the classification error.

4.1   Elucidation of the High Dimensionality and Noise Challenges

In general, supervised learning algorithms try to find a pattern in the dataset that connects the features to the labels. These tools implicitly assume that the "true" connection is the only one. This research questions this assumption, by asking how many patterns, at a given error rate, will be present in a given dataset, just by chance. If there are r such "chance patterns", as well as the one true pattern, then a learner has only 1 chance in r+1 of identifying this correct pattern – i.e., of finding the actual meaningful rule. Notice that, if all we can use is this dataset, there is no way to distinguish these r+1 classifiers (that is, cross-validation will not help, nor with permutation tests, as this pattern applies to the entire dataset.)

To specify our framework, we assume that we are given a dataset D of n instances, each involving p features and we know that the best achievable accuracy is 1-e. We also focus on a given specific set of hypotheses H – e.g., conjunctions, k-DNF, or m-term k-DNF over a subset of k features. We then ask what the expected number of classifiers, from H, that achieve an accuracy of 1-e over D is. We first consider simple

Boolean datasets, and H = Conjunctions over k features. We assume that the dataset D = (X, Y) is generated completely randomly: each $X_{ij} \sim$ Bernoulli (0.5) is drawn independently, and similarly each $Y_i \sim$ Bernoulli (0.5) independent of X and the other Y_j's. In each of these cases, we try to find an upper-bound for the number of matching classifiers involving k relevant features with each problem. Considering the number of instances (n), the number of features (p), the number of relevant features (k), the upper-bound of acceptable error (noise) (e), and the hypothesis space H, we count the number of matching classifiers. Table 2 represents the number of matching classifiers in the Boolean function learning case and we can observe that for Conjunctions:

1. The number of matching classifiers increases in $O(p!)$ considering the number of features (p).
2. The number of matching classifiers increases in $O(2^k)$ considering the number of relevant features (k).
3. The number of matching classifiers increases in $O(2^{ne})$ considering the noise (e) and the number of instances (n).

Given these observations and considering more complex hypotheses such as m-term k-DNFs, it is not surprising at all to come up with cases in which the learning algorithm cannot distinguish the true classifier/pattern from other apparently equally good classifiers/patterns.

Table 2. The upper-bound for the number of matching classifiers given the number of instances (n), the number of features (p), the number of relevant features (k), and the noise (e)

Hypothesis Space	Upper-bound for the Number of Matching Classifiers
Conjunctions	$\dfrac{1}{2^{n-ne}} \dbinom{n}{n-ne} \dbinom{p}{k} 2^k$
k-DNFs	$\dfrac{1}{2^{n-ne}} \dbinom{n}{n-ne} 2^{\binom{p}{k}2^k}$
m-term k-DNFs	$\dfrac{1}{2^{n-ne}} \dbinom{n}{n-ne} \dbinom{2^{\binom{p}{k}2^k}}{m}$

4.2 Concentration on the Label Heterogeneity Challenge

While we try to learn disease-associated patterns from new high dimensional omics datasets, we often ignore the problem of label heterogeneity. As an example reconsider the breast cancer prediction problem: breast cancer is biologically heterogeneous as current molecular classifications based on clinical determinations of steroid hormone receptor (like ER) status, human epidermal growth factor receptor 2 (HER2) status, or proliferation rate status (PR) suggest a minimum of four distinct biological subtypes [8]. Our dataset ignored these sub-classes and merged them into the single label: breast cancer case. We might be able to produce a more accurate predictor if we employed more detailed labelling of these sub-classes, to produce a classifier that could map each subject to a molecular subtype. In this research, we assume the pattern we are looking for is in the form of a m-term k-DNF function matching with Equation 1 and try to design a novel algorithm for learning this function.

References

1. Collins, F.S., Morgan, M., Patrinos, A.: The human genome project: Lessons from large-scale biology. Science 300, 286–290 (2003)
2. Wright, A., Hastie, N.: Genes and Common Diseases. Cambridge University Press, New York (2007)
3. Hastie, T., Tibshirani, R., Friedman, J.: The elements of statistical learning: Data mining, inference, and prediction, 2nd edn. Springer, New York (2009)
4. Hajiloo, M., Damavandi, B., Hooshsadat, M., Sangi, F., Cass, C.E., Mackey, J., Greiner, R., Damaraju, S.: Using genome wide single nucleotide polymorphism data to learn a model for breast cancer prediction. BMC Bioinformatics (in press)
5. Hajiloo, M., Sapkota, Y., Mackey, J.R., Robson, P., Greiner, R., Damaraju, S.: ETHNOPRED: A novel machine learning method for accurate continental and sub-continental ancestry identification and population stratification correction. BMC Bioinformatics 14(1), 61 (2013)
6. Valiant, L.G.: A theory of learnable. Communications of the ACM 27, 1134–1142 (1984)
7. Vapnik, V., Chervonenkis, A.: On the uniform convergence of relative frequencies of events to their probabilities. Theory of Probability and its Applications 16(2), 264–280 (1971)
8. Bertucci, F., Birnbaum, D.: Reasons for breast cancer heterogeneity. Journal of Biology 7(2), 6 (2008)

Shape-Based Analysis for Automatic Segmentation of Arabic Handwritten Text

Amani T. Jamal and Ching Y. Suen

CENPARMI (Centre for Pattern Recognition and Machine Intelligence)
Computer Science and Software Engineering Department, Concordia University
Montreal, Quebec, Canada
{am_jamal,suen}@cenparmi.concordia.ca

Abstract. Text segmentation is an essential pre-processing step for many methods of recognition and for spotting systems as well. There are some characteristics in Arabic that differentiates it from Latin-based scripts. In this thesis proposal, we address the challenges of segmenting offline Arabic handwritten text. Our proposed approach of text segmentaion utilizes the knowledge of Arabic writing. Furthermore, a method for touching segmentation is proposed. To facilitate touching segmentation, a new learning-based baseline estimation method is introduced.

Keywords: Document Analysis, Arabic Handwritten Documents, Text Segmentation, Touching Segmentation, Baseline Estimation.

1 Introduction

Arabic is the mother tongue of more than 300 million people in more than 20 countries. The Arabic script was first documented in 512 AD. More than thirty languages use the Arabic alphabet such as Farsi, Pashtu, Urdu, and Malawi.

The tasks of off-line recognizers are considered difficult since only an image of a script is available. One of the challenges in offline handwriting related systems is the complexity of segmenting text into words. When the writing style is unconstrained, recognition and retrieval of individual components is less reliable. Therefore, they must be grouped into words, before the recognition and spotting stages.

Most of the techniques in handwritten document retrieval and recognition fail if the texts are wrongly segmented into words. Sometimes the cause of failure in Arabic-related methods is the incorrectly segmented text into sub words or Part of Arabic Word (PAW), when PAWs are treated as main units.

Text segmentation into words, specifically in Arabic, faces four main challenges: (1) lack of well defined boundaries between words, (2) touching components, (3) disconnected (broken) components and (4) stop words. These problems have not been solved. Detection and correction of such faults will improve the performance of the recognition and spotting systems. A problem with touching affects the performance of many approaches, such as analytical methods for printed and handwritten documents, semi-holistic techniques, holistic approaches, and word spotting systems.

O. Zaïane and S. Zilles (Eds.): Canadian AI 2013, LNAI 7884, pp. 334–339, 2013.

Many methods have been proposed for segmenting text into words. These methods can be categorized into two approaches: thresholding and classification. A few methods have been applied on Arabic text without considering the uniqueness of this language. In [1], a threshold was determined after measuring the distance using a vertical histogram. The experiments were done on city name images that have a maximum of 3 words per image. The accuracy ranges from 66.67% to 80.34%. In [2], a classification technique was used based on 9 extracted features. The experiment was done on 100 documents with an accuracy of 60%.

1.1   Arabic Characteristics

Twenty-two letters in the Arabic language must be connected on a baseline within a word. The remaining six letters cannot be connected from the left, which we call non-left-connected (NLC) letters. In this way, NLC letters separate a word into several parts depending on how many of those letters are included in a word. Figure 1 shows one word with two PAWs. Some applications use PAW as the main units for recognition or spotting, while others use PAW as distinctive features to improve the accuracy of their systems such as lexicon reduction.

1.2   Challenges

Generally, handwritten texts lack the uniform spacing that is normally found in machine-printed texts. However, in Arabic handwritten text, separation into words is more challenging due to the existence of PAWs. Texts have two types of spacing, intra-word gaps and inter-word gaps. In Arabic, intra-word gaps are the ones between two PAWs, where the word must be disconnected due to NLC letters. This is part of the structure of the language. In Arabic machine-printed text, the inter-word gaps are much larger than intra-word gaps. However, in Arabic handwritten documents, the spacing between the two types is mostly the same, as pointed out in Figure 2.

There are some other issues that add to the complexity of text segmentation. These problems arise from touching and broken PAWs. They appear to be due to the poor printing or scanning, or a writing style [3]. Sometimes adjacent PAWs connect to one another, either between two adjacent words or within a word. Segmentation of such touching is a difficult problem. The difficulty is due to the fact that PAW vary in length (number of letters), consist of dots, contains non-basic characters (additional characters), or have directional markings. More difficulty is added by words having an unknown number of PAWs. Broken and touching PAW lead to unknown or unrecognized connected components (CCs). In other words, touching problems yield to under segmentation. In the case of a broken PAW problem, it is always subject to over segmentation. Figure 3 shows an example of touching and broken PAW.

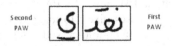

Fig. 1. An Arabic word with two PAWs

Fig. 2. Intra and inter word gaps in Arabic language

Fig. 3. Touching and Broken PAWs

2   Proposed Approach

The focus of this proposal is on segmenting the text into words and PAWs. In addition, a new method for touching PAWs is introduced. The main difference with our segmentation approach from previous methods is utilizing the uniqueness of Arabic writing. To enable touching segmentation, we introduce a learning-based technique for baseline estimation. Our approach for segmentation is a two-stage strategy. In the first stage, referred to as Text Segmentation, the text will be segmented into words and PAWs. In the second stage, named Touching Segmentation, the touching PAWs and words will be segmented. A block diagram of our overall methodology is illustrated in Figure 4.

2.1   Utilizing Knowledge of Arabic Writing

In [1], the authors pointed to the importance of using the language specific knowledge for Arabic text segmentation. In addition, in [4], the authors claim that one of the problems of Arabic text segmentation is the inconsistent spacing between words and PAWs. Our method of text segmentation is linguistically motivated. In the Arabic alphabet, twenty-two letters out of twenty-eight have different shapes when they are written at the end of a word as opposed to the beginning and middle. Therefore, recognizing these shapes can help to identify the end of a word. In addition, there are just fifteen main shapes that can be used to distinguish the end of a word, since the rest of the characters have the same main part but with a different number and/or position of dots. Our touching segmentation approach is based on a study of Arabic handwriting styles. Nineteen letters have either an ascender or descender. After analyzing many Arabic documents and words, it was observed that most of the touching occurs from the overlapping of adjacent PAWs. Writing descenders with a long stroke is a writing style [5].When the last letter of a PAW is a descender, and it encroaches into the adjacent PAW, this causes pixels to touch among the two PAWs. In other words, the touching occurs when a bottom-curved letter at the end of a PAW overlaps another letter. Ascender touching occurs well above the baseline when adjacent PAWs end and start with ascenders.

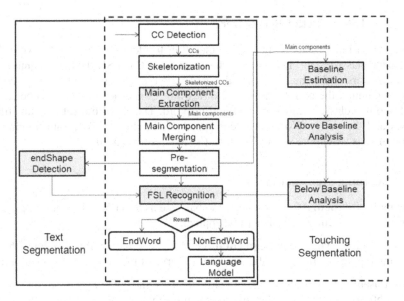

Fig. 4. Overall Methodology

2.2   Text Segmentation

Connected Components Detection. Normally, a PAW is composed of several Connected Components (CCs).Therefore, the first main step in segmentation of an Arabic handwritten word is detecting and labeling its CC. CC analysis is the most efficient approach since the Arabic script consists of several overlapping CC.

Skeletonization. Skeletonization or thinning is the processes of converting an image into a one-pixel thin representation. Skeletonization facilitates feature extraction, analysis and reduces the amount of memory usage.

Main Component Extraction. Morphological Reconstruction that is based on a thin horizontal structuring element located in the middle zone of the image is used to extract the main components of Arabic text. Most threshold methods failed to find the main components.

Broken Component Merging. Broken main components affect our method. We are investigating this issue.

Pre-segmentation. Based on a gap metric, a first rough estimation of segmentation points will be assigned.

EndShape Detection. At this step the main purpose is to detect the last letter of the extracted main component. After extracting some essential points in the skeletonized CC, a recognition-based approach is used to extract the endShape.

EndShape Recognition. At this stage, the extracted endShape of CC will be recognized using a Final Shape letter (FSL) recognizer as described in Section 2.3.

2.3   Final Shape Letter Recognizer

We created a Final Shape Letter (FSL) classifier to identify the end of a word. This recognizer classifies the last part of the main CC. The FSL database contains the shape of letters at the final position. But NLC letter shapes are written the same way at the beginning, the middle or the end of a word. Consequently, FSL can be categorized into two classes: EndWord and notEndWord. The EndWord set contains fifteen classes and the notEndWord set is composed of six classes. To create training and testing models a Support Vector Machine (SVM) has been used.

2.4   Touching Segmentation

Baseline Estimation. After estimating the baseline as described in Section 2.5, above and below the baseline analysis will be applied.

Above Baseline Touching Segmentation. After estimating the top line, the CC above the top line is traced to label the turning curve of the peak as a segment point.

Below Baseline Touching. The CCs below the baseline are detected and their widths and heights are calculated. If they exceed the threshold, the above baseline component will be extracted to be classified using the FSL recognizer. If it belongs to the FSL classes, the last point (pixel) of FSL or the touching point with the baseline is labeled as a segment point.

2.5   Baseline Estimation

The dots and their positions, which are below or above the baseline, provide essential information in Arabic text recognition. Word descenders and ascenders can be identified by the baseline positions. However, for our methodology we will estimate the baseline to facilitate touching segmentation. The proposed methods have not reached satisfactory results for a text line with one or few words. One of the issues that affect single word baseline estimation is words that have letters with long bottom curves [5]. In addition, dots and non-basic letters can affect baseline estimation methods. These issues can be handled by our learning-based approach.

The baseline of an Arabic word consistently lies below the middle line of an image [1]. After analyzing some isolated word images, we observed that the baseline of the words after cropping and normalization was roughly between the 30th and 50th rows of the image. As a consequence, the number of baseline classes at most will be twenty. For supervised learning of baseline positions of handwritten words, we need some handwritten words as training samples with the positions of their baselines. The IFN/ENIT database [6] will be used since they include the position of the baseline in the ground truth. Let $W = \{w_1, w_2, \ldots w_n\}$ be a collection of training word samples. $F = \{f_1, f_2, \ldots f_m\}$ contains baseline-dependent features of a word. $B = \{b_1, b_2, \ldots b_j\}$ are the baseline classes. Our training model consists of P, a set of baseline-dependent features including the position of the baseline (Eqs(1)).

$P = \{(w_i, f_1, f_2, f_3..f_m, b_j) \mid w_i$ is a word, f_1 to fm is a set of features, b_j is the position of a baseline\}   Where $i = 1$ to n, $f_{i1}, f_{i2}, f_{i3}.....f_{im}, 0 \leq b_j \leq 20$.   (1)

2.6   Performance Evaluation

The number of matches between the entities (words and PAWs) detected by the algorithm and the entities in the ground truth will be counted. Three metrics will be used which are detection rate (DR), recognition accuracy (RA) and performance metric, which is based on DR and RA.

3   Conclusion

Segmentation of text into words (Holistic) and word segmentation into PAWs (semi-holistic) are essential preprocessing steps for Arabic applications. New algorithms and techniques that can improve the accuracy of text segmentation were introduced. The knowledge of Arabic writing has been utilized to improve the segmentation. The focus is on automatic segmentation of Arabic handwritten texts into words and PAWs by using shape-based analysis. Furthermore, a method for segmenting touching PAWs was proposed. To facilitate touching segmentation, a new learning-based baseline estimation method was introduced.

References

1. AlKhateeb, J., Ren, J., Ipson, S.S., Jiang, J.: Knowledge-based Baseline Detection and Optimal Thresholding for Words Segmentation in Efficient Pre-processing of Handwritten Arabic Text. In: 5th International Conference on Information Technology: New Generations (ITNG), Las Vegas, USA, pp. 1158–1159 (2008)
2. Srihari, S., Srinivasan, H., Babu, P., Bhole, C.: Spotting Words in Handwritten Arabic Documents. In: Document Recognition and Retrieval XIII (SPIE), San Jose, USA, pp. 606702_1–606702_12 (2006)
3. Elgammal, A., Ismail, M.: A Graph-based Segmentation and Feature Extraction Framework for Arabic Text Recognition. In: 6th International Conference on Document Analysis and Recognition (ICDAR), Seattle, USA, pp. 622–626 (2001)
4. Kchaou, M., Kanoun, S.: Segmentation and Word Spotting Methods for Printed and Handwritten Arabic Texts: A Comparative Study. In: 13th International Conference on Frontiers in Handwriting Recognition (ICFHR), Bari, Italy, pp. 274–279 (2012)
5. Pechwitz, M., Abed, H., Märgner, V.: Handwritten Arabic Word Recognition Using the IFN/ENIT database. In: Guide to OCR for Arabic Scripts, pp. 169–213 (2012)
6. Pechwitz, M., Maddouri, S., Märgner, V., Ellouze, N., Amiri, H.: IFN/ENIT- database of Handwritten Arabic Words. In: International Symposium on Frontiers in Writing and Document (CIFED), Hammamet, Tunisia, pp. 129–136 (2002)

A Probabilistic Framework
for Detecting Unusual Events
in Mobile Sensor Networks

Shehroz S. Khan

David R. Cheriton School of Computer Science
University of Waterloo
Canada
s255khan@uwaterloo.ca

Abstract. The thesis addresses the problem of identification of unusual events in mobile sensor networks. Most existing activity recognition systems based on computer vision devices or wearable sensors attempt to recognize normal activities of daily living, however they may not be well-suited to identify abnormal activities, especially the ones that have not been encountered before. In this thesis, I will study the following research problems:

- Learning classifiers with little or no data from unusual events.
- Low-cost, discriminatory features and most suitable sensors to be used for unusual events detection.
- Learning high level hierarchy of activities to help in detecting transitions to unusual events.
- Incremental / Online learning to update the unusual events classification model dynamically as the new data is received by the sensors.

To tackle these research problems, I propose to use the sequential classifiers such as Hidden Markov Models and its variants.

Keywords: Unusual Event Recognition, Activity Recognition, Hidden Markov Models, Mobile Sensors.

1 Activity Recognition

Activity Recognition [1] studies the actions, behaviours and goals of an agent and attempts to build systems to recognize them. The research in activity recognition has led to the successful realization of intelligent pervasive environments that can provide context, assistance, monitoring and analysis of an agent's activities that are usually backed up by advanced machine learning and vision algorithms. Most of the research in activity recognition is either based on sensors [1] or computer vision [2]. A drawback with sensor based methods is their intrusive nature; a person has to wear sensor based gadgets all the time which may be uncomfortable to carry and he may refuse to wear it [3]. Vision based system works well in an indoor setting, however when a person goes out of the premises these systems cannot provide much help.

O. Zaïane and S. Zilles (Eds.): Canadian AI 2013, LNAI 7884, pp. 340–345, 2013.

2   Unusual Events

A central focus of many of the studies based on Activity Recognition is the detection of usual daily human activities e.g. walking, hand washing, making breakfast etc. However, in many scenarios, detection of unusual activities is of more importance as it may render a person at risk and vulnerable. Consider an activity monitoring system where the normal activities such as walking, sitting, or standing are important to identify, but the more challenging and useful thing to identify is when the person deviates from these normal or relatively safe activities. In this context these *unusual* activities may include incurring a fall or suffering a stroke. A typical activity recognition system may misclassify 'fall' as one of the already existing normal activities because 'fall' may not have occurred earlier. An alternative strategy is to learn the model with specific unusual activities (e.g. fall [4]). However, this may require extensive domain knowledge, an understanding of the activities that may be encountered and sufficient data collection for the type of unusual activity to be modelled. Nonetheless, in the best scenario, these algorithms may only be able to detect the specific unusual activity on which they are trained and cannot be generalized to other types of unusual activities. Another example is the unusual transition within normal activities that may point to an emergency situation. For example, transitioning from lying to running may indicate a danger but a typical activity recognition system would not be able to detect that phenomenon. Moreover, in emergency situations it is important to first identify if an unusual event has occurred and later on efforts can be expanded to find their specific details.

Zhang et al. [5] defines *unusual events* for the audio-visual steams as the ones that are *rare*, *unexpected* and hold *relevance* for a particular task. The rarity of unusual events yet to be observed leads to a lack of sufficient data for training the classification model. More than one type of unusual event may also occur in a data sequence and the unexpectedness of unusual events makes it difficult to model them in advance. Yin et al. [6] mention that due to the scarcity of such activities, *it is a challenging task to design an abnormality detection system that has a low false positive rate.* Collecting abnormal activity data can be cumbersome because it may require the person to actually undergo such unusual events which may be harmful. In addition to very few or no labelled data, the diversity and types of unusual events further makes it difficult to model them efficiently.

3   Mobile Sensors for Activity Recognition

Smartphones are the latest types of mobile phones that are built on a mobile computing platform with embedded operating systems and are capable of more advanced computing abilities than a traditional mobile phone. Most of the modern smartphones are equipped with various sensors, such as accelerometer, gyroscope, compass, proximity sensor, and ambient light sensor. Other types of sensors available are front and back facing cameras, microphone, GPS, WiFi and

Bluetooth radios [7]. In the last few years, there has been considerable amount of research work done for general activity recognition using smartphones. One of the reason for their popularity is that they are non-invasive, easy to carry, work both indoor and outdoor and are equipped with sensors that are useful for activity recognition.

Researchers have used most of the available sensors in a smartphone for activity recognition, and employed many standard classification techniques including the ones based on heuristics. The problem with heuristic method is that they identify activities based on fixed or pre-defined thresholds and lacks generalization. Moreover, it is very difficult to adapt thresholds for new or unseen activities without prior knowledge (which may not be available) and setting new thresholds can conflict with the existing ones and can render the system useless. Standard classification methods such as Decision Trees, Support Vector Machines, Neural Networks, Nearest Neighbours etc are shown to be employed for activity recognition, however the sensor readings from the smartphones are temporal and dynamic in nature, therefore these choices may not be very pertinent. Another problem with employing the standard classification methods is that most of them works for binary/multi class classification scenarios and may not be able to identify new or unusual activities, specially the ones that are not encountered earlier. These algorithms may also not work well when the unusual class of interest is severely under-sampled.

Most of the research on unusual activity recognition using smartphone aims to identify 'falls' and use different smartphone sensors along with various classification methods to achieve this task. These methods assume that data from 'falls' is available. However, falling is a complex phenomenon and classification models based on limited dataset may either over-fit or may not generalize well on new patterns/types of falls. Moreover, these methods cannot identify other types of unusual events such as a stroke.

4   Research Contributions

In this thesis, I will work on developing classification algorithms to identify unusual events in sensor networks, specially on smartphone devices. The major contribution with several proposed tunings are presented below:

– **Learning classification model with little or no data from unusual events.** The thesis will investigate the role of sequential classifiers, such as Hidden Markov Models (HMM), to achieve this task. One of the advantage of the (HMM) is that they can be used for temporal classification by modelling the dynamics in data sequences effectively and consider the history of actions when taking a decision on the current sequence. HMMs can automatically absorb a range of model boundary information for continuous activity recognition [8] in scenarios where activity boundaries are not easily detectable. I propose variants of HMM based approaches to identify unusual events based on mobile sensor data. These approaches use only normal activity data for training and the goal is to identify the deviants. In scenarios when limited unusual events data may be available, techniques based on over-sampling of minority class [9] will be explored and discriminative temporal classifiers such as Conditional Random Fields (CRF) [10] will be implemented and compared with HMMs to study their benefits for the task.

- **Low-cost, discriminatory features and most suitable sensors to be used for unusual events detection.** The proposed sequential classification algorithms are based on low-cost features across different individual and are placement / orientation of sensor independent. Different types of sensors such as accelerometer, gyroscope, pressure sensors etc will be used in this study. I have performed preliminary experiments on a real-world human activity recognition dataset collected with body worn sensors. The proposed classifiers are able to identify artificially generated unusual motion events with good success using only few low-cost features and can generalize over different people and sensor positions [11].

- **Learning high level hierarchy of activities to help in detecting transitions to unusual events.** Hierarchical HMM helps in understanding higher level transition between various activities and their role in identifying unusual events is proposed in the thesis. The different levels of granularity can help in better modelling of unusual events. A proposed hierarchical model is to identify static (e.g. sitting, lying etc) and dynamics activities (e.g. walking, running) and further divide dynamic activities as long term and short term (e.g. jumping) and identify other types of unseen short-term unusual events such as falling or stroke.

- **Incremental / Online learning to update the unusual events classification model dynamically as the new data is received by the sensors.** The real utility of a temporal unusual event detection mechanism can be achieved if new methods can be developed and adapted to support online identification with fast response time in detecting unusual events. In this direction, approaches based on incremental HMM learning will be investigated.

Besides these these major contributions, I will address the following adjoining research issues in this thesis:

- Investigating the role of non-invasive and cheap sensing devices such as a smartphone to detect abnormal activities.
- The computation of features, sampling rate and classification of unusual events on a sensing device directly affects its energy consumption, therefore I plan to address this issue as well in my research. More specifically highly discriminative orientation invariant and computationally efficient low-cost features will be investigated that can provide better recognition rates for temporal unusual events.
- The applicability of classifier ensembles for HMM is not well-studied. It is proposed to learn diverse and accurate HMM based classifiers based on different initial settings and/or different feature subsets and to study their relative advantage over baseline classifier.

4.1   Datasets

To conduct the research proposed in Section 4 and achieve the stated goals, several experiments will be conducted using different types of datasets:

- Simulated Data – Data simulation will be done to accomplish two tasks, namely:
  - To artificially generate different types of unusual events to check the validity of the proposed methods. This method is helpful to work with those activity recognition datasets that have no samples for unusual events. An experiment is described in Khan et al. [11]
  - To artificially over-sample the minority class that represents unusual events / falls. This type of data is useful to test discriminative temporal method such as CRF and is useful to work with datasets that have few samples for unusual events.
- Publicly Available Datasets : This activity involves testing the proposed models on the existing activity recognition datasets. The advantage is that it will be easy to compare the results from the existing methods and understand their relative merits and demerits. The following (non-exhaustive) list of datasets is currently considered:
  - i http://www.kn-s.dlr.de/activity – DLR Human Activity Recognition with Inertial Sensors, contains normal and fall data
  - ii http://archive.ics.uci.edu/ml/datasets/OPPORTUNITY+ Activity+Recognition – Data from Opportunity data challenge, very comprehensive and large, can only be used after careful study about it.
  - iii http://robotics.usc.edu/~harsh/datasets.html - Walking speed data, Robotic Embedded Systems Laboratory, University of Southern California, Los Angeles
  - iv http://architecture.mit.edu/house_n/data/Accelerometer/ BaoIntilleData04.htm – MIT Placelabs data - contains data from 20 ADL - very rich but no abnormal activity.
  - v http://www.hcii-lab.net/data/scutnaa/EN/naa.html – 10 ADLS done by 44 individuals, no abnormal activity, using only one accelerometer worn at different places, Human-Computer Intelligent Interaction Laboratory, South China University of Technology, China
- Dataset Collection by Actors : To collect real-time data, it is proposed to collect data for various ADLs. For testing purposes and trying out over-sampling methods, a limited amount of data for unusual activities such as fall, stroke etc will also be collected. It is important to know that unusual events does not only signify the mentioned defined events, but it can be an unusual activity transition for e.g. a sudden transition from sleeping to running may indicate some type of emergency. These unusual transitions can be inferred from the transition matrix of HMM which should give a probabilistic relationship between various activities. The proposed testbed for dataset collection by actors, will include the development of a mobile application for the Android smartphone platform, which will be developed

using the Google Android API. For testing purposes, the application will give the user the provision to log the activities that will be used to train the classifiers after manual verification and annotation. The data collected through smartphone will be transferred to a server using GSM/Wifi signals at regular interval. The server will perform the offline classification using proposed HMM approaches. Incremental HMM techniques will also be explored that can update the model as the new data arrives in blocks and does not have to re-run from scratch. Once the classifiers are trained, the updated model will be transferred back to the smartphone and for any new activity mobile phone just have to match with the model present in its memory. This process will save the time spent in computation of features, should consume less mobile battery and should not introduce delays in response time.

References

1. Chen, L., Khalil, I.: Activity recognition: Approaches, practices and trends. In: Activity Recognition in Pervasive Intelligent Environments, pp. 1–32. Atlantis Press (2011)
2. Fiore, L., Fehr, D., Bodor, R., Drenner, A., Somasundaram, G., Papanikolopoulos, N.: Multi-camera human activity monitoring. Journal of Intelligent and Robotic Systems 52(1), 5–43 (2008)
3. Schulze, B., Floeck, M., Litz, L.: Concept and design of a video monitoring system for activity recognition and fall detection. In: Mokhtari, M., Khalil, I., Bauchet, J., Zhang, D., Nugent, C. (eds.) ICOST 2009. LNCS, vol. 5597, pp. 182–189. Springer, Heidelberg (2009)
4. Yu, X.: Approaches and principles of fall detection for elderly and patient. In: IEEE 10th International Conference on ehealth Networking Applications and Services, HealthCom 2008, pp. 42–47 (2008)
5. Zhang, D., Gatica-Perez, D., Bengio, S., McCowan, I.: Semi-supervised adapted HMMs for unusual event detection. In: CVPR (1), pp. 611–618 (2005)
6. Yin, J., Yang, Q., Pan, J.J.: Sensor-based abnormal human-activity detection. IEEE Trans. Knowl. Data Eng. 20(8), 1082–1090 (2008)
7. Lane, N., Miluzzo, E., Lu, H., Peebles, D., Choudhury, T., Campbell, A.: A survey of mobile phone sensing. IEEE Communications Magazine 48(9), 140–150 (2010)
8. Yang, J., Xu, Y.: Hidden markov model for gesture recognition. Technical Report CMU-RI-TR-94-10, Robotics Institute, Pittsburgh, PA (May 1994)
9. Bowyer, K.W., Chawla, N.V., Hall, L.O., Kegelmeyer, W.P.: SMOTE: Synthetic minority over-sampling technique. CoRR abs/1106.1813 (2011)
10. Lafferty, J.D., McCallum, A., Pereira, F.C.N.: Conditional random fields: Probabilistic models for segmenting and labeling sequence data. In: Brodley, C.E., Danyluk, A.P. (eds.) ICML, pp. 282–289. Morgan Kaufmann (2001)
11. Khan, S.S., Karg, M.E., Hoey, J., Kulic, D.: Towards the detection of unusual temporal events during activities using hmms. In: Proceedings of the 2012 ACM Conference on Ubiquitous Computing, UbiComp 2012, pp. 1075–1084. ACM, New York (2012)

Intelligent Tutoring Systems Measuring Student's Effort During Assessment

Peter Lach

University of Regina, Regina SK S4S 0A2, Canada
lach200p@cs.uregina.ca

Abstract. Development of eye tracking technology brings new opportunities for use in computer science and e-learning. This paper will present a possible way of how an eye tracker can be used with intelligent tutoring system to enhance learning experience of students. The task of an intelligent tutoring system discussed here is to recognize the degree of students effort when answering questions and then respond with appropriate feedback to motivate student.

Keywords: Intelligent Tutoring Systems, eye tracking, pupil dilatation, mental workload.

1 Introduction

One of the problems affecting learning process in Intelligent Tutoring Systems is for the tutor accurately estimate student's knowledge state. In this paper, I present research on exploring eye tracking technology which is used to modify student model in Intelligent Tutoring Systems during assessment. In particular, I will discuss pupillary response during student's interaction with learning environment. Pupil size is known to correlate well with the mental workload for discrete, non-interactive tasks [2]. The magnitude of the pupillary dilation appears to be a function of processing load, or the mental effort required to perform the cognitive task[1]. Although there are many other techniques how to asses mental workload such as physiological measures (pupil size, heart rate, EEG) or subjective ratings, pupil size is most suited for assessment student mental workload because it doesn't disrupts student ongoing activities, provides real-time information, and is less intrusive than other physiological measures[1]. The purpose of this paper is to describe my ongoing research aimed to develop an Intelligent Tutoring System which utilizes eye tracking data in real-time to get information about students mental workload during answering questions. This information is then used by system to evaluate students knowledge state and adapt learning environment to student's current needs. The paper is structured as follows. Section 2 will introduce basic concepts from areas such Intelligent Tutoring Systems, Eye Tracking Technology and Index of Cognitive Activity. After this, section 3 discuss new Intelligent Tutoring System developed for this research. Focus is on both Intelligent Tutoring Systems and Eye Tracking. Section 4, at last, will draw some conclusions and provides some directions for future work.

O. Zaïane and S. Zilles (Eds.): Canadian AI 2013, LNAI 7884, pp. 346–351, 2013.

2   Theoretical Background

2.1   Intelligent Tutoring Systems

Intelligent Tutoring Systems (ITSs) are any computer systems that can be used in learning and contains intelligence. ITSs show a vast variety of ways to conceptualize, design and develop tutorial services that support learning[3]. It is an outgrowth of the earlier computer aided instruction (CAI) model. Traditional CAI presents instructional materials in a rigid tree structure to guide the students from one content page to another. This early CAI systems did not support any adaptation to student needs and were unable to provide the individualized attention that a human instructor can provide. ITSs are intended to take some of the part of a human teacher or tutor, and to take at least some of the initiative in an educational dialogue. ITSs are typically separated into several different parts, where each part plays an individual role. The basic four-component architecture of ITSs is made of domain model, student model, tutoring model and user interface component.

Domain Model. Also known as expert knowledge contains the concepts, rules and problem-solving strategies of domain to be learned. This module can fulfil multiple roles: as a source of knowledge being taught or as an assessment module of students knowledge state.

Student Model. This module stores information which are specific to each individual student. Student model is viewed as a dynamic model that implements several functions[3]. According to Wenger [4] student model has the following three tasks:

1. It must gather implicit and explicit data about student.
2. It must use this data to create a representation of student's knowledge state.
3. It must be able to asses the student's knowledge level.

Tutoring Model. Tutoring model represents the teacher and takes input from domain and student models. It makes decisions about tutoring strategies and actions. This model is reflected in different forms of interaction with student: Socratic dialogs, hints or feedback[3].

User Interface Component. This component control the way how the knowledge is presented to the student. It also provide an interface for interaction between ITSs and student.

2.2   Eye Tracking Technology

Eye movement can be divided into two components: fixations, where eye is almost still looking at one point, and saccades which represent a fast eye movement between two fixations. We know also other eye events which are used in research such as blinks and pupil dilatation. All these movements can be tracked with any currently available eye tracking technology. Current eye trackers are very expensive and therefore they still remain only as research tool. There are already few existing projects using eye tracker in learning environment. Among the first projects which include eye tracking technology in e-learning is AdeLE(Adaptive E-Learning through Eye Tracking)[5]. Goal of this project is to develop a e-learning system which uses real-time eye tracking data to support adaptive teaching and learning. In the system described here [7], eye tracking data were used by emphatic software agent to identify learners focus and interest areas on screen. The character agents are eye-aware because they use eye movements, pupil dilation, and changes in overall eye position to make inferences about the state of the learner and to guide his behaviour. User studies showed that this type of agent could have beneficial effects on learners motivation and concentration during learning. e5Learning is another learning environment using eye tracking technology[6]. This system uses rectangular areas on screen to track users gaze position. Each of these Regions of Interest (ROI) have associated time threshold predefine by the author of the course. System then tracks each of these regions whether or how those areas have been accessed by the user. System developed in this project[8] detects student disengagement and boredom during learning process. System tracks student eye gaze patterns. When student looked away from screen for certain period of time tutor assumed that the student was disengaged. Research of Intelligent Tutoring Systems using eye tracking technology is very large and is not limited to this projects mentioned before. These examples illustrate how and eye tracking technology can be used to obtain low-level mechanic account of cognitive processes during learning. However none of those examples explored further pupillary response as a measure of mental workload.

2.3   Index of Cognitive Activity

Although dilation of the pupil in response to increased attention was first observed early in 20th century[9], the first systematic study of the phenomenon appears to have been that of Hess and Polt[10]. Under conditions of constant illumination and accommodation, pupil size has been observed to vary systematically in relation to a variety of physiological and psychological factors, including non-visual stimulation, habituation, fatigue, sexual and political preference, and level of mental effort[11]. The magnitude of the pupillary dilation appears to be a function of processing load, or the mental effort required to perform the cognitive task[1]. Beatty[12] has shown that task-evoked pupillary response uniquely reflects the momentary level of processing load and is not an artefact of non-cognitive confounding factors. The Index of Cognitive Activity (ICA) uses the signal processing techniques of wavelet analysis to detect small but reliable increases in pupil size while minimizing the impact of changes in light[13]. The Index is computed for each second of a task. It is essentially the ratio of the number of unusual increases in pupil size observed in one second over the theoretical maximum of such increases for one second. The Index is scaled by the hyperbolic tangent function so that its values lie between 0 and 1. Typically, the values for all seconds are averaged to produce a single over-all Index value for any given task[14]. The ICA has several advantages over other techniques that measure changes in pupil dilation. First, it does not require averaging over trials or over individuals. Second, it can be applied to a signal of any length.

3 Method

3.1 System Architecture

Tutoring System Component. To simplify the task of developing an ITS, I restricted the scope of the problem as follows: firstly the system is build to teach student about basic concepts of data structures, secondly, only 6 data structures are covered in this system (array, string, list, queue, stack, tree). These topics include basic concepts of data structure manipulations e.g. add new element and remove element from data structure. System developed here is designed to adapt general framework of ITS using four component architecture. Domain knowledge is coded using XML and stored outside of the system. This domain is represent in tree structure where each data structure type represents new topic. I have split this topic further down to each individual concept. For example, data structure 'array' is a topic which has multiple concepts e.g. 'add' or 'remove' elements. Each concept is then divided into pages which are then presented to student. Each page holds small peace of information related to selected concept and topic. Student model in this system is represent by set of separate statistical measurements and stored in multiple text files. Student knowledge state is represented using Bayesian Network which is used by Tutoring module to make decision about next tutoring steps. Tutoring module in this system is represented by set of 'if-then' rules. Communication between tutor and student is managed by user interface module which in this system is developed using Simple Direct Media Layer library.

Effort Assessment Component. This component is designed to identify how much effort was given by student when answering question during assessment. Input from eye tracking device is used to create reading pattern. During reading process of required concepts system records how much of required text has been covered by student. This will create understanding if student has required knowledge for answering question. Other input from eye tracker is used during assessment process. System records pupil dilatation during question answer to create ICA. Both reading patter and ICA are then used in Effort Assessment Component to create better believe about students newly gained knowledge. Output from this component is also used to create systems feedback to students response.

3.2 Eye Tracking Device

I have used SR Research EyeLink II eye tracking device for this research. The EyeLink II system consists of three miniature cameras mounted on a headband.

Two eye cameras allow binocular eye tracking or easy selection of the subject's dominant eye. Each camera has built-in illuminators, digitally corrected for even lighting of the entire field of view. Together with digital compensation for changes in ambient lighting, this results in exceptionally stable pupil acquisition. EyeLink II has the highest resolution (noise limited at $< 0.01^o$) and fastest data rate (500 samples per second). Configuration of the system consists of host pc which is controlling EyeLink II. Headband and monitor markers are connected to host pc which process all data generate by eye tracker. Display pc which is hosting the Intelligent Tutoring System application is connected to host pc using PCI card to retrieve real-time data from subject.

3.3 Procedure

Before student start to use this system we need to perform calibration of eye tracking device. After this student has access to ITS application. System will present student with topics related to data structures which student can choose from. Student doesn't need to follow any hierarchy when selecting topics. System will automatically generate sequence of required concepts for selected topic using prerequisites chain. Before the concept is presented to student system will perform drift correction to create any errors in eye tracking data which could be caused by shifting of headband. System then presents concept to student using text and picture. During this step, system records data from eye tracking device and creates map of accessed parts. This will generate reading patter which will be used later by Effort Assessment Component. When student finished reading concepts, he/she can access assessment page where student is required to answer questions related to concept just covered. Before accessing the assessment page system will perform again drift correction but it will also record pupil dilatation data to create base ICA index. After drift correction step student is presented with question and possible answers. At this point system starts to record eye movement and pupil dilatation. When student answers question, system stops recording data and evaluate the following: correct answer, required concept covered, increase in mental workload. Evaluation of those factors will update systems believe about student knowledge state and generate feedback which is then presented to student. In case that student answered correctly and system detected increase in mental workload during assessment and student also finished reading required concept, system will understand the response as that the student have required knowledge about concept. If student's answer is correct but student did not finished reading required concept and also system did not detected any increase in mental workload, system will understand this as case where student could guest the answer and will re-assess the student by asking another question from the same concept. An example feedback to student with incorrect answer but covered required concept and low mental workload would be "Wrong, your answer is incorrect! You need to try think harder about the answer".

4   Discussion

System presented in this paper is a real working learning environment using eye tracking device to identify student effort during assessment. I need to perform user study to see how effective this system can be when interacting with students. But I believe that by getting information about student's eye response system can interact with student in more efficient way and provide more appropriate feedback. In this paper I have summarized my ongoing work in area of Intelligent Tutoring Systems and my thesis research.

References

1. Iqbal, S.T., Zheng, X.S., Bailey, B.P.: Task-evoked pupillary response to mental workload in human-computer interaction. Ext. Abstracts CHI 2004, pp. 1477–1480. ACM Press (2004)
2. Juris, M., Velden, M.: The Pupillary Response to Mental Overload. Physiological Psychology 5(4), 421–424 (1977)
3. Nkambou, R., Bourdeau, J., Mizoguchi, R.: Advances in Intelligent Tutoring Systems. Springer, Heidelberg (2010) 978-3-642-14362-5
4. Wenger, E.: Artificial Intelligence and Tutoring Systems. Computational approaches to the communication of knowledge (1987)
5. Pivec, M., Trummer, C., Pripfl, J.: Eye-Tracking Adaptable e-Learning and Content Authoring Support. Informatica 30, 83–86 (2006)
6. Calvi, C., Porta, M., Sacchi, D.: e5Learning, an E-Learning Environment Based on Eye Tracking. In: Proceedings of the 8th IEEE International Conference on Advanced Learning Technologies (ICALT 2008), Santander, Spain, July 1-5, pp. 376–380 (2008)
7. Wang, H., Chignell, M., Ishizuka, M.: Empathic Tutoring Software Agents Using Real-time Eye Tracking. In: Proc. of the Eye Tracking Research and Applications Symposium, ETRA 2006, pp. 73–78 (2006)
8. D'Mello, S., Olney, A., Williams, C., Hays, P.: Gaze tutor: A gaze-reactive intelligent tutoring system. International Journal of Human-Computer Studies 70(5), 377–398 (2012)
9. Lowenstein, O.: Experimentelle Beitrage zur Lehre von den Katatonischen Pupillenveranderungen. Monatschrift far Psychiatrie und Neurologie 47, 194–215 (1920)
10. Hass, E.H., Polt, J.M.: Pupil size in relation to mental activity during simple problem-solving. Science 143, 1190–1192 (1964)
11. Hoeks, B., Levelt, W.J.M.: Pupillary dilation as a measure of attention: A quantitative system analysis. Behavior Research Methods, Instruments and Computers 25(1), 16–26 (1993)
12. Beatty, J.: Task-Evoked Pupillary Responses, Processing Load, and the Structure of Processing Resources. Psychological Bulletin 91(2), 276–292 (1982)
13. Marshall, S.: The Index of Cognitive Activity: Measuring Cognitive Workload. In: Proceedings of IEEE Conference on Human Factors and Power Plants, pp. 7–9 (2002)
14. Bartels, M., Marshall, S.: Measuring cognitive workload across different eye tracking hardware platforms. In: Proc. of ETRA 2012, pp. 161–164 (2012)

Sparse Representation for Machine Learning

Yifeng Li

School of Computer Science, University of Windsor,
401 Sunset Avenue, Windsor, Ontario, N9B 3P4, Canada
li11112c@uwindsor.ca, yifeng.li.cn@gmail.com

Abstract. Sparse representation is a parsimonious principle that a signal can be approximated by a sparse superposition of basis functions. The main topic of my thesis research is to apply this principle in the machine learning fields including classification, feature extraction, feature selection, and optimization.

Keywords: sparse representation, machine learning, classification, feature extraction, feature selection, optimization.

1 Introduction

Sparse representation (SR) is a principle that a signal can be approximated by a sparse linear combination of dictionary atoms [2]. It can be formulated as $b = x_1 a_1 + \cdots + x_k a_k + \epsilon = Ax + \epsilon$, where $A = [a_1, \cdots, a_k]$ is called a *dictionary*, a_i is called a *dictionary atom* or *basis vector*, x is a sparse *coefficient vector*, and ϵ is an error term. Sparse representation involves sparse coding and dictionary learning. Given a new signal b and dictionary A, learning the sparse coefficient x is termed *sparse coding*. Given training data D, learning the dictionary A is called *dictionary learning*.

For understanding SR, an example of l_1-regularized SR is given in the following from a Bayesian perspective (more details can be found in [13]). Suppose each atom a_i is normally distributed with zero mean and diagonal covariance, x follows a Laplace distribution with zero mean and diagonal covariance, and the error ϵ follows a Gaussian distribution with zero mean and diagonal covariance. First, we fix the dictionary A to learn x. Its maximum *a posteriori* (MAP) estimation can be formulated as

$$\min_{x} f(x) = \frac{1}{2}\|b - Ax\|_2^2 + \lambda\|x\|_1, \tag{1}$$

where $\lambda \geq 0$ is a scalar to balance the trade-off between reconstructive error and sparsity. This model is called l_1-*least-squares* (l_1LS) sparse coding. In regularization theory, it is known as a l_1-*regularized* model. Equation (1) coincides with the well-known *LASSO* [14]. Second, The l_1-regularized dictionary learning model can be expressed as

$$\min_{A,Y} f(A,Y) = \frac{1}{2}\|D - AY\|_F^2 + \frac{\alpha}{2}\sum_{i=1}^{k}\|a_i\|_2^2 + \lambda\sum_{i=1}^{n}\|y_i\|_1, \tag{2}$$

where $\alpha \geq 0$ controls the scale of the dictionary atoms.

O. Zaïane and S. Zilles (Eds.): Canadian AI 2013, LNAI 7884, pp. 352–357, 2013.

It has been reported that SR is very robust to noise and redundancy in the data [3]. The main problem I am addressing in my doctoral research is to apply the SR principle in machine learning. Since machine learning is a wide area, I focus on feature extraction, feature selection, and classification in my dissertation. I categorize the implementations of the SR principle into two groups – i) the methods using sparse coding only, ii) and the methods using both sparse coding and dictionary learning. In the subsequent sections, the problem in each group is defined and the existing solutions are surveyed. The optimization issue is also addressed. I describe my current solutions and mention future works to be completed in my thesis. My methods have been applied in various high-throughput genomic data analysis. However, due to page limit, I omit this part in this paper. Interested readers are referred to [1, 12, 13]. Hereafter, I denote the training data by $D \in \mathbb{R}^{m \times n}$ where m and n are the numbers of features and samples, respectively. The class labels are in the column vector $c \in \{1, 2, \cdots, C\}^n$ where C is the number of classes. A set of p new samples is represented with $B \in \mathbb{R}^{m \times p}$.

2   Sparse Coding for Classification

2.1   Problem Statement

Sparse coding classification methods are based on the assumption that a new sample can be approximated by a sparse superposition of all training samples. Given the training data $\{D, c\}$, in order to predict the class label of a new sample b using sparse coding, the sparse coefficient x must be obtained first by optimizing a model, and then the class label of b is predicted by defining a decision function $g(b|x, D, c) \in \{1, 2, \cdots, C\}$.

2.2   Existing Solutions

Basis pursuit (equivalent to Equation (1)) has been applied to face recognition in [15]. First, the sparse code is learned by basis pursuit. Next, *nearest subspace* (NS) rule is used as a decision function. The NS rule is defined as $g(b) = \arg\min_{1 \leq i \leq C} r_i(b)$, where $r_i(b)$ is the regression residual corresponding to the i-th class: $r_i(b) = \|b - A\delta_i(x)\|_2^2$, where $A = D$, $\delta_i(x) : \mathbb{R}^n \to \mathbb{R}^n$ returns the coefficients for class i. Its j-th element is given by x_j, if atom a_j is in class i, otherwise 0.

In [16], a kernel extension of a l_1-model is proposed, it is equivalent to $\min_x f(x) = \frac{1}{2}\|b' - A'x\|_2^2 + \lambda\|x\|_1$, where $b' = (\phi(A))^\mathsf{T}\phi(b)$ and $A' = (\phi(A))^\mathsf{T}\phi(A)$. $\phi(\cdot)$ is a function that maps a sample from input space into high-dimensional feature space. The essence of their idea is to first map all samples in high-dimensional feature space, and then project them onto n-dimensional space by the transformation matrix $\phi(A)$. In the n-dimensional space, basis pursuit is applied.

2.3   My Contributions

Instead of using the l_1-regularized model, I propose the following non-negative sparse coding for classification [11]:

$$\min_x f(x) = \frac{1}{2}\|b - Ax\|_2^2, \quad x \geq 0. \tag{3}$$

This is inspired by *non-negative matrix factorization* (NMF). In usual circumstance, the optimal solution to Equation (3) is very sparse. The relation between non-negativity and sparsity can be explained by either the active-set theory in optimization, or a Bernoulli prior in Bayesian inference [13]. Combining the l_1-norm and non-negativity I obtain the l_1-non-negative sparse coding model [9, 13]:

$$\min_{x} f(x) = \frac{1}{2}\|b - Ax\|_2^2 + \lambda^{\mathrm{T}}x, \quad x \geq 0, \tag{4}$$

where $\lambda = \{\lambda\}^n$. In sparse coding, I name the training samples corresponding to nonzero coefficients the *support atoms*. The rational of using non-negative sparse coding is that a unknown sample resides in the conical region of the active atoms. The minimum cone of a unknown sample may be well explained by its vertices (that is the active atoms). The classification methods of using the above two models are called *non-negative least squares* (NNLS) and l_1NNLS approaches. I propose the *k-nearest neighbor* (*k*-NN) based decision rule, in [13], which can take less time than the NS rule, but obtain similar accuracy. I have demonstrated that NNLS requires very few training samples in order to obtain significant accuracy. Through strict statistical comparison, it has also been shown that NNLS has a performance comparable to that of SVM.

I have extended the l_1LS, NNLS, and l_1NNLS models to kernel versions by applying the dimension-free property in sparse coding. My rational is in the following. Since least squares optimization is a specific *quadratic programming* (QP) problem, we can reformulate Equation (1) to a l_1-*regularized QP* (l_1QP) problem:

$$\min_{x} f(x) = \frac{1}{2}x^{\mathrm{T}}Hx + g^{\mathrm{T}}x + \lambda\|x\|_1, \tag{5}$$

where $H = (\phi(A))^{\mathrm{T}}\phi(A)$, and $g = -(\phi(A))^{\mathrm{T}}\phi(b)$. Similarly, the non-negative and l_1-non-negative models can be reformulated to the following non-negative QP problem:

$$\min_{x} \frac{1}{2}x^{\mathrm{T}}Hx + g^{\mathrm{T}}x, \quad \text{s.t. } x \geq 0, \tag{6}$$

where $g = -(\phi(A))^{\mathrm{T}}\phi(b)$ for NNLS, and $g = \lambda - (\phi(A))^{\mathrm{T}}\phi(b)$ for l_1NNLS. Thus the optimization of sparse coding models is *dimension-free*. Via replacing inner products with kernel matrices, we can easily obtain the kernel sparse coding. It has been reported that my kernel sparse coding based classifier can obtain good performance [9, 13].

2.4   Future Works

First, the learning bound of sparse coding approaches will be studied under the statistical learning theory. Qualitatively speaking, the first term in Equations (1), (3), and (4) aims to minimize the empirical error, while the sparsity-inducing term is to reduce the *Vapnik-Chervonenkis dimension*. Second, the choice of an appropriate kernel is crucial in order to obtain good classification performance. Thus my future work in this direction will be focused on kernel learning for space coding approaches.

3 Dictionary Learning for Feature Extraction

3.1 Problem Statement

The sparse coding based approach is an instance-based learning. For each new sample, a large QP needs to be solved, it is hence inefficient for large-scale data. We thus need to learn a dictionary to capture the main latent patterns. For classification, dictionary learning is a scheme of dimension reduction. The classification involves three phases. First, a dictionary A is learned from training data D and possibly c, that is solving the matrix decomposition $D \approx AY$. Columns of Y are the images of training samples in the feature space. Second, a classifier g is trained over Y and c. Third, the images (denoted by X) of the new samples in the feature space are obtained as well by solving the sparse coding $B \approx AX$, and their class labels are predicted by the classifier $g(X)$.

3.2 Existing Solutions

We can view NMF as a model of unsupervised dictionary learning. It has been used for clustering and feature extraction before my study. For instance, it has been applied to reduce the dimensionality of gene expression data [7]. New samples are usually projected into the feature space by applying pseudo-inverse $X = A^{\dagger}B$. The drawback of this is that the non-negative constraint of X is violated. A kernel solution to a l_1-model was proposed in [4]. It is inefficient because the sparse code of each sample is updated separately and the dictionary atoms are not well-represented in the feature space.

3.3 My Contributions

I present a fast generic unsupervised dictionary learning framework in [13] and [8]. It solves the following two generic models:

$$\min_{A,Y} \frac{1}{2}\|D - AY\|_F^2 + \lambda\|Y\|_1 \quad \text{s.t. } \|a_i\|_2 = 1; \text{ if } t = \text{true}, Y \geq 0, \quad (7)$$

$$\min_{A,Y} \frac{1}{2}\|D - AY\|_F^2 + \frac{\alpha}{2}\sum_{i=1}^{k}\|a_i\|_2^2 + \lambda\sum_{i=1}^{n}\|y_i\|_1 \quad \text{s.t. if } t = \text{true}, Y \geq 0, \quad (8)$$

where t indicates if non-negative constraint should apply on Y. The advantages of this framework are that i) A can be updated analytically; ii) columns of Y are updated in a parallel fashion; and iii) inner products among training data and dictionary are only required in optimization rather than the original data. The inner product $A_\phi^{\mathsf{T}}A_\phi$, rather than the intractable A_ϕ, is iteratively updated for nonlinear kernel. I also propose a supervised dictionary learning method in [10], where I reveal that the sparse coding of a new sample must be consistent with the dictionary learning model in training phase.

3.4 Future Works

First, unlike PCA and ICA, SR can learn non-orthogonal and redundant basis vectors. Independent basis vectors are selected during the sparse coding of a signal. Hence it is interesting to investigate how accuracy changes with the number of basis vectors. Second, I plan to enforce sparsity on dictionary as well which is useful for variable selection. Third, spurred by *Bayesian factor regression modeling*, I plan to design a supervised dictionary learning model that combines dictionary learning and Bayesian regression. Finally, kernel supervised dictionary learning models will also be addressed.

4   Optimization for Sparse Representation

4.1   Problem Statement

Fast sparse coding algorithm is crucial in sparse coding and dictionary learning. Unfortunately, as in Equations (5) and (6), sparse coding is a large-scale QP problem. Moreover, the l_1-regularized models are non-smooth. Therefore, solving this QP problem efficiently for huge amount of data is an important topic in sparse representation.

4.2   Existing Solutions

There are two typical sparse coding algorithms for the l_1-regularized model. One is the interior-point method [6], and another the is proximal method [5]. The former approximates the non-smooth l_1-norm by a smooth function. The later is a first-order approach. It has been shown that first-order methods are efficient for non-smooth problems.

4.3   My Contributions

I proposed to use active-set algorithms for various sparse coding models in [8, 13]. I applied the following three properties. First, the optimization is dimension-free, therefore the input of my algorithms are inner products. Second, the active-set method is usually quite efficient for small and medium-sized problems. It thus makes dictionary learning very fast. Third, there are many common but expensive computations among the sparse coding of different signals using active-set method. My algorithms hence allow the sparse coding of multiple signals to share common computations in a parallel fashion.

4.4   Future Works

Inspired by the optimization of SVM, I am working on a *decomposition method* for large-scale sparse coding. The basic idea is in fact an implementation of the block-coordinate-descent scheme. In each iteration, a few coefficients violating the *Karush-Kuhn-Tucker* (KKT) conditions are selected in the working set, and the rest are fixed. Only the coefficients in the working set are updated by a fast QP solver. This procedure iterates until no coefficient violates the KKT conditions. *Sequential minimal optimization* (SMO) is the extreme case of the decomposition method for SVM. I am devising SMO for large-scale sparse coding.

5   Conclusions

The main topic of my thesis dissertation is to devising learning methods which apply the principle of sparse representation. The problems or challenges, and current solutions are presented in this paper. The future works mentioned above will be finalized and included in my dissertation. Meanwhile, I am developing two open-source toolboxes [1,12] including the implementations of low level optimizations and high level machine learning applications. The purpose is to serve the machine learning community and receive constructive suggestions for my study.

Acknowledgments. I greatly thank my advisor, Dr. Alioune Ngom, and all professors for their helps in my study including Dr. Luis Rueda, Dr. B. John Oommen, Dr. Richard Caron, and Dr. Michael Ochs. My research is supported by IEEE CIS Summer Research Grant 2010, OGS Scholarship 2011-2013, NSERC Grants #RGPIN228117-2006 and #RGPIN228117-2011, CFI Grant #9263, and scholarships from University of Windsor.

References

1. The sparse representation toolbox in matlab,
   http://cs.uwindsor.ca/~li11112c/sr
2. Bruckstein, A.M., Donoho, D.L., Elad, M.: From sparse solutions of systems of equations to sparse modeling of signals and images. SIAM Review 51(1), 34–81 (2009)
3. Elad, M.: Sparse and Redundant Representations. Springer, New York (2010)
4. Gao, S., Tsang, I.W.-H., Chia, L.-T.: Kernel sparse representation for image classification and face recognition. In: Daniilidis, K., Maragos, P., Paragios, N. (eds.) ECCV 2010, Part IV. LNCS, vol. 6314, pp. 1–14. Springer, Heidelberg (2010)
5. Jenatton, R., Mairal, J., Obozinski, G., Bach, F.: Proximal methods for hierarchical sparse coding. JMLR 12(2011), 2297–2334 (2011)
6. Kim, S.J., Koh, K., Lustig, M., Boyd, S., Gorinevsky, D.: An interior-point method for large-scale $l1$-regularized least squares. J-STSP 1(4), 606–617 (2007)
7. Li, Y., Ngom, A.: Non-negative matrix and tensor factorization based classification of clinical microarray gene expression data. In: BIBM, pp. 438–443. IEEE Press, Piscataway (2010)
8. Li, Y., Ngom, A.: Fast kernel sparse representation approaches for classification. In: ICDM, pp. 966–971. IEEE Press, Piscataway (2012)
9. Li, Y., Ngom, A.: Fast sparse representation approaches for the classification of high-dimensional biological data. In: BIBM, pp. 306–311. IEEE Press, Piscataway (2012)
10. Li, Y., Ngom, A.: Supervised dictionary learning via non-negative matrix factorization for classification. In: ICMLA, pp. 439–443. IEEE Press, Piscataway (2012)
11. Li, Y., Ngom, A.: Classification approach based on non-negative least squares. Neurocomputing (in press, 2013)
12. Li, Y., Ngom, A.: The non-negative matrix factorization toolbox for biological data mining. BMC Source Code for Biology and Medicine (2013),
    http://cs.uwindsor.ca/~li11112c/nmf (under revision)
13. Li, Y., Ngom, A.: Sparse representation approaches for the classification of high-dimensional biological data. BMC Systems Biology (in press, 2013)
14. Tibshirani, R.: Regression shrinkage and selection via the lasso. Journal of the Royal Statistical Society. Series B (Methodological) 58(1), 267–288 (1996)
15. Wright, J., Yang, A., Ganesh, A., Sastry, S.S., Ma, Y.: Robust face recognition via sparse representation. TPAMI 31(2), 210–227 (2009)
16. Yin, J., Liu, X., Jin, Z., Yang, W.: Kernel sparse representation based classification. Neurocmputing 77, 120–128 (2012)

Extracting Information-Rich Part of Texts Using Text Denoising

Rushdi Shams

Department of Computer Science
University of Western Ontario
London, ON N6A 5B7,Canada
rshams@uwo.ca

Abstract. The aim of this paper is to report on a novel text reduction technique, called *Text Denoising*, that highlights information-rich
content when processing a large volume of text data, especially from the
biomedical domain. The core feature of the technique, the text readability
index, embodies the hypothesis that complex text is more information-
rich than the rest. When applied on tasks like biomedical relation bearing
text extraction, keyphrase indexing and extracting sentences describing
protein interactions, it is evident that the reduced set of text produced
by text denoising is more information-rich than the rest.

1 Introduction

Often, to test a method's scalability as well as its performance across genres of
texts, there is a need to process large volumes of text data in many disciplines
of NLP, be it textual relation extraction, summarization or meta-tagging. It
has been reported by many researchers [11][17] that machine learning as well as
rule-based approaches show improvements over their benchmarks with increased
training data. However, the use of large volume of data can create several bottlenecks. One is technical—processing large data, like that from biomedical texts,
slows down many algorithms; another is even more important—algorithms can
exhibit a decreased accuracy because of the noise, which are irrelevant or redundant data for a given classification task, added by information-poor parts of
texts.

There are several statistics, like word-level feature *tf–idf* and sentence-level
feature *sentence position*, that help identify information richness. Although the
degree of use shows their popularity, these features have some serious limitations.
For example, *tf–idf* computes document similarity directly in the word-count
space which may be slow for large vocabularies and sentence position is useful
for summarization but is superficial in relation extraction. In other words, they
are either task-specific and/or domain-specific measures.

Text readability has multivariate features that consider many attributes like
length of paragraph, words and sentences, and number of polysyllabic and mono-
syllabic words. In this paper, I report a text reduction technique called *Text
Denoising* that reduces text data based on text readability, especially from the

O. Zaïane and S. Zilles (Eds.): Canadian AI 2013, LNAI 7884, pp. 358–363, 2013.

biomedical domain, to that which is more information-rich by removing most of the noise. The reduced text is also expected to be task-independent and informative enough to improve accuracy of NLP tools across disciplines.

2   Proposed Method

Among text readability scores, the following five measures are considered as yardsticks— Fog Index (hereinafter, FI) [7], Flesch reading ease score (FRES) [5], Smog Index [9], Forcast Index [1], and Flesch-Kincaid readability index (FKRI) [8]. The choice of using text readability as an *information richness statistic* is motivated by the results of an experiment by Duff and Kabance [3]. In their experiment, a passage with no more than two phrases were converted into primer prose and FI was applied to test its readability. They found that the score was low (i.e., the prose was extremely easy to read). The authors concluded that easy texts obscure the relationships and ideas as they de-emphasize both. In contrast, difficult texts emphasize relationships and ideas yielding low readability. I suggest that the describing of biomedical relations, meta information, etc. lengthens sentences as well as increases the use of polysyllabic words which are the two principal components of many of the readability indexes.

Both rule-based and machine learning-based versions of *Text Denoising* are based on this principle that use text readability as a key feature and applies it at the sentence-level to identify those sentences within a text, called denoised text, where content information, such as biomedical relations, is more likely to occur. The rest of the text is called noise text. I am interested to observe the effect of using text denoising on different tasks and genres of text.

3   Text Denoising on Relation Extraction

I developed a corpus of 24 texts that describe four pairs of related MeSH C and MeSH D concepts reported by Perez *et al.* [12]. I applied the rule-based version of text denoising on these texts to extract related biomedical concepts. The only rule I set for this task was to extract 30% of the low-readability sentences from the texts according to their FI score. This threshold is termed as the *denoising threshold* and the texts extracted are called *denoised texts*; the rest is called *noise text*. This threshold point was set heurisiticaly considering the stability in the frequency of appearance of the related concepts in the corpus. Other than 30%, the results with different denoising thresholds ranging from 10% to 50%, however, was not satisfactory. I ranked the pairs of concepts present in the denoised texts using their frequency. Most of the concept pairs with higher ranks, however, did not contain any semantic relations according to UMLS semantic relation network. Therefore, I re-ranked the pairs according to their positive predictive value (PPV) (similar to *precision* measure used in information retrieval evaluation tasks) and sensitivity. The pairs of concepts found from this re-ranking showed a convincing accuracy of 75% (ratio of semantically related concepts to total) against the output of the UMLS semantic relation network. Table 1 shows an output from a paper on one

Table 1. Extracted related concepts for a paper on Ischemia and Glutamate

Rank	Related Concepts	Semantic Relation
1	Ischemia-Glutamate	Yes
2	Levels-Ischemia	No
3	Levels-Glutamate	Yes
4	Glutamate-Neurons	Yes
5	10min-Ischemia	Yes
6	Glutamate-CA4	Yes
7	Increase-Glutamate	Yes
8	10min-Glutamate	No
9	Ischemia-5min	Yes
9	Glutamate-5min	No

of the four pairs of concepts. Of note, I found that the noise texts did not have any related biomedical concepts. The detailed experimental setup and results are reported by Shams and Mercer [13].

Later, I performed an experiment with four other readability scores mentioned in Section 2 on the same corpus. A comparative result showed that FI outperformed the other indexes by extracting more meaningful relations [14]. I also analyzed the performance of the indexes considering the performance of FI as a benchmark. Table 2a and 2b show that the SMOG index is a close second to FI followed by FKRI, while FRES and FORCAST performed poorly. It can also be noted that the SMOG index, like FI, uses the core measure of *complex words* which reveals the fact that the measure of complex word fits best for text denoising and biomedical relation extraction.

4   Text Denoising on Keyphrase Extraction

I investigated the usability of denoised texts as training data for machine learning-based keyphrase indexers called KEA [17], KEA++ [11] and Maui [10]. I applied the indexers with their classifiers induced from denoised training data on three datasets, namely FAO-780, CERN-290 and NLM-500. These datasets are composed of texts from the domains of agriculture, physics and biomedical science. I

Table 2. (a) Micro-average and (b) macro-average precision, recall and F-Score of the indexes on biomedical relation extraction

(a)				(b)			
Score	Precision	Recall	F-Score	Score	Precision	Recall	F-Score
SMOG	95.83	82.14	88.46	SMOG	96.88	82.60	89.16
FKRI	88.89	82.76	85.71	FKRI	89.73	82.60	86.01
FRES	82.61	65.52	73.08	FRES	80.83	65.63	72.44
FORCAST	81.82	62.07	70.59	FORCAST	77.88	61.61	68.72

Table 3. F-Scores of the keyphrase indexers with text denoising and its benchmark on three datasets

(a) Performance of KEA

Classifier	FAO-780		CERN-290		NLM-500	
	F-Score	t-value	F-Score	t-value	F-Score	t-value
with Text Denoising	23.03	5.07	14.73	3.42	14.60	4.14
Benchmark	20.76		12.29		12.21	

(b) Performance of KEA++

Classifier	FAO-780		CERN-290		NLM-500	
	F-Score	t-value	F-Score	t-value	F-Score	t-value
with Text Denoising	27.98	3.78	23.28	2.40	20.15	6.38
Benchmark	25.19		21.04		17.91	

(c) Performance of Maui

Classifier	FAO-780		CERN-290		NLM-500	
	F-Score	t-value	F-Score	t-value	F-Score	t-value
with Text Denoising	31.87	2.76	24.42	2.26	31.50	3.52
Benchmark	31.86		24.92		31.13	

compared the result with their benchmark performances that were achieved by using the full-text training data. Convincingly, in a 10-fold cross validation experiment, both KEA and KEA++, with their classifiers induced from denoised training data, outperformed their respective benchmark F-scores [15]. Maui, on the other hand, had mixed results and its denoised text induced classifier performs comparably with its benchmark [16]. The F-Scores are listed in Table 3a, 3b and 3c where a *t-value* greater than or equal to 2.26 indicates the statistical significance of the results at 95% confidence. Of note, unlike the fixed denoising threshold of 30% for relation extraction, I found that to get bias-free classifiers for the indexers, the denoising threshold point needed to be varied (usually between 30%–70%) for different genres of texts. This outcome confirms that the rule to decide the amount of text to be extracted from texts substantially depends for different writing styles.

5   Text Denoising on Extracting Protein Relations Bearing Sentences

In an attempt to eliminate the denoising threshold which depends on writing style (Section 4), I decided to develop a machine learning version of text denoising. The classification task in hand was to annotate sentences of a set of texts with either *positive* or *negative* labels based on the presence of protein interactions. The feature set chosen is composed of 35 features like various parameters of readability indexes, term frequency, inverse sentence frequency, biomedical

Table 4. Performance of text denoising on extracting protein relations bearing sentences against the gold standard

Dataset	Precision	Recall	F-Score
BioNLP	82.5	87.8	85.1
BioDRB	84.7	91.1	87.8
FetchProt	90.8	89.2	90

named entity, verbs and acronyms, stopwords, semantic words, and sentence positions. After applying a series of well known classifiers like Bayesian classifiers, Random Forest, SVM, AdaBoost, and Bagging, the classifier that performed best was chosen which is Bagging stacked with Random Forest. The corpora used for this experiment are BioNLP, BioDRB and FetchProt that contain over 85, 000 sentences. Two automated tools called RelEx [6] and WRelEx [4] are used to assign binary labels to each sentence of these corpora depending on the presence of any protein relations. Having realized after this assignment that the classes are negatively skewed (almost doubled the positive labels), synthetic positive samples are produced using SMOTE [2] where the minority class is over-sampled by taking each minority class sample and introducing synthetic examples along the line segments joining any/all of the k, which is five in our setup, minority class nearest neighbors. From initial results, I found that many features were highly correlated with each other but had low correlation with the class. Therefore, I used a *wrapper* method to select a set of bias-free features. However, I observed that this set of features varies for different corpora. Table 4 shows the precision, recall and F-Score of text denoising in a 10-fold cross validation setup, considering the highly agreed upon annotation of RelEx and WRelEx as the gold standard. It can be noted that the outcome of this experiment without using SMOTE was not satisfactory as the F-scores were under 80%.

6   Conclusion and Future Work

The proposed text denoising method performed much the same on several tasks and kinds of texts: the reduction of texts according to the readability improved relation mining, keyphrase indexing and extracting sentences that describe protein relations. This result strongly suggests that sentences that are difficult to read are more information-rich than the rest. The effect of text denoising is yet to be examined for text categorization and summarization. I am currently investigating the effect of readability on e-mail spam detection. The results so far are interesting as I am labelling spam and ham based on the readability of e-mail text content only (i.e., without looking at the mail header). Also, I intend to train benchmark summarizers with denoised texts and see how they perform against gold standard summaries.

References

1. Caylor, J.S., Stitch, T.G., Fox, L.C., Ford, J.P.: Methodologies for determining reading requirements of military occupational specialities. Technical Report 73-5, Human Resources Research Organization, Alexandria, VA (1973)
2. Chawla, N.V., Bowyer, K.W., Hall, L.O., Kegelmeyer, W.P.: Smote: Synthetic minority over-sampling technique. Journal of Artificial Intelligence Research 16, 321–357 (2002)
3. Duffy, T.M., Kabance, P.: Testing a readable writing approach to text revision. Journal of Educational Psychology 74, 733–748 (1982)
4. Faiz, S.I.: Discovering higher order relations from biomedical text. Master's thesis, Department of Computer Science, The University of Western Ontario, Canada (2012)
5. Flesch, R.: A new readability yardstick. Journal of Applied Psychology 32, 221–233 (1948)
6. Fundel, K., Küffner, R., Zimmer, R.: Relex - relation extraction using dependency parse trees. BMC Bioinformatics 23(3), 365–371 (2007)
7. Gunning, R.: Fog index after twenty years. Journal of Business Communication 6(3), 3–13 (1969)
8. Kincaid, J.P., Fishburne, R.P., Rogers, R.L., Chissom, B.S.: Derivation of new readability formulas (automated readability index, fog count, and flesch reading ease formula) for navy enlisted personnel. Research Branch Report 8-75, Chief of Naval Technical Writing: Naval Air Station Memphis (1975)
9. McLaughlin, G.H.: Smog grading – a new redability formula. Journal of Reading 12(8), 639–46 (1969)
10. Medelyan, O.: Human-competitive automatic topic indexing. PhD thesis, University of Waikato, New Zealand (2009)
11. Medelyan, O., Witten, I.: Domain-independent automatic keyphrase indexing with small training sets. Journal of the American Society for Information Science and Technology (JASIST) 59(7), 1026–1040 (2008)
12. Perez-Iratxeta, C., Bork, P., Andrade, M.: Literature and genome data mining for prioritizing disease-associated genes. In: Eisenhaber, F. (ed.) Discovering Biomolecular Mechanisms with Computational Biology. Molecular Biology Intelligence Unit, pp. 74–81. Springer (2006)
13. Shams, R., Mercer, R.E.: Extracting connected concepts from biomedical texts using fog index. Elsevier Procedia - Social and Behavioral Sciences 27, 70–76 (2011)
14. Shams, R., Mercer, R.E.: Evaluating core measures of text denoising for biomedical relation mining. In: 3rd International Workshop on Global Collaboration of Information Schools (WIS 2012), Taipei, Taiwan (2012)
15. Shams, R., Mercer, R.E.: Improving supervised keyphrase indexer classification of keyphrases with text denoising. In: Chen, H.-H., Chowdhury, G. (eds.) ICADL 2012. LNCS, vol. 7634, pp. 77–86. Springer, Heidelberg (2012)
16. Shams, R., Mercer, R.E.: Investigating keyphrase indexing with text denoising. In: 12th ACM/IEEE Joint Conference on Digital Libraries (JCDL 2012), pp. 263–266. ACM (2012)
17. Witten, I., Paynter, G., Frank, E., Gutwin, C., Nevill-Manning, C.: KEA: Practical automatic keyphrase extraction. In: Proceedings of the 4th ACM Conference on Digital Libraries, Berkeley, CA, USA, pp. 254–255 (1999)

A Novel Content Based Methodology
for a Large Scale Multimodal Biometric System

Madeena Sultana

Dept. of Computer Science, University of Calgary, Calgary, AB, Canada
msdeena@ucalgary.ca

Abstract. Recently, Content Based Image Retrieval (CBIR) system has drawn enormous attention of researchers because of its efficiency in recognizing images from large databases as well as growing demand from real world applications. According to many, biometrics recognition is one of the most potential applications of CBIR. However, no research work has been published up to date on content based multimodal biometric systems. In this proposal, a content based multimodal biometric system, where color, texture, and shape features are combined to enhance the recognition accuracy of the system, is proposed. The preliminary result of the proposed content based feature fusion method for face recognition demonstrates its potential to boost up the recognition performance of a large scale multimodal biometric system.

Keywords: Content based image retrieval system, multimodal biometrics, face recognition, visual image features.

1 Problem Statement and Motivation

Because of the increasing demand of browsing and recognising digital images from large databases in different applications, Content Based Image Retrieval (CBIR) [9], where images are retrieved based on their visual contents, has been considered as a hot topic of research in last few years [10, 11]. Generally, CBIR exploits low level image features such as color, texture, or shape through low dimensions and fast feature extraction methods. Consequently, CBIR systems require less memory and computation time than traditional methods. One of the most potential applications of CBIR, according to many researchers [4, 15], is a large scale biometric system. Nevertheless, very few research works have been published on the content based biometric systems, despite growing demands in both research and government spheres. Biometric security is of utmost importance to Canada, since as early as this year, many Canadians will be issued the new ePassport (Biometric Passport) [5]. Jason Kenney, Citizenship and Immigration Minister of Canada, said that the safety and security of Canadians can be protected through the use of biometrics and it will "strengthen and modernize" Canada's immigration system as well [6]. Recent reports on security of traditional passwords also point out how easy it is nowadays to break majority of "strong" passwords, and inform of a new generation of biometric pass-

O. Zaïane and S. Zilles (Eds.): AI 2013, LNAI 7884, pp. 364–369, 2013.

words based on multiple biometric sources [8]. While scientific research revealed that multimodal biometric enhances the recognition accuracy significantly over single biometric and provides reliable means of authentication [7, 8], there has been no research up to date on content based multimodal biometric system. This has motivated me to develop the proposed content based feature combination method for large scale multimodal biometric system. Despite the recognition enhancement, multimodal biometric system faces a number of challenges. Multiple biometric traits increase the number of high dimensional features that consequently raise the memory requirements as well as computation time. Situation becomes even worse while dealing with large databases containing bulk amount of high resolution images. I propose to address these problems through a novel combination of dimensionality-reduction and weighted fusion methods. In summary, the main objective of my research is to enhance the performance of a large scale multimodal biometric system through the fusion of multiple low level features of different biometric traits.

2   Proposed Plan of Research

I propose a novel content based image retrieval approach based on the weighted features of color, texture, and shape to be used as the backbone of my multi-modal recognition system. The three well-known, simple, and fast feature extraction algorithms: color histogram [12], Gabor filter [13], and pseudo Zernike moments [14] will be used to enhance the accuracy and efficiency of the proposed retrieval system. Color histogram method will extract the color features, Gabor filter will be used next to extract the texture, and pseudo Zernike moments [14] will be applied to extract the shape attribute of an image. The weighted color, texture, and shape features will be combined as a single descriptor to reduce feature dimensionality and achieve reduction in required storage. Finally, classification will be accomplished by the Support Vector Machine (SVM). The robustness and accuracy of an image recognition system using weighted combination of color and shape features from large databases has been demonstrated in my previous work [2, 3]. To adopt the specific image retrieval system [2] for biometric recognition, texture feature is included to boost the recognition accuracy. Moreover, color histogram, texture, and pseudo Zernike moments are simple and fast features to extract from large databases, rather than features obtained by traditional Eigen vector methods. Thus, the novelty of this research lies in the hybridization of the three content based low level features for multimodal biometrics. Applying the multimodal technique in feature level, the proposed content based methodology can enhance the interclass variability and thus improve the recognition and performance of the system. Furthermore, the main drawbacks of traditional multimodal systems working on large datasets, such as memory requirements and computational complexity, are overcome by the choice of features and the weighted feature fusion technique. In addition, the fast GPU implementations of these methods are already tested in a different context of image retrieval [3]. Therefore, my proposed method can be implemented both in CPU and GPU to enable fast real-time recognition on massive data sets.

3   Progress to Date

Initially, I applied my weighted fusion technique for face recognition, to verify the recognition performance of combined low level color, texture, and shape features. In this case color feature is extracted by color histogram, texture is computed by Gabor filter, and affine moment invariants are considered as shape attributes. Experimental results and discussions are presented in the following two subsections [1]:

3.1   Experimental Results

Recognition performance is evaluated by creating following three databases from standard AT&T [16] and AR [17] datasets. In AR dataset, each image can be denoted as I_{ij} where subscript i indicates the subject and j indicates different pose of that subject.

1. **Grayscale database:** Among the 40 subjects of AT&T dataset, 20 subjects were chosen randomly for this database. Therefore, it contains 200 images in total with 10 images per person with variation of pose, expression, rotation, and time.
2. **Color database:** From the cropped images of AR dataset, I have chosen 20 subjects (i=1, 2, 3, ..., 20) randomly from both sessions having varying pose and expressions. Therefore, this database contains total 160 images, composed by a subset D_{color}= $\{I_{i1}, I_{i2}, I_{i3}, I_{i4}, I_{i14}, I_{i15}, I_{i16}, I_{i17}\}$ of AR dataset having different pose and expressions e.g. natural expression $\{I_{i1}, I_{i14}\}$, smile $\{I_{i2}, I_{i15}\}$, anger $\{I_{i3}, I_{i16}\}$, scream $\{I_{i4}, I_{i17}\}$ in two different sessions.
3. **Critical query database:** This database was created to evaluate my system in critical conditions such as persons wearing sunglasses under different illumination and time $\{I_{i8}\text{-}I_{i10}, I_{i21}\text{-}I_{i23}\}$. For this purpose, I have created a query database containing total 120 images, $Q_{critical}$= $\{I_{i8}, I_{i9}, I_{i10}, I_{i21}, I_{i22}, I_{i23}\}$ from AR dataset where i=1, 2, 3, ..., 20. The query images from this database are matched to my color database (D_{color}).

Since, I applied content based features and feature extraction techniques for face recognition, primarily I computed the average precision and recall, the standard quantitative measures for verifying the effectiveness of any CBIR system. The precision and recall are defined as follows [19]:

$$Precision = \frac{I_N}{R} = \frac{\text{True Positives}}{\text{True Positive+False Positives}} \tag{1}$$

$$Recall = \frac{I_N}{T} = \frac{\text{True Positives}}{\text{True Positive+False Negatives}}, \tag{2}$$

where I_N is the number of images retrieved that are most relevant to query, T is the total number of similar images in the database, and R is the total number of retrieved images. Different researchers use different values for R and T. I considered T=10 for grayscale database and T=8 for color database, since former database contains 10 different images per person and latter database contains 8 images per person, respectively. The 5 topmost images retrieved from database are considered as the search result of any query image, so $R=5$. F_β score is also computed to evaluate the efficiency of the proposed retrieval system [20]:

$$F_\beta = (1 + \beta^2) \times \frac{Precision \times Recall}{(\beta^2 \times Precision) + Recall} \tag{3}$$

Since the number of retrieved images is a fixed value ($R=5$), precision result better reflects the performance of the proposed method. Therefore, to emphasize precision value, I computed F_β, where $\beta=0.5$. Table 1, Table 2, and Table 3 summarize the recognition performance of different methods in different databases.

Table 1. Performance comparison of different methods for grayscale database

Method	Avg. Precision	Avg. Recall	$F_{0.5}$ Score
Color histogram method [12]	0.95	0.48	0.79
Affine moment invariant method [18]	0.49	0.24	0.41
Gabor filter method [13]	0.94	0.47	0.78
Proposed method	**0.98**	**0.49**	**0.82**

Table 2. Performance comparison of different methods for color database

Method	Avg. Precision	Avg. Recall	$F_{0.5}$ Score
Color histogram method [12]	0.95	0.59	0.85
Affine moment invariant method [18]	0.72	0.45	0.64
Gabor filter method [13]	0.97	0.60	0.86
Proposed method	**0.99**	**0.62**	**0.88**

Table 3. Performance comparison of different methods for critical query database

Method	Avg. Precision	Avg. Recall	$F_{0.5}$ Score
Color histogram method [12]	0.65	0.41	0.58
Affine moment invariant method [18]	0.48	0.30	0.43
Gabor filter method [13]	0.93	0.58	0.83
Proposed method	**0.94**	**0.59**	**0.84**

3.2   Analysis of Experimental Outcomes

The high precision, recall, and F-score of the proposed method confirm that combined low level content based features can efficiently recognize face images from standard face databases. In addition, the proposed method can recognize person from both grayscale and color images having different pose, expression, sessions, alignment, facial details, illumination variation, and some occlusions (e.g. sunglasses). Moreover, comparative analyses in Table 1, Table 2, and Table 3 demonstrate that the proposed fusion of features enhances the recognition performance over single feature based methods. Another significant advantage of the proposed method is its tuneable weight parameters. In critical query database, the performance of color histogram significantly degraded because of illumination change and occlusion. However, this does not deteriorate the performance of the proposed method since higher weight is assigned to the texture feature for this database.

4   Future Work

Although the overall performance of our content based face recognition method was satisfactory, the recognition rate of affine moment invariants could be better. Therefore, instead of affine moment invariants, I will apply pseudo Zernike moments [14] to extract the shape feature from multiple biometric traits. Next, the recognition accuracy of the proposed method will be verified in terms of False Acceptance Rate (FAR) and False Rejection Rate (FRR). A SVM based classification technique is needed to be implemented to identify person from multiple biometric traits. In addition, performance of the proposed method will be compared to other traditional and advanced methods. Finally, a weight learning system will be developed to adjust the weights automatically and then the performance of the proposed method will be evaluated using large scale database. A GPU based parallel matching technique will be included to achieve additional computational efficiency as well.

5   Contributions

The contributions of the proposed research work can be summarized as follows: 1) Enhancing accuracy and robustness of the system by using multiple biometric traits. 2) Reducing memory and computation complexity of large-scale multimodal biometric system by applying novel consolidation of low level features. 3) Employing weighted combination of content based image features for the first time for multimodal biometric systems. 4) Enabling real-time computation by parallel processing in CPU and GPU. 5) Developing complete software system that has a very high potential for commercialization.

Acknowledgement. The author is thankful to her supervisor Dr. Marina Gavrilova and the anonymous reviewers for their helpful comments.

References

1. Sultana, M., Gavrilova, M.: A Content Based Feature Combination Method for Face Recognition. In: 8th International Conf. on Computer Recognition Systems (CORES), Poland, May 27-29 (in press, 2013)
2. Sultana, M., Uddin, M.S.: Trademark Recognition using a Weighted Combination of Different Image Features. Int. J. of Computer Theory and Engineering (IJCTE) 4(6), 1035–1038 (2012)

3. Sultana, M., Mamun, N.M., Uddin, M.S., Ali, M.: A GPU Based Efficient Trademark Retrieval Technique using a Weighted Combination of Multiple Image Features. In: IEEE Conference on Communication, Science & Information Engg. (CCSIE), London, UK, pp. 83–88 (2011)
4. Choraś, R.S.: Image Feature Extraction Techniques and Their Applications for CBIR and Biometric Systems. Int. J. of Biology and Biomedical Engineering 1(1), 6–16 (2007)
5. Passport Canada,
   http://www.ppt.gc.ca/support/faq.aspx?lang=eng&id=q810
   (last accessed on March 05, 2013)
6. CICS News: Nationals of 29 Countries to Require Biometrics to Enter Canada (December 10, 2012), http://www.cicsnews.com/?p=2570 (last accessed on March 5, 2012)
7. Monwar, M.M., Gavrilova, M.L.: Multimodal Biometric System Using Rank Level Fusion Approach. IEEE Trans. on System, Man and Cybernetics—PART B 39(4), 867–878 (2009)
8. Yampolskiy, R., Gavrilova, M.: Artimetrics: Biometrics for Artificial Entities. IEEE Robotics and Automation, Magazine 19(4), 48–58 (2012)
9. Kato, T.: Database Architecture for Content Based Image Retrieval. In: Image Storage and Retrieval Systems, pp. 112–123 (1992)
10. Eitza, M., Hildebranda, K., Boubekeurb, T., Alexaa, M.: An Evaluation of Descriptors for Large-scale Image Retrieval from Sketched Feature Lines. Computers & Graphics 34(5), 482–498 (2010)
11. Arampatzis, A., Zagoris, K., Chatzichristofis, S.A.: Dynamic Two-stage Image Retrieval from Large Multimedia Databases. Information Processing & Management 49(1), 274–285 (2013)
12. Swain, M.J., Ballard, D.H.: Color Indexing. Int. J. of Computer Vision 7, 11–32 (1991)
13. Daugman, J.G.: Uncertainty Relations for Resolution in Space, Spatial Frequency, and Orientation Optimized by Two-Dimensional Visual Cortical Filters. Journal of the Optical Society of America A 2, 1160–1169 (1985)
14. Chong, C.-W., Mukundan, R., Raveendran, P.: An Efficient Algorithm for Fast Computation of Pseudo-Zernike Moments. Int. J. Pattern Recogn. Artif. Int. 17(6), 1011–1023 (2003)
15. Datta, R., Joshi, D., Li, J., Wang, J.Z.: Image Retrieval: Ideas, Influences, and Trends of the New Age. ACM Computing Surveys 40(2), Article 5, 60 pages (2008)
16. AT&T Lab. Cambridge, http://www.cl.cam.ac.uk/research/dtg/attarchive/facedatabase.html (last accessed on March 03, 2013)
17. Martinezand, A.M., Avinash, C.K.: PCA versus LDA. IEEE Trans. on Pattern Analysis and Machine Intell. 23(2), 228–233 (2001)
18. Flusser, J., Suk, T.: Rotation Moment Invariants for Recognition of Symmetric Objects. IEEE Trans. Image Proc. 15, 3784–3790 (2006)
19. Viitaniemi, V., Laaksonen, J.: Evaluating the Performance in Automatic Image Annotation: Example Case by Adaptive Fusion of Global Image Features. Signal Process. Image Commun. 22(6), 557–568 (2007)
20. Chinchor, N.: MUC-4 Evaluation Metrics. In: Fourth Message Understanding Conference, pp. 22–29 (1992)

Author Index